Water Supply System Reliability, Safety and Risk Modelling & Assessment, Volume II

Water Supply System Reliability, Safety and Risk Modelling & Assessment, Volume II

Guest Editors

Katarzyna Pietrucha-Urbanik
Janusz Rak

MDPI

Basel • Beijing • Wuhan • Barcelona • Belgrade • Novi Sad • Cluj • Manchester

Guest Editors

Katarzyna Pietrucha-Urbanik
Department of Water Supply
and Sewerage Systems
Rzeszow University of
Technology
Rzeszow
Poland

Janusz Rak
Department of Water Supply
and Sewerage Systems
Rzeszow University of
Technology
Rzeszow
Poland

Editorial Office
MDPI AG
Grosspeteranlage 5
4052 Basel, Switzerland

This is a reprint of the Special Issue, published open access by the journal *Water* (ISSN 2073-4441), freely accessible at: https://www.mdpi.com/journal/water/special_issues/54F659246O.

For citation purposes, cite each article independently as indicated on the article page online and as indicated below:

Lastname, A.A.; Lastname, B.B. Article Title. *Journal Name* **Year**, *Volume Number*, Page Range.

ISBN 978-3-7258-7040-0 (Hbk)
ISBN 978-3-7258-7041-7 (PDF)
https://doi.org/10.3390/books978-3-7258-7041-7

Contents

About the Editors

Katarzyna Pietrucha-Urbanik

Katarzyna Pietrucha-Urbanik, DSc, PhD, Eng., is an Associate Professor at the Department of Water Supply and Sewerage Systems, Faculty of Civil, Environmental Engineering and Architecture, Rzeszow University of Technology (Poland). Her research focuses on the reliability and safety of municipal water systems, failure risk analysis, risk assessments for consumers, and the resilience of water distribution infrastructure.

Janusz Rak

Janusz Rak, PhD, DSc, Eng., is a Professor at the Department of Water Supply and Sewerage Systems, Faculty of Civil, Environmental Engineering and Architecture, Rzeszow University of Technology (Poland). His research interests include critical infrastructure protection, reliability and safety of water supply systems, failure modelling and risk analysis, and reliability-based risk assessment to support resilient operation of water utilities.

Preface

Water supply systems must deliver potable water on demand under increasing pressure from ageing assets, climate variability, and growing expectations for service continuity. This Reprint brings together recent advances in reliability engineering, risk analysis, and data-driven modelling applied to water supply and distribution networks. The contributions illustrate how integrated assessment from human reliability and operational risk to leakage detection and life-cycle energy analysis can support safer, more resilient and sustainable water services.

Katarzyna Pietrucha-Urbanik and Janusz Rak
Guest Editors

water

MDPI

Editorial

Reliability, Safety and Risk Modelling in Water Supply Systems for Climate-Resilient Infrastructure

Katarzyna Pietrucha-Urbanik * and Janusz Rak

Department of Water Supply and Sewerage Systems, Faculty of Civil, Environmental Engineering and Architecture, Rzeszow University of Technology, Al. Powstancow Warszawy 6, 35-029 Rzeszow, Poland; rakjan@prz.edu.pl

* Correspondence: kpiet@prz.edu.pl

Modern water supply systems comprise complex technical infrastructure, reacting to extreme user demands and ever-growing environmental pressure. The complexity of distribution networks, the unpredictability of failures, and the shortages and contamination of water mean that the reliability and safety of water supply infrastructure remain key challenges for cities and regions around the world. The second volume of this Special Issue of *Water*, devoted to reliability, safety, and risk modelling in water supply systems, shows how the application of a wide range of approaches utilising advanced data analysis methods to innovative treatment technologies can contribute to the reliability and resilience of these crucial systems. In light of accelerating climate variability and increasing hydrological extremes, strengthening the climate resilience of water supply systems has become an essential dimension of reliability and risk assessment, linking traditional engineering approaches with adaptive and forward-looking strategies.

The first group of contributions focuses on machine learning approaches and digital risk modelling. The two opening papers examine leakage risk in water distribution networks. Wu and co-authors (Contributions 1 and 2) proposed a multi-component attention-based solution that integrates a conditional GAN with data-balancing techniques. Tested on data from the B2 and B3 districts in Zhengzhou, the model improved the recall rate for detecting leak-prone pipeline segments by more than 40% and demonstrated that increasing representative data points significantly enhances predictive performance. The same team later developed a model using graph neural networks (GNN), integrating network topology, GIS variables, and historical failure records. The model identified key leak-related factors such as pipe age, material, and previous failures, outperforming established machine learning approaches. Together, these papers illustrate a global shift toward leveraging operational data and artificial intelligence to improve failure prediction in critical infrastructure.

Medeiros and co-authors (Contribution 3) further show that predictive models can be developed even under limited data availability. Using historical failure data from the Brazilian utility CAGEPA and multilayer perceptron neural networks, the authors predicted with over 80% accuracy when the next failure would occur in specific network segments within a 90-day horizon. By distinguishing between new and recurring events and incorporating variables such as terrain elevation and the history of previous breaks, the study highlights the importance of systematic data collection and analytics-driven maintenance planning.

Zhang and colleagues (Contribution 4) extended digital assessment to fire protection systems, developing an IoT-based platform for intelligent testing of fire pumps. Sensors monitor flow, pressure, power, and efficiency, transmitting data to an analytical module.

The use of the discrete Fréchet distance to compare performance curves enables automatic detection of pump anomalies, while optimisation through genetic algorithms enhances discriminatory capability. This system demonstrates how IoT-supported diagnostics can increase the operational reliability of firefighting installations.

A second thematic group addresses energy-efficient design and strategic management of distribution systems. Ryu and Lee (Contribution 5) developed an improved life cycle energy analysis (LCEA) model for optimal WDS design, incorporating new pipe maintenance criteria and a resilience index based on nodal pressure. A revised unit energy formulation and modernised pipe replacement standards reduced total energy consumption by approximately 35% compared with models focused solely on cost minimisation. This demonstrates that energy-efficient design can simultaneously reduce emissions, improve hydraulic conditions, and increase resilience to failures.

Large-scale infrastructure interventions were explored in a case study by Song and co-authors (Contribution 6), analysing how future water transfer projects will reshape the supply network in the Erhai Lake region by 2035. Using landscape ecology methods combined with network theory, the authors assessed the connectivity of 215 nodes and 216 links in the 2020 network and predicted increases of 122 nodes and 163 links due to planned projects. These changes substantially strengthen overall system resilience. Increasing the number of nodes and decentralising flows enhances tolerance to failures, underscoring the importance of resource redistribution in mitigating climate-driven water inequalities.

Environmental factors and water quality also play a fundamental role in system reliability. Shrestha and colleagues(Contribution 7) evaluated the removal of iron and manganese in groundwater using dual-media filters filled with manganese-oxide-coated sand and ceramic. Traditional aeration and rapid sand filtration were insufficient for manganese removal, whereas the coated ceramic medium reduced concentrations from 1.10 mg/L to below 0.01 mg/L. Increasing pH from 7.5 to 9.0 improved stability. These findings are vital for regions affected by metal-rich groundwater, given the health risks associated with chronic manganese exposure.

Amin and co-authors (Contribution 8) analysed pipe scale deposition in Surakarta, Indonesia, revealing that groundwater-supplied zones exhibited higher dissolved solids, hardness, and alkalinity than surface water zones. Although traditional indices suggested moderate calcium carbonate scaling potential, deposit analysis showed a dominance of manganese (50–80%) and iron (46–64%). The deposits likely developed through oxidation by manganese-oxidising bacteria, exacerbated by the absence of groundwater chlorination. With an estimated deposition rate of 1660 kg per year, the study demonstrates that scaling indices alone are insufficient; mineralogical analysis is essential for accurate risk assessment.

Tahraoui and colleagues (Contribution 9) addressed polluted surface water sources, conducting a year-long evaluation of coagulation–flocculation treatment with aluminium sulphate. Frequent sampling and jar tests combined with SVM modelling enabled the determination of optimal coagulant dosage. Follow-up analyses after one year confirmed consistently high removal efficiencies, supporting the need for continuous improvement of treatment technologies, especially where surface water degradation poses major health risks.

Because water supply systems are anthropotechnical systems, technical performance is inseparable from human operation. Żywiec, Tchórzewska-Cieślak, and Sokolan (Contribution 10) analysed operator behaviour in water treatment plants using the Fuzzy-Bayes CREAM method. Estimated human error probabilities ranged from 0.0005 to 0.0746, depending on the process. The study emphasises that risk assessments incorporating human factors more accurately reflect operational conditions and that decision-support tools must accompany increasing automation to avoid new vulnerabilities.

Taken together, the contributions in this volume demonstrate that improving the reliability and safety of water supply systems requires an integrated, interdisciplinary approach. Combining engineering analysis, machine learning, IoT-based diagnostics, innovative filtration materials, and recognition of human factors allows water infrastructure to be managed more effectively.

Several common conclusions emerge:

- The reliability of predictive models depends on high-quality data and the integration of GIS, monitoring, and historical failure records;
- Energy-optimised design incorporating life cycle analysis and resilience indices yields ecological and operational benefits;
- Climate change and uneven water availability necessitate large transfer projects supported by system-level analyses;
- Effective treatment of iron, manganese, and other contaminants is essential for safeguarding public health;
- Human reliability assessment and operator training must accompany the digitalisation of water systems.

Looking ahead, a promising direction is the development of integrated smart water platforms combining IoT sensing, predictive analytics, and water quality monitoring. In the context of rapid urbanisation and climate change, ensuring safe and sustainable water supplies becomes a global priority. Solutions based on artificial intelligence, including digital twins enabling real-time simulation, together with strengthened cybersecurity frameworks, will be indispensable for future-ready water infrastructure.

Author Contributions: All authors equally contributed to the development of this manuscript. All authors have read and agreed to the published version of the manuscript.

Funding: This research received no external funding.

Acknowledgments: This Special Issue brings together the dedicated work of researchers advancing knowledge in water resources, environmental management, and sustainable infrastructure. We extend our sincere thanks to all authors and reviewers whose contributions shaped this collection. Their efforts have enriched the scientific discourse within this Special Issue, and we look forward to continued collaboration in pursuing innovative solutions for water, environmental, and sustainability challenges.

Conflicts of Interest: The authors declare no conflicts of interest.

List of Contributions:

1. Wu, W.; Zhang, J.; Kang, Y.; Tang, Z.; Pan, X.; Liu, N. Harnessing Risks with Data: A Leakage Assessment Framework for WDN Using Multi-Attention Mechanisms and Conditional GAN-Based Data Balancing. *Water* **2024**, *16*, 3329. https://doi.org/10.3390/w16223329.
2. Wu, W.; Pan, X.; Kang, Y.; Xu, Y.; Han, L. To Feel the Spatial: Graph Neural Network-Based Method for Leakage Risk Assessment in Water Distribution Networks. *Water* **2024**, *16*, 2017. https://doi.org/10.3390/w16142017.
3. Medeiros, V.d.S.; Santos, M.D.d.; Brito, A.V. Case Study for Predicting Failures in Water Supply Networks Using Neural Networks. *Water* **2024**, *16*, 1455. https://doi.org/10.3390/w16101455.
4. Zhang, S.; Li, Y.; Chen, X.; Zhou, R.; Wu, Z.; Zarhmouti, T. A Novel IoT-Based Performance Testing Method and System for Fire Pumps. *Water* **2024**, *16*, 792. https://doi.org/10.3390/w16050792.
5. Ryu, Y.M.; Lee, E.H. Optimal Design of Water Distribution System Using Improved Life Cycle Energy Analysis: Development of Optimal Improvement Period and Unit Energy Formula. *Water* **2024**, *16*, 3300. https://doi.org/10.3390/w16223300.
6. Song, K.; Jiang, X.; Wang, T.; Yan, D.; Xu, H.; Wu, Z. The Impact of Large-Scale Water Diversion Projects on the Water Supply Network: A Case Study in Southwest China. *Water* **2024**, *16*, 357. https://doi.org/10.3390/w16020357.

7. Shrestha, A.M.; Kazama, S.; Sawangjang, B.; Takizawa, S. Improvement of Removal Rates for Iron and Manganese in Groundwater Using Dual-Media Filters Filled with Manganese-Oxide-Coated Sand and Ceramic in Nepal. *Water* **2024**, *16*, 2450. https://doi.org/10.3390/w16172450.
8. Amin, S.; Kazama, S.; Sawangjang, B.; Takizawa, S. Causes and Effects of Scale Deposition in Water Supply Pipelines in Surakarta City, Indonesia. *Water* **2024**, *16*, 2275. https://doi.org/10.3390/w16162275.
9. Tahraoui, H.; Toumi, S.; Boudoukhani, M.; Touzout, N.; Sid, A.N.E.H.; Amrane, A.; Belhadj, A.; Hadjadj, M.; Laichi, Y.; Aboumustapha, M.; et al. Evaluating the Effectiveness of Coagulation–Flocculation Treatment Using Aluminum Sulfate on a Polluted Surface Water Source: A Year-Long Study. *Water* **2024**, *16*, 400. https://doi.org/10.3390/w16030400.
10. Żywiec, J.; Tchórzewska-Cieślak, B.; Sokolan, K. Assessment of Human Errors in the Operation of the Water Treatment Plant. *Water* **2024**, *16*, 2399. https://doi.org/10.3390/w16172399.

water

MDPI

Article

Harnessing Risks with Data: A Leakage Assessment Framework for WDN Using Multi-Attention Mechanisms and Conditional GAN-Based Data Balancing

Wenhong Wu [1,2], Jiahao Zhang [1,2], Yunkai Kang [1,2,*], Zhengju Tang [3], Xinyu Pan [1,2] and Ning Liu [3]

1 School of Information Engineering, North China University of Water Resources and Electric Power, Zhengzhou 450046, China; wuwenhong@ncwu.edu.cn (W.W.); x20231090801@stu.ncwu.edu.cn (J.Z.); pan19760928@gmail.com (X.P.)

2 Henan Province Water Distribution Network Intelligent Management Engineering Research Center, Zhengzhou 450046, China

3 Zhengzhou Water Affairs Group Co., Ltd., Zhengzhou 450007, China; czyml@163.com (Z.T.); leovanci@gmail.com (N.L.)

* Correspondence: kykinfdarksoul@gmail.com

Abstract: Assessing leakage risks in water distribution networks (WDNs) and implementing preventive monitoring for high-risk pipelines has become a widely accepted approach for leakage control. However, existing methods face significant data barriers between Geographic Information System (GIS) and leakage prediction systems. These barriers hinder traditional pipeline risk assessment methods, particularly when addressing challenges such as data imbalance, poor model interpretability, and lack of intuitive prediction results. To overcome these limitations, this study proposes a leakage assessment framework for water distribution networks based on multiple attention mechanisms and a generative model-based data balancing method. Extensive comparative experiments were conducted using water distribution network data from B2 and B3 District Metered Areas in Zhengzhou. The results show that the proposed model, optimized with a balanced data method, achieved a 40.76% improvement in the recall rate for leakage segment assessments, outperforming the second-best model using the same strategy by 1.7%. Furthermore, the strategy effectively enhanced the performance of all models, further proving that incorporating more valid data contributes to improved assessment results. This study comprehensively demonstrates the application of data-driven models in the field of "smart water management", providing practical guidance and reference cases for advancing the development of intelligent water infrastructure.

Keywords: water distribution network; risk assessment; data-driven model; smart water management

1. Introduction

Water distribution networks (WDNs) are vital links connecting water sources to end users, playing a crucial role in residents' daily lives and urban development. However, many cities face severe leakage issues in their water distribution systems. These problems not only result in the waste of water resources but may also lead to service disruptions, water quality contamination, damage to the structural integrity of the network, and increased maintenance costs. Therefore, the occurrence of leakage incidents is a key indicator of the overall normal pipeline system and a core factor in assessing the risk level of water distribution networks. On the foundation of high-standard urban water facility planning and construction, establishing an objective and scientific evaluation system for the normal water pipelines is essential for enhancing the operation and maintenance of these systems [1,2]. This will not only extend the lifespan of the network and reduce operating costs but also significantly improve water management efficiency, ensuring water supply safety and sustainable development [3].

Over the past decade, industry and academia have introduced various methods for the controllable risk assessment of water pipelines. A commonly used approach is leakage risk assessment based on the Analytic Hierarchy Process (AHP) with fuzzy mathematics. For example, H. Fares et al. [4] utilized a Hierarchical Fuzzy Expert System (HFES) to assess the influence of 16 pipeline risk factors on failure probability, concluding that pipe age and material are key risk factors. Similarly, I.A. Shaban et al. [5] constructed a leakage factor network through literature analysis, identifying that the type of pipeline connection significantly impacts leakage. They combined expert opinions to assign weights to risk factors and quantify pipeline risk in the evaluation process [6,7]. However, while these methods provide quantitative assessments of pipeline leakage risk, the subjectivity introduced by expert knowledge limits their applicability across different regions and systems.

With the increasing pace and quality of digital transformation in WDNs, data-driven models have emerged as essential tools for optimizing operations and maintenance strategies [8]. A typical framework for data-driven risk assessment involves systematically collecting and cleaning Geographic Information System (GIS) data related to pipeline networks, selecting leakage risk factors as independent variables for model training, and formulating assessment metrics that balance engineering requirements [9]. For example, Li et al. (2020) [10] incorporated spatial clustering patterns of pipe bursts into their model, using the Extreme Gradient Boosting (XGBoost) algorithm to assess pipeline network risks. Zhou et al. (2021) [11] similarly employed XGBoost, selecting parameters such as pipe material, diameter, length, age, and location to assess the leakage risk of a city's water network. Fan et al. (2022) [12] further advanced the field by proposing a data-driven machine learning model that integrates various factors—including engineering, geological, climatic, and socio-economic variables—and applied five machine learning algorithms to refine the risk assessment process. These methods enable probabilistic mapping of risk factors across different feature spaces in sample data [13,14], reducing the subjectivity of traditional weight-based evaluation methods and enhancing the reliability of assessment results.

Recent studies have shown that leakage data is significantly scarcer compared to non-leakage data. Training a model with an imbalanced dataset often leads to overfitting. For example, Zhou et al. [11] found that based on pipeline repair records, the number of non-leaking pipelines far exceeds that of leaking pipelines. As a result, models trained on such imbalanced datasets tend to classify most pipelines as low-risk. However, addressing dataset imbalance can significantly enhance model performance without altering the core machine-learning process for leak prediction. Liu et al. [15], for instance, encountered severe data imbalance in their pipeline dataset and compared the performance of Random Forest and Logistic Regression models after applying undersampling, oversampling, and Synthetic Minority Over-sampling Technique (SMOTE) [16]. The results showed that SMOTE effectively improved model performance. These studies highlight the importance of incorporating data balancing techniques into the workflow of data-driven models. However, most of the current methods rely on interpolation techniques, which, while enriching the leakage samples, fail to account for the specific characteristics of the pipelines.

Despite the success of data-driven pipeline risk assessments across various regions and data quality levels, several challenges persist. These include limited data interoperability between GIS, smart water platforms, and leakage prediction systems, inadequate explanation of assessment results, and the absence of a standardized assessment framework. Most critically, data imbalance remains a significant challenge, as it directly impacts model performance. Existing modeling approaches often struggle to capture the unique characteristics of real-world pipeline data.

To address these shortcomings, this study develops a systematic leakage risk assessment framework for urban water distribution networks based on pipeline data from the B2 and B3 metering areas in Zhengzhou City, Central China. The architecture of this system, shown in Figure 1, comprises five components: the construction of a pipeline feature database, an enhanced data scheme for leakage samples, risk feature interaction learning model training, Shapley Additive Explanations (SHAP) risk factor analysis, and a leakage risk assessment visualization platform. This system bridges the information gap between existing systems, focusing on resolving the issue of data imbalance in leakage samples. It also integrates data and pipeline assets from management units to promote the renewal and renovation of underground networks, providing a practical paradigm for the sustainable development of water supply systems and efficient water resource management.

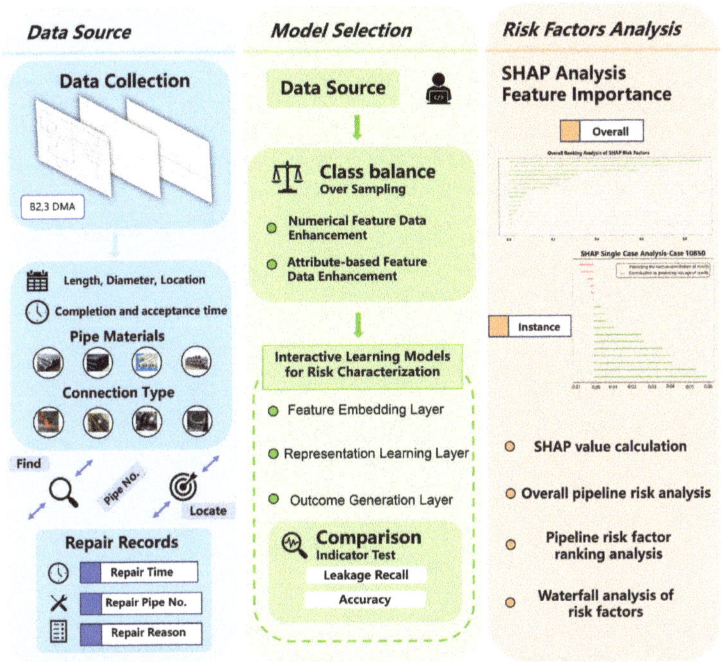

Figure 1. Framework.

In Section 2, this paper presents the modeling approach for predicting pipeline leakage probability, including strategies for balancing leakage samples and the principles of an attention mechanism-based prediction model. Section 3 outlines the metrics used to evaluate the model and analyze pipeline risk factors, introducing a leakage risk assessment platform developed based on this research. Finally, Section 4 summarizes the key contributions of this study and discusses future research directions.

2. Research Methodology

2.1. Construction of Pipeline Characterization Database Based on Data Governance

Data quality is critical to the performance of data-driven models, making robust data governance essential for improving data quality, particularly when dealing with missing records and diverse data sources. This study focused on governing pipeline data from the B2 and B3 metering areas in Zhengzhou, Central China, to ensure comprehensive coverage of pipeline feature diversity. The B2 and B3 metering areas have a combined water supply capacity of approximately 260,000 m^3/day, with a total pipeline length of 1922 km (320,713 pipelines, of which 243,682 are recorded in the GIS database). Of these, the B2 metering area accounts for 46%, while the B3 metering area makes up 52% (The data for this study were sourced from the GIS. However, 2% of the dataset lacked clear classification into either B2 or B3 categories, so these data were excluded from the analysis.). It should be noted that this study data retrieval is based on the time of entry into the system, and some pipeline records indicate completion dates before the year 2000. This effort yielded 320,713 fundamental pipeline records, 3055 leakage point entries, and 1964 maintenance records. The data were sourced from three distinct departments: (1) leakage point measurements provided by the leakage detection department, (2) pipeline maintenance records from manual maintenance enterprises, and (3) GIS and District Metered Areas (DMA) data related to the B2 and B3 DMA from local water companies, covering the period from 2000 to 2023.

To ensure the high availability of the database, the research used the pipeline section number as the primary keyword to link the three data sources. The paper first established a comprehensive pipeline database by adhering to the "Trinity" data positioning principles. Subsequently, this research conducted a prudent screening of pipeline risk factors with reference to comprehensive simulations, field experiments, historical data analysis [16], and industry-standard approaches. As outlined in Table 1, these factors are categorized into three main groups: (1) physical and pipeline-specific factors, (2) geographic-related factors, and (3) process-oriented factors.

Table 1. Higher completeness pipeline risk factors.

Type	Risk Factors	Feature Type	Data Sets	Data Completeness Rate
Physical and pipeline-specific factors	Nominal diameter (cm)	numerical	[15, 2200]	99.9%
	Pipe age (month)	numerical	[1, 69]	86.4%
	Pipe material	categorical	{PE pipe, PPR pipe, UPVC pipe, Galvanized pipe, Steel pipe, Steel-plastic pipe, Grey cast iron, Ductile iron, Concrete pipe}	97.3%
	Pipe Section Length (m)	numerical	[0.0067, 193.4854]	99.9%
Geographic-related factors	Location of road	categorical	{Road name}	83.8%
	Pipe type	categorical	{Municipal administration, Courtyard}	61.1%
Process-related factors	Connection type	categorical	{Socket type rubber ring interface, Hot-melt, Threaded, Socketed asbestos cement joints, Welded, Flanged, Jointed asbestos cement}	95.2%

Based on the screened risk factors, the data undergoes an additional round of refinement, where information for each pipeline unit is consolidated according to its respective pipe section number. This process leads to the construction of a comprehensive WDN pipeline characterization database. Considering the general characteristics of machine learning models, the terms "pipeline features" and "pipeline risk factors" are used interchangeably in the remainder of this paper.

2.2. GAN Data Enhancement Scheme for Pipeline Feature Specialization

The scarcity of leakage records in real-world scenarios leads to highly imbalanced datasets, limiting model reliability and cross-regional transferability, especially in supervised learning [17]. This research analysis shows common samples outnumber leakage samples by over a hundredfold, indicating an extreme imbalance. A specialized data augmentation method is needed to generate samples similar to the leaking pipeline and prevent the model generalization capability from degrading.

This research utilizes Conditional Tabular Generative Adversarial Networks (CTGAN) to model the distribution of leakage data and generate synthetic samples [18]. This approach accounts for the unique characteristics of pipeline risk factors, including numerical features such as pipe age and length and sparse, high-dimensional categorical features like pipe material and connection type [19]. Overall these data are mixed type data. However, traditional interpolation-based augmentation methods, such as the SMOTE, cannot handle categorical features represented by one-hot vectors. Therefore, this research implements a tailored approach by modeling numerical and categorical features separately and introducing conditional vectors, represented as masks, to enforce necessary constraints during the Generative Adversarial Networks (GAN) model's training phase. This method applies random sampling to the leakage data based on a learned mixture Gaussian model and then recombines all sampled features into a new set of pipeline characteristics. The pipeline features will be richer by supplementing the features which be generated by this method.

When dealing with numerical features of pipelines, the process is as follows:

(1) First, a mixture of Gaussian distribution patterns needs to be learned for each numerical type feature C_i, and the learned Gaussian mixture $\mathbb{P}_{C_i}(c_{i,j})$ is shown in Equation (1):

$$\mathbb{P}_{C_i}(c_{i,j}) = \sum_{k=1}^{n} \mu_k \mathcal{N}(c_{i,j}; \eta_k, \varphi_k) \quad (1)$$

where C_i is all numerical features. $c_{i,j}$ is the j numerical feature. μ_k is the weight of the learned k mode. η_k is the Gaussian distributed mean of the learned k mode and φ_k is the standard deviation of the k mode.

(2) Second, the kind of pattern that generates the data is included as part of the condition vector and constrains the range of values generated.

(3) Finally, the probability weights are then assigned to the patterns learned for each feature value $c_{i,j}$, and the pattern classification to which the generated sample belongs is selected based on the weights. Finally the pattern class scalar is generated to represent the values of numerical features $\alpha_{i,j}$, as shown in Equation (2):

$$\alpha_{i,j} = \frac{c_{i,j} - \eta_k}{4\varphi_k} \quad (2)$$

For categorical features, the research goal is to constrain the generator result $\widehat{d_1}, \cdots, \widehat{d_{N_d}}$. And as far as possible, satisfy the $\mathcal{P}_G(row|D_{i^*} = k^*) \approx \mathcal{P}(row|D_{i^*} = k^*)$. The specific process is as follows:

(1) For the attribute class feature D_i in row i, it generates the zero-padded mask vector m_i, where the k_{th} element of m_i represents the mask of category k.

(2) For each category k, based on the number of times the category appears in the column D_i, n (where the total number of categories is N), the probability function is calculated as $P(D_i = k) = \frac{n}{N}$.

(3) The category k^* is then selected according to the probability function and the element corresponding to k^* in the mask vector m_i is updated to 1.

(4) All the generated attribute class feature masks are spliced to form the final discrete quantity generation result $d_{1,i}, \cdots, d_{N_d,i}$.

The virtual leakage sample generated by the above two processes and thus the final generator can be represented as:

$$\widehat{r_j} = \alpha_{1,j} \oplus \beta_{1,j} \oplus \cdots \alpha_{N_c,j} \oplus \beta_{N_c,j} \oplus d_{1,j} \oplus \cdots \oplus d_{N_d,j} \quad (3)$$

where $\widehat{r_j}$ is the final generated j pipeline sample feature. c is the dimension of the numerical feature. α_{N_c} is the numerical type feature of this generated sample. β_{N_c} is the modal quantity of the numerical type feature of this generated sample, which is uniquely heat-coded. If the calculated value belongs to the learned pattern 3 (The variational Gaussian mixing model VGM mentions three modes [20]), then we can get $\beta_{i,j} = [0, 0, 1]$.

As shown in Figure 2 (Where $\oplus$ is concatenation. In the "Mode of Learning" section at the center of the figure, the dashed line represents the Gaussian mixture mode, while the solid line represents the mode of all data features.), this research approach uses a generalized GAN framework comprising a generator and a discriminator [19]. To quantify the discrepancy between the synthetic pipeline features $\widehat{r_j}$ generated by the generator and the authentic pipeline features r_j, this study employs the cross-entropy loss function as our primary metric. This choice is grounded in its ability to precisely capture the distance between the model's output and the true classification labels, thereby ensuring a robust assessment of the generated data's fidelity. The formula for the cross-entropy loss function is presented as follows:

$$L(r_j, \widehat{r_j}) = -[r_j \cdot log(\widehat{r_j}) + (1 - r_j) \cdot log(1 - \widehat{r_j})] \quad (4)$$

where $\widehat{r_j}$ is the virtual pipeline feature, r_j is the real pipeline feature, and $L(r_j, \widehat{r_j})$ is the cross-entropy loss function, which is smaller as the prediction gets closer to the real label.

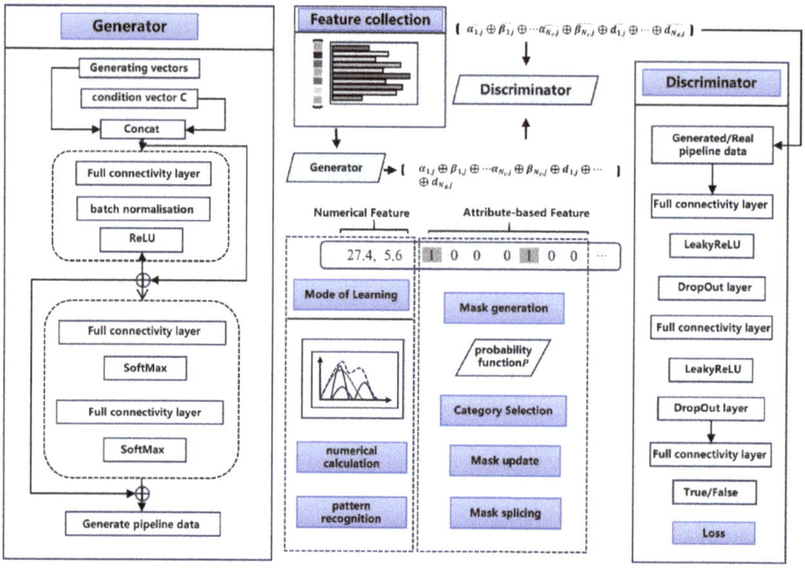

Figure 2. Data Enhancement Program.

2.3. *Interactive Learning Model of Risk Factors Based on Multiple Attention Mechanisms*

Current data-driven modeling methodologies often struggle to adequately capture the intricate interplay among risk factors, especially when dealing with attributes such as pipeline material and segment connection methods. This challenge leads to insufficient learning capabilities for high-dimensional characterizations [21].

To address this issue, this research optimized the leakage prediction architecture by breaking down the overall task into three objectives:

(1) Unify numerical and categorical features into a common space.
(2) Learn representations within this space that can clearly distinguish the risk patterns of leaking pipelines from normal pipelines.
(3) Learn an appropriate mapping to output the pipeline representations as leakage probabilities.

These objectives are achieved sequentially, and the process can be described using the following equations:

$$\{X_1, X_2 \ldots, X_M\} \overset{f_1(\cdot)}{\Rightarrow} \{E_1, E_2 \ldots, E_D\} \overset{f_2(\cdot)}{\Rightarrow} \{R_1, R_2 \ldots, R_D\} \overset{f_3(\cdot)}{\Rightarrow} \widehat{Y} \quad (5)$$

where X denotes the features that the pipeline has, M denotes the number of features, E denotes the embedded representation of the pipeline in embedding space which has the D dimensions, R represents the final learned pipeline representation in that space and $\widehat{Y}$ denotes the result of predicting pipeline leakage.

To address the above three objectives, the study designs the following three model layers concerning the commonly used Multi-Head Self-Attention Block (MHSA Block) and Residual Connection [22,23] in deep learning:

(1) Feature Embedding Layer: This layer transforms both numerical and attribute-based features into a unified dimensional space, denoted $f_1(\cdot)$;
(2) Representation Learning Layer: Here, pipeline features are fused to generate higher-order interactions among risk factors, denoted $f_2(\cdot)$;
(3) Outcome Generation Layer: Leveraging residual connections, this layer fosters information flow between layers, enabling efficient risk probability mapping while preserving critical information throughout the network, denoted $f_3(\cdot)$.

The overall architecture of the model can be depicted in Figure 3.

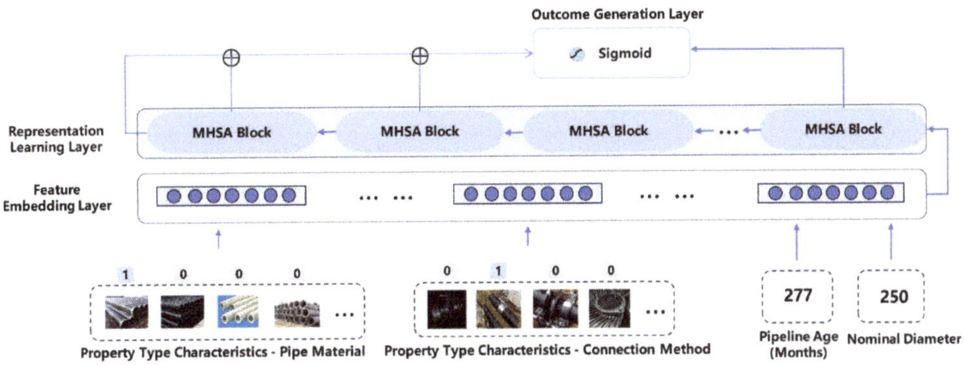

Figure 3. Pipeline Risk Prediction Modeling Framework.

The processed batch of pipeline features is first passed through $f_1(\cdot)$ to obtain an initial representation e of the embedding space, and then e is passed through $f_2(\cdot)$ (where the multi-head attention modules [24,25] are stacked within $f_2(\cdot)$) in order to obtain higher-order representations that contain different combinations of risk feature interactions [26]. The first step of $f_2(\cdot)$ is to compute the attention scores $\alpha_{m,k}^{(h)}$ of the pipelines m, k obtained by random sampling within each attention head h, which denotes the similarity between the two in this embedding space, and can be described by Equations (6) and (7):

$$\alpha_{m,k}^{(h)} = \frac{exp\left(\psi^{(h)}(e_m, e_k)\right)}{\sum_{l=1}^{M} exp(\psi^{(h)}(e_m, e_k))} \tag{6}$$

$$\psi^{(h)}(e_m, e_k) = \left\langle w_{Query}^{(h)} e_m \middle| w_{Key}^{(h)} e_k \right\rangle \tag{7}$$

where the inner product function $\psi^{(h)}(\cdot,\cdot)$ is used to compute the similarity between two pipelines in separate attention head h. $w_{Query}^{(h)}, w_{Key}^{(h)}$ is the weight matrix.

The second step of $f_2(\cdot)$ is to cumulatively aggregate features between different pipelines within the same batch guided by the attention scores and to achieve splicing aggregation between attention heads, which is shown in Equations (8) and (9):

$$\tilde{r}_m^{(h)} = \sum_{k=1}^{M} \alpha_{m,k}^{(h)} \left(w_{Value}^{(h)} e_k\right) \tag{8}$$

$$\tilde{r}_m = \tilde{r}_m^{(1)} \oplus \tilde{r}_m^{(2)} \oplus \cdots \oplus \tilde{r}_m^{(H)} \tag{9}$$

where $\tilde{r}_m^{(h)}$ is the new representation learned by every block, and $\tilde{r}_m$ is the higher-order interaction representation. Then, we splice the computational results of all head attention mechanisms to finally obtain the pipeline higher-order interaction representation $\tilde{r}_m$. where $\oplus$ denotes the cascade operation and H denotes the number of attention heads.

In order to preserve the previously learned features, the $f_3(\cdot)$ aggregates the results of the $f_2(\cdot)$ through residual concatenation and outputs the final binary classification prediction vectors using the Sigmoid function, which is shown in Equations (10) and (11):

$$\widehat{r_m} = ReLU\left(\widetilde{r_m} + W_{Res}r_m\right) \tag{10}$$

$$\widehat{y} = sigmoid\left(w^T(\widehat{r_1} \oplus \widehat{r_2} \oplus \widehat{r_3} \ldots \widehat{r_M}) + b\right) \tag{11}$$

The result generation layer aggregates the results of the multi-layer representation learning layer through residual linkage. And output the final binary prediction vector using the Sigmoid function. The process is shown below:

(1) 0 to 0.5: The pipeline is deemed to be in a normal state, suggesting a low likelihood of experiencing a leakage accident.
(2) 0.5 to 0.75: The pipeline is classified as in average condition, indicating an elevated risk of leakage occurrence.
(3) 0.75 to 1: The pipeline is identified as risky, signaling a high probability of imminent leakage.

These thresholds enable a nuanced pipeline safety assessment, allowing for targeted interventions and preventive measures.

2.4. Methodology for Analyzing Pipeline Risk Factors Based on SHAP Values

For non-expert statisticians in pipeline operations and maintenance (O&M), relying solely on quantitative leakage probabilities presents challenges in informing repair decisions. Although data-driven models provide probabilistic risk assessments, they often lack detailed analysis of individual risk factors, limiting their ability to offer specific, actionable insights for the targeted maintenance and repair of pipelines.

To address this issue and improve the model's interpretability, this study incorporates SHAP (Shapley Additive Explanations). SHAP utilizes the Shapley value concept from game theory [27], quantifying each feature's contribution to the model's final prediction. SHAP assigns an importance value to each feature for a given prediction [28], illustrating the expected change in the model's output when the attribute values of that feature are adjusted. This approach captures the influence of individual risk factors and is mathematically formalized as follows:

$$\phi_i(x) = \sum_{S \in 1,\ldots,p/i} \frac{|S|!(p - |S| - 1)!}{p!}[f(x_S \cup i) - f(x_S)] \tag{12}$$

where x_S denotes the input after removing features from the feature set S in x. p denotes the number of features in the input. $|S|$ denotes the size of S. $f(x_S \cup i)$ denotes the model output containing feature i. $f(x_S)$ denotes the model output without feature i.

The results of SHAP analysis are generally visualized through a bar chart that illustrates the overall contribution of each risk factor to the final prediction. In this chart, the order and width of the bars reflect the relative influence of different features. SHAP's visual approach offers a clear and intuitive way to assess the impact of individual risk factors on the model's predictions. This enables practitioners to develop targeted pipeline maintenance and repair plans based on a detailed understanding of the risk factors.

3. Results

3.1. Analysis of Pipeline Risk Prediction Model Performance

In this section, the effectiveness of the data augmentation method and data-driven model is validated through comparative experiments on the constructed pipeline feature database. Two main experiments are conducted: (1) Visualizing the distribution of synthetic data versus original data in high-dimensional space to observe their spatial relationships and (2) Comparing the accuracy of leakage detection between the proposed and commonly used data-driven models. Higher accuracy in identifying leakage pipelines indicates more reliable probabilistic mapping results. The pipeline feature database is split into two sets for these comparisons, with 80% used for model training and 20% for testing.

The performance of the data augmentation methods is evaluated using TSNE (t-distributed Stochastic Neighbor Embedding) and KDE (Kernel Density Estimation) dimensionality reduction. TSNE converts high-dimensional similarity into low-dimensional clustering, making similar pipeline samples cluster in lower dimensions [29]. KDE, a non-parametric method, estimates the probability density function and visualizes data density as a histogram [30]. Combining TSNE and KDE allows us to understand the distribution of data points after dimensionality reduction. As shown in Figure 4, similar points in high-dimensional space form distinct clusters in the low-dimensional space. The traditional SMOTE and optimized GAN methods effectively generate clear cluster boundaries and balance sample proportions. Compared to the existing SMOTE interpolation method, the approach used here produces a more uniform sample distribution, reflecting the richness of the generated data. Additionally, the closer the generated samples are to the original samples in the TSNE diagram, the higher their similarity. These similar samples carry additional information, which helps refine the model's decision boundaries during training, thereby enhancing model quality while balancing leakage and normal pipeline samples.

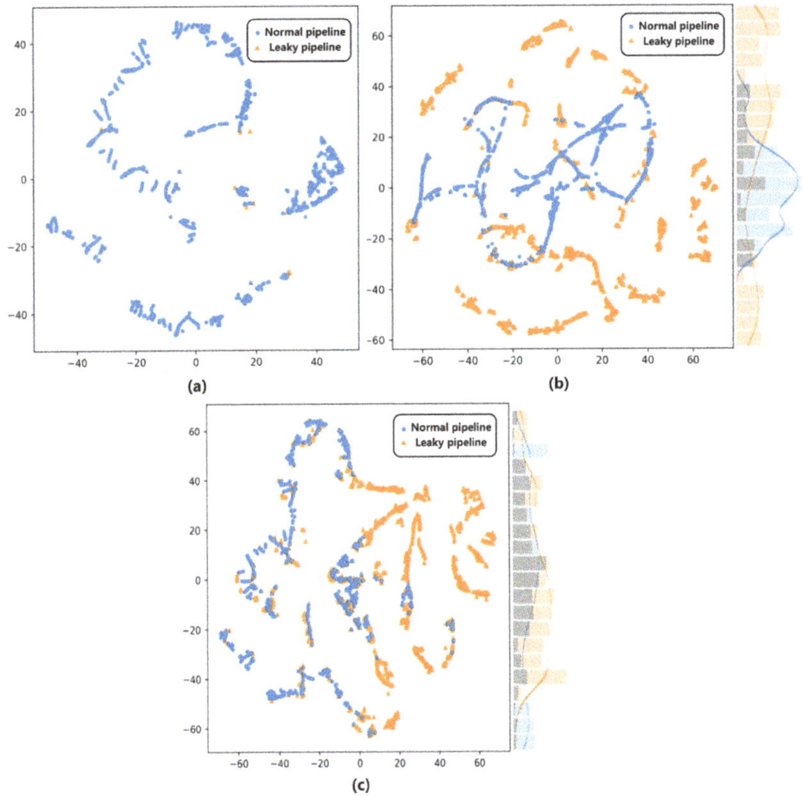

Figure 4. Enhanced sample balance results for leakage sample data: (**a**) Original Data; (**b**) SMOTE Data Enhancement Method; (**c**) Conditional GAN Data Augmentation Method. (The horizontal and vertical axes represent the two-dimensional vector values obtained by dimensionality reduction of the high-dimensional representation of the samples, serving only as markers).

The outcomes of the comparative experiments are summarized in Table 2, which details the assessment accuracy of various data-driven models for detecting leaky pipelines and evaluating overall pipeline segments, with different proportions of synthetic data incorporated into the model training process. Given the potential severity of underreporting leakage incidents, this research aims to find a method that can maintain sensitivity in identifying leaking pipelines without compromising the accuracy of recognizing non-leaking pipelines. The experimental results show that the model employed in this case, which utilizes CTGAN and an attention mechanism, achieved the best performance in testing. Compared to commonly used data-driven models, this approach improved the detection accuracy of leaky pipelines by 2.30% and the overall pipeline recognition accuracy by 1.84%. What's more, the Leakage-Recall and Accuracy metrics provide crucial insights into the model's performance in these areas.

Table 2. Comparison of experimental results.

Model Name	Enhancing Data Participation Rates	Method of Parameterization	Leakage Pipe Segment Assessment Accuracy	Overall Pipe Segment Assessment Accuracy
This case	0% 30% 50%	Grid search	34.88% 61.28% 75.64%	88.96% 87.68% 89.64%
Random forest	0% 30% 50%	Grid search	27.45% 48.63% 73.94%	88.02% 85.98% 84.32%
Decision tree	0% 30% 50%	Grid search	29.43% 32.78% 64.32%	79.82% 78.37% 81.66%
XGBoost	0% 30% 50%	Grid search	31.64% 52.47% 69.27%	82.72% 81.53% 83.37%
AdaBoost	0% 30% 50%	Grid search	32.73% 57.84% 71.62%	84.28% 83.62% 85.72%
BP neural network	0% 30% 50%	Grid search	26.69% 46.29% 62.79%	78.43% 76.82% 77.54%
LightGBM	0% 30% 50%	Grid search	28.73% 53.38% 67.94%	81.93% 81.23% 82.47%

The evaluation metrics are defined using the following equations based on a confusion matrix:

- True Normal (TN) refers to samples accurately classified as non-leaky pipelines.
- True Leakage (TL) denotes samples correctly identified as leaky pipelines.
- False Normal (FN) represents pipelines mistakenly classified as intact despite actual leakage.
- False Leakage (FL) indicates pipelines incorrectly predicted as leaky when functioning properly.

Additional evaluation metrics are derived from this confusion matrix to comprehensively assess the model's effectiveness.

$$\textit{Leakage-Recall} = \frac{TN}{FL + TN} \tag{13}$$

$$\textit{Overall-Accuracy} = \frac{TN + TL}{TN + TL + FL + FN} \tag{14}$$

In addition to model selection for the overall framework, the results of the comparative experiments identified several key trends. Notably, leakage assessment accuracy across all models significantly improved with increasing data participation rates. For instance, the proposed model's accuracy for leakage segment assessments increased from 34.88% at 0% participation to 75.64% at 50%, representing a notable rise of 40.76%. Similarly, other models, including Random Forest and Decision Tree, exhibited enhanced accuracy with higher data participation, underscoring the value of additional data. However, overall pipe segment assessment accuracy remained relatively stable despite increased participation rates, with the proposed model's accuracy fluctuating between 87.68% and 89.64%. While Decision Tree and BP Neural Network showed slightly lower overall accuracy, they exhibited a comparable stabilizing trend, indicating effective data augmentation for leaky pipeline samples. By balancing the dataset, this research ensured that the models learned the leak pipeline risk characterization patterns, benefiting from increased sample diversity. Nevertheless, in some cases, prediction accuracy decreased with rising generated data participation, suggesting that not all generated samples enhance model performance, thereby necessitating quality screening of generated data.

3.2. SHAP Analysis of Pipeline Risk Factors

The SHAP method leverages model interpretability to quantify the impact of various risk factors on pipeline leakage incidents. In this section, the reliability of the SHAP analysis results is verified by comparing them with heuristic data analysis of risk factors and the trend observed in the pipeline model predictions for the study area. As shown in Figure 5, the pipeline's age and material, particularly grey cast iron, which is prone to oxidation, contributed most significantly to the model's predictive outcomes, making them the highest risk factors for pipelines. In the overall ranking of risk factors, grey cast iron, pipe age, and connection types such as bell-and-spigot asbestos-cement joints had the greatest impact. Additionally, pipeline diameter was shown to be a significant factor, consistent with common engineering knowledge that leakages are more concentrated in small-diameter, short-length pipes, particularly DN100 pipes.

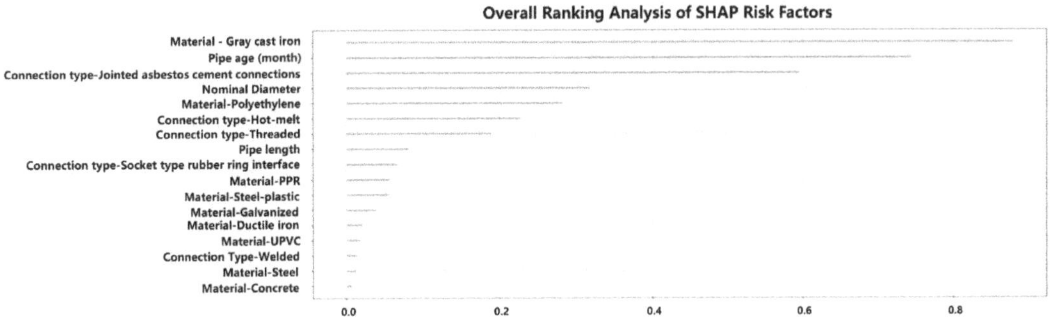

Figure 5. SHAP analysis results: Overall Ranking Analysis of SHAP Risk Factors.

To validate the reliability of the aforementioned overall pipeline analysis results, this paper also conducted separate studies in Case 10,850 and Case 486. The results of the Case 10,850 SHAP analysis are shown in Figure 6. In Case 10,850, the SHAP analysis for leakage prediction similarly highlighted the contribution of grey cast iron and bell-and-spigot asbestos-cement joints, which aligns with the analysis of overall pipeline risk factors. It is also worth noting that grey cast iron pipelines are commonly connected using bell-and-spigot or threaded joints, both of which are prone to leakage if sealing is inadequate or if temperature changes cause joint loosening. This explains why threaded and bell-and-spigot joints are identified as risk factors for leakage in this case.

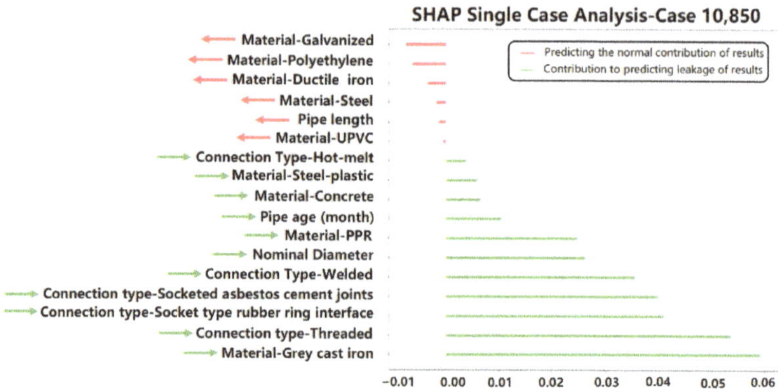

Figure 6. SHAP analysis results: SHAP Single Case Analysis-Case 10850.

The results of the Case 486 SHAP analysis are shown in Figure 7. In Case 486, the SHAP analysis showed that pipeline age and diameter contributed the most to leakage prediction, again consistent with the global analysis of pipeline risk factors. This highlights that as pipelines age, material degradation becomes a more prominent issue. Additionally, older pipelines may lack modern corrosion protection measures, making them more susceptible to leakage. Therefore, this study recommends replacing pipelines that have exceeded their service life or are made from outdated materials. Overall, the SHAP analysis of these two individual cases is consistent with the global analysis, indicating that pipeline age, diameter, and material are critical factors influencing pipeline leakage.

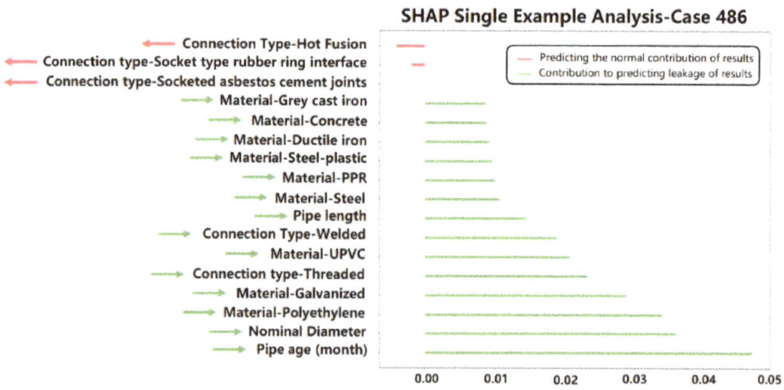

Figure 7. SHAP analysis results: SHAP Single Example Analysis-Case 486.

Specifically, the analysis of leakage frequency by pipeline material is detailed in Table 3, the source of data is the leaky pipeline obtained in Section 2.1. The number of leaks per kilometer is calculated by Equation (15). The results reveal that the leakage rates per kilometer for PE and grey cast iron pipes are significantly higher than those of other materials. The number of predicted hazardous pipelines for PE and grey cast iron pipes also exceeds that of other materials. For instance, PE pipes exhibit a leakage frequency of 5.72 per kilometer, with 598 leaks and 4396 predicted hazardous pipelines, indicating a prominent leakage issue and considerable future maintenance and management challenges. Grey cast iron pipes have a similar total leakage count of 598, with a leakage frequency of 2.67 per kilometer and 6229 predicted hazardous pipelines, highlighting a severe risk of leakage in the coming years. In contrast, despite having the longest total length (1094.37 km), ductile iron pipes show a leakage rate of only 0.17 per kilometer, with a total of 186 leaks and just 463 predicted hazardous pipelines, demonstrating high stability and low risk. Other materials, such as PPR and steel-plastic pipes, display lower leakage frequencies and predicted hazardous pipelines, suggesting their overall reliability and manageable future risks. Overall, PE and grey cast iron pipes exhibit the most severe leakage issues and should be prioritized for future maintenance (consistent with SHAP analysis). Ductile iron pipes show excellent performance with minimal future leakage risk. PPR and steel-plastic pipes also perform reliably and at a low risk.

$$Leak\ per\ kilometer = \frac{Number\ of\ leaks}{Total\ length} \tag{15}$$

where *Leak per kilometer* denotes the number of leaks per kilometer recorded from 2021 to 2023. *Number of leaks* denotes the number of leaks in the pipeline. *Total length* the total length of the pipeline.

Table 3. Enlightening data analysis results for pipeline materials.

Pipe Material	Total Length (km)	Number of Leaks	Recorded from 2021 to 2023 Leaks per Kilometer (Times/km)	The Number of Hazardous Pipelines Predicted
PE pipe	104.61	598	5.72	4396
PPR pipe	157.83	298	1.89	440
UPVC pipe	32.03	56	1.75	611
Galvanized pipe	21.64	43	1.99	800
Steel pipe	26.34	12	0.46	4
Steel-plastic pipe	169.81	149	0.88	114
Grey cast iron	224.10	598	2.67	6229
Ball-milled cast iron	1094.37	186	0.17	463
Concrete pipe	22.82	24	1.05	379

The analysis demonstrates that the model and SHAP analysis effectively differentiate the impact of pipeline materials (including connection methods) on leakage risk.

The results of the enlightening data analysis regarding pipeline age are illustrated in Figure 8. Pipelines older than 15 years exhibit a higher frequency of leaks per kilometer compared to those with a shorter age. The distribution violin plot comparing leakage and normal pipeline age distributions indicates that the proportion of leaks increases significantly with the pipeline age. According to the model's predictions, the percentage of pipelines classified as hazardous is 35.2% for those older than 30 years, 20.4% for those aged 20–30 years, and 39.1% for those aged 15–20 years. This shows a strong correlation between pipeline age and leakage risk, which aligns with the SHAP analysis indicating that the risk of leakage increases with pipeline age [31]. In addition, it is noteworthy that the pipeline leak rate declines after 20 years, indicating that leaks tend to occur shortly after installation. This suggests a higher likelihood of failures during the pipeline's early service life or wear-in period, following a characteristic "bathtub curve" failure pattern [12,32]. This phenomenon aligns with findings from previous studies.

3.3. Leakage Risk Assessment Visualization Platform

Implementing the visualization of model prediction results and reliably monitoring the overall pipeline condition are critical to addressing information silos in modern water systems. This study developed a pipeline leakage risk assessment visualization platform by integrating and mapping data from pipeline feature databases, GIS-based spatial information, and model predictions, with a detailed GIS map of pipeline status at its core. The platform's interface and key functions are shown in Figure 9. It allows for network-wide, category-specific, region-specific, and individual pipeline status queries, as well as statistical feedback. By offering four levels of pipeline network visualization, the platform equips O&M personnel with flexible tools for querying leakage risk and efficiently identifying high-risk pipelines. When a pipeline is classified as poor or dangerous, it indicates a high likelihood of leakage, prompting increased inspection frequency by O&M personnel.

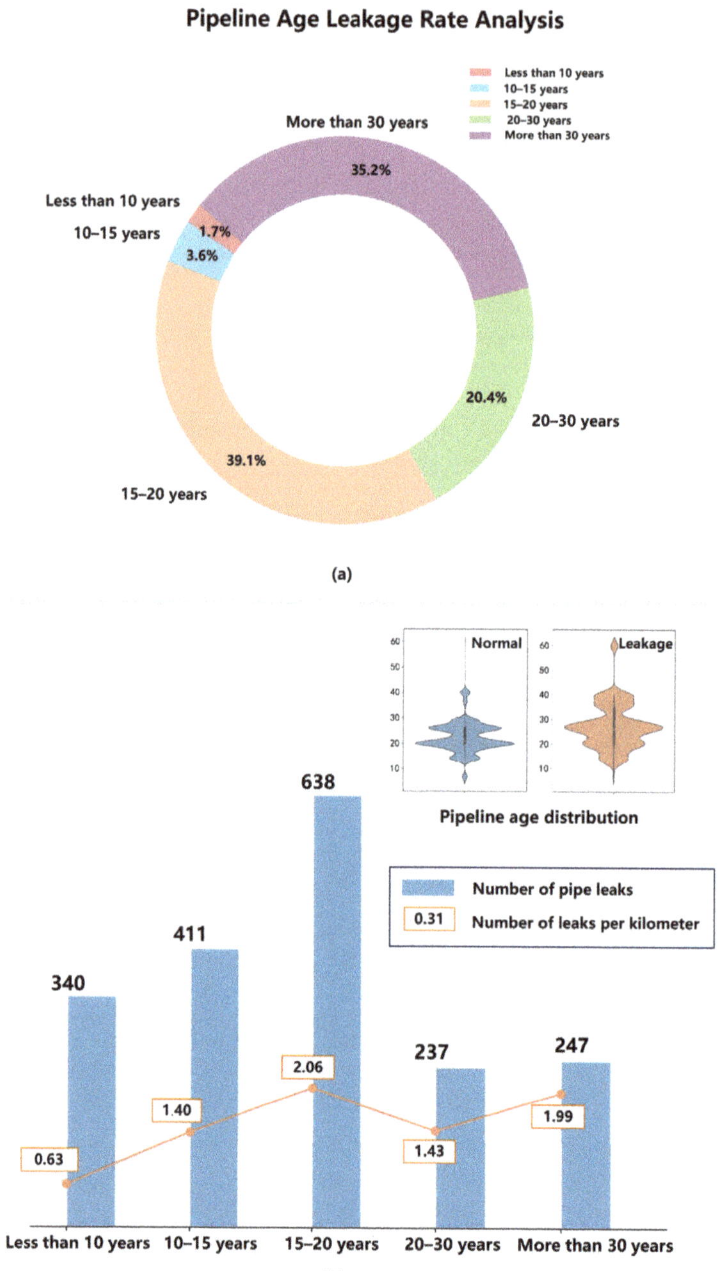

Figure 8. Pipeline age leakage rate analysis: (**a**) Age distribution of pipelines assessed as a level of risk, (**b**) Leakage ratios for pipelines of different ages.

Figure 9. Visualization Platform: (**a**) Leakage Point; (**b**) Pipeline Location; (**c**) B2, B3 DMA location; (**d**) Total Pipeline leakage risk status statistics; (**e**) Classified pipeline leakage risk status statistics and positioning; (**f**) Individual pipeline leakage risk status information; (**g**) Region-specific leakage risk status statistics.

This research applied the model to all pipelines in the B2 and B3 DMAs. After excluding 73,324 pipelines with missing features (22.86% of the total), the model classified 202,630 pipelines as normal (63.18%), 31,323 as poor (9.77%), and 13,436 as hazardous (4.19%). Overall, the pipeline risk in this region is low, with leakage-prone pipelines exhibiting some spatial clustering. Hazardous pipelines display significant variations in key factors such as length, age, and diameter. SHAP analysis of these factors provides actionable insights for pipeline O&M personnel. The recommended action plans are as follows:

- For pipelines predicted to be risky due to age, material, or connection method, immediate replacement is advised.
- If the predicted risk is due to pipe length or if the pipeline is categorized as poor, its location should be recorded, and the frequency of leakage detection should be increased.
- For pipelines categorized as poor or where length is a significant risk factor, their locations should be recorded, and the frequency of leakage detection should be increased.

From a macro perspective, based on recommendations from relevant planning documents in the sector regarding pipeline leakage rates, we suggest developing repair plans for high-risk pipelines such as PE and grey cast iron pipes. These should be progressively replaced with ductile iron and steel pipes that meet safety and accident rate standards. Utilizing the visualization platform, O&M units can coordinate the replacement of high-risk pipelines in concentrated areas, minimizing disruptions to daily life.

4. Conclusions

This study utilizes the WDN pipeline data from Zhengzhou, central China, to demonstrate the application of a data-driven pipeline leakage risk assessment model. It presents a comprehensive multi-model approach, offering a reliable and practical example for constructing a water distribution network leakage risk assessment system. The main conclusions are as follows:

(1) The data-driven model predictions serve as the basis for pipeline leakage risk assessment, eliminating the reliance on expert judgment or manually assigned risk levels. Knowledge is derived from patterns in the pipeline data, making this approach more objective and stable compared to traditional methods.

(2) The case study provides a detailed analysis of pipeline attributes and addresses sample imbalance using conditional vector optimization adversarial neural networks. This method balances the leaking and non-leaking pipelines in the database and explores the relationships between risk factors. By applying a multi-attention mechanism, higher-order risk feature combinations are created, improving the model's ability to identify leaking pipelines and significantly enhancing performance. The optimized model achieved a 40.76% improvement in recall for leakage segment assessments, outperforming the second-best model by 1.7%.

(3) The introduction of the SHAP explanatory model allows for visual analysis of risk factors at both the overall pipeline level and the individual pipeline level. This improves the model's explanatory power and provides specific guidance for practical application. SHAP analysis reveals that pipeline age, diameter, material, and connection type as key factors influencing pipeline leakage, aligning with previous research findings. First, for older pipelines, the materials may have reached the end of their service life or reflect outdated manufacturing techniques, making them more prone to leakage. Regarding pipeline diameter, small-diameter and short-length pipelines, particularly DN100, are more susceptible to leaks. In terms of material, grey cast iron is the most significant contributor to leakage due to its susceptibility to oxidation; it was commonly used in early WDN, alongside galvanized pipes, where manufacturing processes were less advanced and service life longer. Additionally, the connection types associated with grey cast iron pipes, typically socket and threaded joints, can be prone to leakage if not properly sealed or if they loosen due to factors like temperature fluctuations. This accounts for the higher risk of leakage associated with threaded and socketed connections in this context. The next most significant contributors were PE and PPR pipes, whose susceptibility to leakage is influenced by connection type as well as environmental factors and construction quality. In terms of connection types, threaded and socketed interfaces also showed high contributions in the SHAP analysis. Pipelines with threaded connections are generally small-diameter PE and PPR pipes, where high pressure at the joints may lead to fractures, indirectly confirming

the tendency of PE and PPR pipes to leak. Socketed connections, typically used with grey cast iron and PE pipes, are also vulnerable to leakage due to construction quality and temperature variations, verifying that both connection type and pipe material impact leakage risk.

(4) Using the model prediction results as a benchmark, this case breaks through the data barriers of the established GIS system, smart water platform and leakage prediction system, and builds a visualization platform integrating risky pipeline query and risky pipeline positioning functions, which provides intuitive and scientific guidance for pipeline maintenance personnel to repair the pipeline network.

This research also observed that using generative models, such as GANs, for data augmentation significantly influenced the performance of certain machine learning models, likely due to variations in the quality of generated data and potential issues with model fit. To address this, future work will focus on developing evaluation metrics for generative samples to ensure data quality. Additionally, the research area for this research is the B2 and B3 DMA in Zhengzhou, Central China, which present limitations in pipeline feature diversity and do not account for the topological relationships between pipelines. In the future work, it is anticipated that this research method could be applied to other areas, with the design of a spatial awareness module to encode inter-pipeline relationships. Finally, to further investigate the time-varying effects of pipeline risk factors and to provide constructive guidance on optimal pipeline replacement timing, we propose introducing a survival analysis approach, taking pipeline age at the time of leakage as the study objective, and utilizing an optimized Survshap method for more in-depth analysis.

Author Contributions: Conceptualization, W.W.; methodology, W.W. and Y.K.; formal analysis, J.Z.; resources, W.W. and Z.T.; data curation, X.P. and N.L.; writing—original draft preparation, J.Z. and Y.K.; writing—review and editing, J.Z. and Y.K.; visualization, J.Z., N.L. and X.P.; supervision, Y.K., Z.T. and N.L.; project administration, W.W. All authors have read and agreed to the published version of the manuscript.

Funding: This research received no external funding.

Data Availability Statement: Data available on request due to restrictions, e.g., privacy or ethical.

Acknowledgments: Thanks to Zhengzhou Water Affairs Group Co., Ltd. for the data support.

Conflicts of Interest: The authors Zhengju Tang and Ning Liu are employed by Zhengzhou Water Group Co., Ltd. The remaining authors declare that the research was conducted in the absence of any commercial or financial relationships that could be construed as a potential conflict of interest.

References

1. Shao, Y.; Zhou, J.; Wang, L.; Wang, P.; Lv, S. New Mode of Informatization Management about Drainage Maintenance Based on Full Life Cycle Management Strategy. *China Water Wastewater* **2024**, *40*, 17–21.
2. Ruan, J. Analysis on state assessment and risk alert of water supply pipe network. *Water Purif. Technol.* **2020**, *39*, 107–110. [CrossRef]
3. Wu, W.; Pan, X.; Kang, Y.; Xu, Y.; Han, L. To Feel the Spatial: Graph Neural Network-Based Method for Leakage Risk Assessment in Water Distribution Networks. *Water* **2024**, *16*, 2017. [CrossRef]
4. Fares, H.; Zayed, T. Hierarchical Fuzzy Expert System for Risk of Failure of Water Mains. *J. Pipeline Syst. Eng. Pract.* **2010**, *1*, 53–62. [CrossRef]
5. Shaban, I.A. Systematic and Scientometric Analyses of Predictors for Modelling Water Pipes Deterioration. *Autom. Constr.* **2023**, *149*, 104710. [CrossRef]
6. Yan, M.; Wang, H.; Liu, Z.; Gong, W.; Dai, X. Establishment and study on the evaluation index system for operational efficiency of urban drainage system. *Chin. J. Environ. Eng.* **2023**, *17*, 3124–3136.
7. Qiao, Q.; Xu, Z.; Wang, Y. Safety assessment method of water distribution network based on the combination of AHP-entropy weighting method. *City Town Water Supply* **2022**, 50–57. [CrossRef]
8. Zhu, P. Neural Network Based Pipe Bursting Risk Assessment Model for Water Supply Pipe Networks. Master's Thesis, Shanghai University of Finance and Economics, Shanghai, China, 2023. [CrossRef]
9. Harvey, R.; McBean, E.A.; Gharabaghi, B. Predicting the Timing of Water Main Failure Using Artificial Neural Networks. *J. Water Resour. Plann. Manag.* **2014**, *140*, 425–434. [CrossRef]

10. Li, C. Risk Assessment of Water Distribution Network Pipeline Failures Based on Spatial Aggregation Attributes and XGBoost Algorithm. Master's Thesis, Zhejiang University, Zhejiang, China, 2021.
11. Zhou, H.; Jiang, S.; Zhou, Z.; Ying, S.; Xu, X.; Weng, X. Exploration of health assessment and model restoration for urban water supply network. *Water Supply Drain.* **2021**, *57*, 491–494.
12. Fan, X.; Wang, X.; Zhang, X.; Yu, X. Machine learning based water pipe failure prediction: The effects of engineering, geology, climate and socio-economic factors. *Reliab. Eng. Syst. Saf.* **2022**, *219*, 108185. [CrossRef]
13. Wang, Y.; Qiao, Q.; Han, B. Application practice of data-driven safety assessment model of water supply pipeline network. *Water Waste Water Eng.* **2023**, *49*, 140–147.
14. Shi, L.; Zhou, F. Technical verification and assessment of urban water supply system risk identification and emergency capability. *Water Waste Water Eng.* **2021**, *57*, 9–15+23.
15. Liu, W.; Wang, B.; Song, Z. Failure prediction of municipal water pipes using machine learning algorithms. *Water Resour. Manag.* **2022**, *36*, 1271–1285. [CrossRef]
16. Chawla, N.V.; Bowyer, K.W.; Hall, L.O.; Kegelmeyer, W.P. SMOTE: Synthetic Minority Over-Sampling Technique. *J. Artif. Intell. Res.* **2002**, *16*, 321–357. [CrossRef]
17. Chang, T. Methodology and Application of Pipe Condition Assessment in Urban Water Distribution System. Master's Thesis, Tsinghua University, Beijing, China, 2018.
18. Xu, L.; Skoularidou, M.; Cuesta-Infante, A.; Veeramachaneni, K. Modeling Tabular Data Using Conditional GAN. In Proceedings of the 33rd International Conference on Neural Information Processing Systems, Vancouver, BC, Canada, 8–14 December 2019.
19. Song, Z.; Jiang, X.; Yu, W.; Qian, J.; Song, Z. Data Augmentation Based on Active Virtual Sample Selection for Bearing Fault Diagnosis. In Proceedings of the 2023 CAA Symposium on Fault Detection, Supervision and Safety for Technical Processes (SAFEPROCESS), Yibin, China, 22–24 September 2023; pp. 1–6.
20. Crowley, J.L. *Pattern Recognition and Machine Learning*; Springer: New York, NY, USA, 2006.
21. Lin, Z.; Khetan, A.; Fanti, G.; Oh, S.P. The power of two samples in generative adversarial networks. *arXiv* **2017**, arXiv:1712.04086. [CrossRef]
22. Michel, P.; Levy, O.; Neubig, G. Are Sixteen Heads Really Better than One? In Proceedings of the 33rd International Conference on Neural Information Processing Systems, Vancouver, BC, Canada, 8–14 December 2019.
23. He, K.; Zhang, X.; Ren, S.; Sun, J. Deep Residual Learning for Image Recognition. In Proceedings of the IEEE Conference on Computer Vision and Pattern Recognition, Online, 5 July 2016; pp. 770–778.
24. Song, W.; Shi, C.; Xiao, Z.; Duan, Z.; Xu, Y.; Zhang, M.; Tang, J. AutoInt: Automatic Feature Interaction Learning via Self-Attentive Neural Networks. In Proceedings of the 28th ACM International Conference on Information and Knowledge Management, Beijing, China, 3–7 November 2019; pp. 1161–1170.
25. Vaswani, A.; Shazeer, N.; Parmar, N.; Uszkoreit, J.; Jones, L.; Gomez, A.N.; Kaiser, Ł.; Polosukhin, I. Attention Is All You Need. In Proceedings of the 31st International Conference on Neural Information Processing Systems, Long Beach, CA, USA, 4–9 December 2017.
26. Miller, A.; Fisch, A.; Dodge, J.; Karimi, A.-H.; Bordes, A.; Weston, J. Key-Value Memory Networks for Directly Reading Documents. *arXiv* **2016**, arXiv:1606.03126.
27. Van den Broeck, G.; Lykov, A.; Schleich, M.; Suciu, D. On the tractability of SHAP explanations. *J. Artif. Intell. Res.* **2022**, *74*, 851–886. [CrossRef]
28. Lundberg, S.M.; Lee, S.-I. A Unified Approach to Interpreting Model Predictions. *arXiv* **2017**, arXiv:1705.07874.
29. Van der Maaten, L.; Hinton, G. Visualizing data using t-SNE. *J. Mach. Learn. Res.* **2008**, *9*, 579–2605.
30. Kalinic, M.; Krisp, J.M. Kernel Density Estimation (KDE) vs. Hot-Spot Analysis—Detecting Criminal Hot Spots in the City of San Francisco. In Proceedings of the 21st Conference on Geo-information Science, Lund, Sweden, 12–15 June 2018.
31. Xu, Q.; Chen, Q.; Li, W.; Ma, J. Pipe Break Prediction Based on Evolutionary Data-Driven Methods with Brief Recorded Data. *Reliab. Eng. Syst. Saf.* **2011**, *96*, 942–948. [CrossRef]
32. Singh, A.; Adachi, S. Bathtub curves and pipe prioritization based on failure rate. *Built Environ. Proj. Asset Manag.* **2013**, *3*, 105–122. [CrossRef]

water

MDPI

Article

Optimal Design of Water Distribution System Using Improved Life Cycle Energy Analysis: Development of Optimal Improvement Period and Unit Energy Formula

Yong min Ryu [1] and Eui Hoon Lee [2,*]

[1] Department of Civil Engineering, Chungbuk National University, Cheongju 28644, Republic of Korea; rmfl45@naver.com
[2] School of Civil Engineering, Chungbuk National University, Cheongju 28644, Republic of Korea
* Correspondence: hydrohydro@chungbuk.ac.kr; Tel.: +82-043-261-2407

Abstract: Water distribution systems (WDSs) are crucial for providing clean drinking water, requiring an efficient design to minimize costs and energy usage. This study introduces an enhanced life cycle energy analysis (LCEA) model for an optimal WDS design, incorporating novel criteria for pipe maintenance and a new resilience index based on nodal pressure. The improved LCEA model features a revised unit energy formula and sets standards for pipe rehabilitation and replacement based on regional regulations. Applied to South Korea's Goyang network, the model reduces energy expenditure by approximately 35% compared to the cost-based design. Unlike the cost-based design, the energy-based design achieves results that can relatively reduce energy when designing water distribution networks by considering recovered energy. This allows designers to propose designs that consume relatively less energy. Analysis using the new resilience index shows that the energy-based design outperforms the cost-based design in terms of pressure and service under most pipe failure scenarios. The implementation of the improved LCEA in real-world pipe networks, including Goyang, promises a practical life cycle-based optimal design.

Keywords: water distribution system; life cycle energy analysis; energy consumption; unit energy formula; new resilience index

1. Introduction

The supply of uncontaminated drinking water to consumers is essential for human life, and a water distribution system (WDS) is critical infrastructure to fulfill this necessity. The design of a WDS is complex because of its operational conditions and related uncertainties [1,2]. The design process involves determining several variables such as the pipe diameter, the pipe length between nodes, the pipe material, pumps, and tanks [3–6]. Various metaheuristic optimization algorithms have been used to derive optimal WDS designs [4,7–10]. Most of these studies have focused on minimizing the cost of designing a WDS [11–13]. Cost minimization studies aim to derive optimal designs based on the initial construction cost of a WDS but do not consider factors such as energy consumption during network construction. However, to construct a large-scale social infrastructure (such as a WDS), an optimal design should consider the entire life cycle, including maintenance and decommissioning.

The life cycle includes the entire process from fabrication to the disposal of any item or facility [14,15]. The major stages to be considered during the WDS life cycle are fabrication, maintenance, and disposal. While the fabrication and disposal stages analyze the static condition of the entire facility, the maintenance stage considers the aging and destruction of pipes over time. Shamir and Howard (1979) confirmed that pipe destruction occurs exponentially and proposed a failure rate formula [16]. Sharp and Walski (1988) derived a simple formula for pipe aging over time using the Hazen–Williams and Darcy–Weisbach equations [17]. Mononobe (1960) proposed a destruction estimation formula based on existing WDS data [18]. Shamir and Howard (1979) proposed a method for planning pipe replacement cycles in a WDS based on the degree of destruction, and Male et al. (1990) suggested a cost-effective method for renewal and replacement cycles in a WDS [16,19]. Studies on the maintenance stages of WDSs and those considering environmental and economic factors based on WDS life cycle analysis have also been conducted [16,19]. Subsequently, the life cycle energy analysis (LCEA) model has been applied to WDSs [12,20]. Representative WDS optimal design studies using LCEA include the study by Filion et al. (2004) and the study by Lee et al. (2015) [14,21].

In this study, a new LCEA was proposed that improved the shortcomings of Filion et al. (2004) and Lee et al. (2015), which are representative LCEA studies [14,21]. In the case of the newly proposed LCEA, to apply it without modifying the basic data of the pipe network to be applied, the C value of the existing pipe network was used, and the C value of the newly buried pipe was equally applied during rehabilitation and replacement. To enhance applicability, a method was employed to calculate rehabilitation and replacement times, considering regional characteristics based on local regulations and pipe C values. In addition, to improve the accuracy of the unit energy calculation formula in the process of improving applicability, a new optimal unit energy calculation formula was proposed by analyzing various types of trend lines and R^2 according to the trend lines. Additionally, a novel resilience index was formulated and applied to assess the performance of the proposed water distribution system design using the new LCEA model. The new resilience index, developed based on Todini's (2000) and Cimellaro et al.'s (2016) resilience indices primarily used in water distribution systems (WDSs), was introduced [22,23]. The previously proposed resilience index integrated factors like the affected population in a system failure, the tank capacity in the WDS, and water quality through calculations [22,23]. The new resilience index focuses on the need to restore water pressure at each node to meet WDS standards for a return to the original state after a failure. The new resilience index presented in this study employs an approach that assesses the overall system performance by considering hydraulic pressure at each node during failure events. Through the utilization of this new resilience index, a comparative analysis was conducted between the cost-based optimal design plan and the energy-based optimal design plan.

2. Methodology

2.1. Overview

This study consisted of four main parts. First, metaheuristic optimization algorithms (MOAs) were selected for the optimal WDS design. Second, optimal improvements and renewal timing based on the aging of the WDS were established. Third, the LCEA model was analyzed using a new unit energy calculation method. Finally, a new resilience index was developed and applied based on the pressure at each node in the water pipe network. The optimal design search method using the improved LCEA method used in this study is shown in Figure 1.

According to Figure 1, the first step is to establish the optimal improvement and regeneration period according to the aging of the WDS. The WDS consists of multiple pipes. For each pipe, the optimal diameter is selected from among various types of diameters. In order to establish the optimal rehabilitation, repair, and replacement period, it is necessary to calculate the degree of aging according to the diameter for each pipe, and the criteria are presented based on this. The second step is to select MOAs. MOAs have been developed in the past, but each has its own advantages and disadvantages. For example, there are MOAs with a small number of parameters and high usability and MOAs with a large number of parameters and internal operators but low usability but good performance. Among the MOAs developed previously, MOAs that showed good performance in the WDS optimal design were selected and applied to the WDS optimal design through parameter sensitivity analysis. The third step is to develop a new unit energy calculation method. In order to improve the accuracy of the existing unit energy calculation method, various unit energy calculation formulas were established, and the optimal unit energy calculation formula was proposed through accuracy analysis. Finally, a new resilience index was developed. Based on the optimal design produced through the proposed method, a new resilience index was developed, applied, and evaluated to analyze the resilience of the design when a disaster occurs.

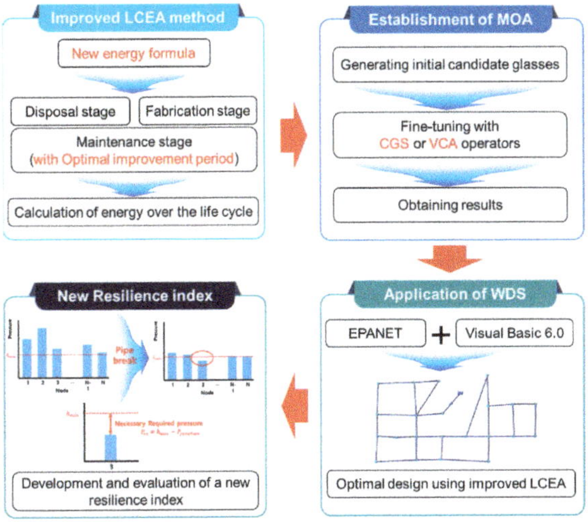

Figure 1. Optimal design search method using improved LCEA model.

2.2. Metaheuristic Optimization Algorithm for Optimal Water Distribution System Design

The optimal design of a WDS consists of deriving an optimal design plan according to the designer's purpose while satisfying the user's requirements such as appropriate water quality and demand. Traditionally, this approach has relied on trial and error based on user experience, with additional mathematical techniques applied [24]. Since 1981, research has been conducted using linear and nonlinear functions to design WDSs to minimize the cost of branched pipe networks [25–27]. However, in the case of linear and nonlinear functions, an optimal design could not be derived due to nonlinear elements in the simulation process and dependence on the initial solution group [6,28,29]. To overcome the limitations of these mathematical techniques, metaheuristic algorithms have been used for WDS optimization [11,30–32]. Among various metaheuristic algorithms, the modified hybrid vision correction algorithm (MHVCA) has demonstrated superior performance in minimizing costs for WDS optimization [10].

In this study, among various metaheuristic optimization algorithms, the modified hybrid vision correction algorithm (MHVCA) proposed by Ryu and Lee (2023) was used [10]. The MHVCA is a metaheuristic optimization algorithm that consists of six parameters that must be set by the user. According to Ryu and Lee (2023), the MHVCA showed good results in the optimal design of six WDSs (Goyang network, Hanoi network, Pescara Network, Zhejiang network, Modena network, and Balerma network) and showed better results than metaheuristic optimization algorithms such as Harmony search and the Genetic algorithm, which have fewer parameters than the MHVCA [10]. In this study, the effectiveness of the MHVCA in WDS optimization was used to explore the optimal design of WDSs to minimize energy consumption. Table 1 provides a description of each important internal operator and parameter used in the MHVCA.

According to Table 1, the CGS, CGSR, and DR are probability parameters for operator selection, while the CF, AF, MTF, and MHR are parameters utilized to fine-tune the solution during iterative calculations. The optimal solution search process of the MHVCA is as follows in Figure 2.

Table 1. Important internal operators and parameters in the MHVCA [10].

Operators (Full Name)	Description	
CGS (Centralized Global Search)	An operator that proceeds with a search after setting a new search range using the optimal value of the current iteration and the median value of the search range	
MTF (Modulation Transfer Function)	A method for adjusting the decision variable based on the distance between the best value among the existing optimal solutions and the new solution using an operator that mimics lens brightness adjustment in the process of manufacturing glasses	
Parameters (Full name)	**Description**	**Value of range (General value)**
MHR (Modified Hybrid Rate)	A parameter used in the process of searching for a new solution, a probability parameter that determines whether fine-tuning (search method using CF, AF, and MTF) is executed	0~0.36 (Self-adaptive)
CGSR (Centralized Global Search Rate)	Probability parameter that determines whether CGS is executed	0~1 (Self-adaptive)
CF (Compression Factor)	A method of reducing the range of region search as the number of iterations increases with an operator that mimics the compression process in the process of manufacturing glasses	0~100 (30)
AF (Astigmatic Correction)	A method of setting and searching the range of region search based on the angle of the astigmatism axis set by the user with an operator that mimics the astigmatism correction process in the process of manufacturing glasses	0~180 (45)
DR (Division Rate)	A probability parameter that determines the direction to be searched within the search range during the process of searching for a new solution	0~1 (0.1)

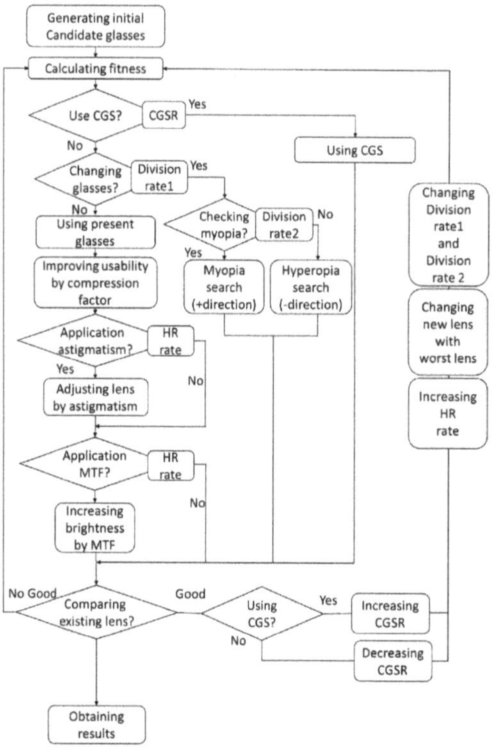

Figure 2. Flowchart of MHVCA.

According to Figure 2, the MHVCA does not search for the number of solution combinations equal to the number of parent generations like the Genetic algorithm in one optimal solution search, but it instead searches for one solution combination (glasses in MHVCA terminology).

2.3. Establishment of Optimal Improvement Period

In the pursuit of cost minimization, the optimal WDS design requires alternatives with minimal initial installation costs. However, to analyze the WDS life cycle, it is essential to consider the energy required for maintenance over time after pipeline installation. Therefore, to simulate the maintenance of a WDS, it is crucial to proactively model the aging of pipes over time. Lee et al. (2015) quantified the Hazen–Williams coefficient (C) based on the equation proposed by Mononobe (1960) and the coefficient proposed by Baek (2002), which is expressed as follows [14,18,33]:

$$C_y = C_0\left(1 - \frac{(0.0961659\text{D} + 1.15507)\sqrt{y}}{D}\right)^{0.723076D^{-0.0660117}} \tag{1}$$

where C_y denotes the C value over time, C_0 denotes the initial C value, D denotes the diameter of the pipe (mm), and y denotes the year after burial or replacement.

A previous study used Equation (1) to propose the rehabilitation and replacement period for pipes, which were suggested based on the value of the Hazen–Williams coefficient (C). When C was ≤ 90, the pipe was rehabilitated; after two rehabilitations, if C was $\leq$80, the pipe was replaced. It was assumed that C recovered to 90–110 after rehabilitation and to 130 after replacement because of the installation of new pipes. However, in this study, the rehabilitation and replacement periods of pipes were not set uniformly based on a fixed value of C. Instead, a method was proposed to consider the Hazen–Williams coefficient (C) of the WDS under consideration. The service life of the water supply and sewage facilities was specified based on the regulations of the Office of Waterworks in South Korea. The WDS used in this study was the Goyang network. In order to estimate the head loss of a pipe, the Darcy–Weisbach or Hazen–Williams formula is used. In the case of the Goyang network, which is the water pipe network applied in this study, the head loss was estimated using the Hazen–Williams formula as a result of analyzing the initially proposed literature and the literature in which the optimal design was conducted [4,10,13,34,35]. Therefore, in this study, the Hazen–Williams formula was also used to conduct the optimal design. The material of pipe that constitutes the Goyang network is assumed to be made of steel according to the existing literature, and the useful service life of the pipe is set to 30 years [10,13,34,35]. Therefore, based on the diameter of the pipes, rehabilitation and replacement criteria were established using the C value after 30 years (useful life for the service of steel pipes). Figure 3 shows the variation in C values by diameter over time.

As shown in Figure 3, after 30 years, the C values for each pipe range from approximately 60 to 68. To establish criteria for rehabilitation and replacement, the average C value for pipes after 30 years was set to about 65 in this study. Furthermore, based on the existing replacement criteria, replacement was performed after two rehabilitation sessions. To determine the recovered C values following rehabilitation and replacement, an approach similar to that proposed in previous studies was adopted. Similarly to previous research, it was assumed that after rehabilitation, the C value of pipes recovered to approximately 85% of their initial C values (130 out of 110).

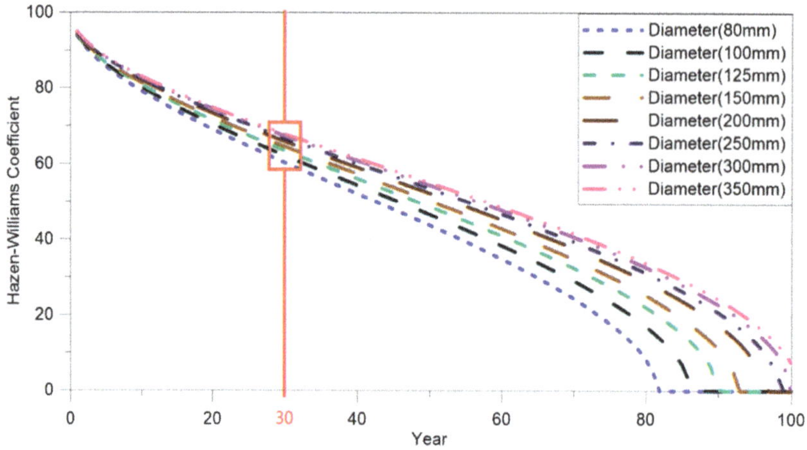

Figure 3. Convergence curve of C by diameter over time (red box is distribution of C according to diameter after 30 years.).

2.4. The LCEA Model Using a New Unit Energy Equation

The proposed LCEA model can be divided into three main phases—Phase 1 determines the rehabilitation period of each pipe using information on the WDS to be optimized (pipe diameters) and regulations specific to the region; Phase 2 calculates the unit energy for each pipe using the WDS information to be optimized (nodes, pipes, pressures, etc.); Phase 3 is further divided into three stages (fabrication, maintenance, and disposal). During the fabrication stage, the fabrication energy of all pipes in the WDS can be calculated using the unit energy. The maintenance stage involves estimating the maintenance energy over the intended life cycle by considering pipe destruction simulations and the rehabilitation period determined in Phase 1. Finally, the disposal stage involves calculating the disposal energy for all pipes in the WDS using unit energy. The improved LCEA model was simulated using Visual Basic 6.0 and EPANET 2.0 (US EPA, 2000) [36]. In this study, Demand-Driven Analysis (DDA) and Pressure-Driven Analysis (PDA) were used in the process of applying EPANET to optimally design the WDS. DDA was used to simulate normal situations for optimal design, and PDA was used to simulate abnormal situations for resilience index analysis [10,37–40]. This study calculated the resilience index based on the optimal design and optimal design plan. In the process of conducting the optimal design, the optimal design plan was derived using the DDA method. PDA was used to calculate the disaster resilience index of each optimal design plan. The demand at each node of the WDS to be applied was set as a fixed value, and the temporal distribution pattern of demand was not applied. The optimal WDS design using the LCEA model can be used to calculate the annual energy consumption. Using this, the year with the lowest energy consumption can be set to be the optimal life cycle. The objective function of the optimal WDS design using the LCEA model can be expressed as follows:

$$E_{all} = \left[E_{fab} + (E_{main} - E_{rec}) + E_{dis} + Penalty\right]/LC \quad (2)$$

where E_{all} denotes the total annual energy (GJ/year), E_{fab} denotes the fabrication energy (GJ/year), E_{main} denotes the maintenance energy (GJ/year), E_{rec} denotes the recycle energy (GJ/year), E_{dis} denotes the disposal energy, *Penalty* denotes the penalty function, and *LC* denotes the optimal life cycle (year). The penalty function can be used to exclude design proposals that do not satisfy the minimum required hydraulic pressure of each node during the process of optimal WDS design. On the one hand, if the minimum required water pressure is satisfied, the penalty is zero; on the other hand, if the minimum required water pressure is not satisfied, a penalty is assigned, as follows:

$$Penalty = \sum_{jc=1}^{tot} 10^{20} \times (h_{minP} - h_{jc}) + 10^{7}, (if\ h_{jc} < h_{minP}) \\ Penalty = 0, (if\ h_{jc} > h_{minP}) \quad (3)$$

where tot denotes the total number of nodes in the WDS, h_{minP} denotes the minimum water pressure of the nodes in the WDS, and h_{jc} denotes the water pressure of jc.

2.4.1. New Unit Energy Equation

For an optimal WDS design using the LCEA model, the energy generated at each stage (fabrication, maintenance, and disposal) must be analyzed, which is based on the length of the pipe. To calculate the unit energy based on diameter, it is essential to ensure accuracy by building on the previously suggested values. In this study, various formulas for unit energy calculations were expressed using trend equations, and an optimal formula for the unit energy calculation was proposed. The trend equation was established using previously studied data, and the diameter data of the Goyang pipe network were utilized to analyze the possibility of additional use. Table 2 presents unit energy equations using existing unit energy equations and various proposed equations and is a table showing the R^2 value for each equation.

Table 2. Previously proposed unit energy formula and various other formulas.

	Previous Function [14]	**R^2**
Fabrication	$e_{fab} = (4.206 \times D^{1.9959}) \times Conv$	0.9962
Disposal	$e_{dis} = (0.2974 \times D^{2.0248}) \times Conv$	0.9952
	Linear function	**R^2**
Fabrication	$e_{fab} = (27.053 \times D - 35.659) \times Conv$	0.9788
Disposal	$e_{dis} = (1.9904 \times D - 2.6401) \times Conv$	0.9789
	Log function	**R^2**
Fabrication	$e_{fab} = (78.805 \times \ln(D) - 31.34) \times Conv$	0.9298
Disposal	$e_{dis} = (5.8014 \times \ln(D) - 2.326) \times Conv$	0.9308
	Exponential function	**R^2**
Fabrication	$e_{fab} = (4.2905 \times D^{1.9677}) \times Conv$	0.9992
Disposal	$e_{dis} = (0.3035 \times D^{1.9927}) \times Conv$	0.9989
	2nd-order polynomial function	**R^2**
Fabrication	$e_{fab} = (4.3669 \times D^2 - 2.0038 \times D + 3.1002) \times Conv$	0.9962
Disposal	$e_{dis} = (0.3116 \times D^2 - 0.0831 \times D + 0.1258) \times Conv$	0.9952
	3rd-order polynomial function	**R^2**
Fabrication	$e_{fab} = (2.1678 \times D^3 - 19.535 \times D^2 + 65.506 \times D - 58.322) \times Conv$	0.9963
Disposal	$e_{dis} = (0.1769 \times D^3 - 1.476 \times D^2 + 5.427 \times D - 4.8875) \times Conv$	0.9979
	4th-order polynomial function	**R^2**
Fabrication	$e_{fab} = (1.2748 \times D^4 - 15.24 \times D^3 + 62.27 \times D^2 - 107.33 \times D + 64.24) \times Conv$	1.0000
Disposal	$e_{dis} = (0.1116 \times D^4 - 1.3474 \times D^3 + 5.9498 \times D^2 - 9.7072 \times D + 5.8445) \times Conv$	1.0000

In this table, D denotes the pipe diameter (m), and $Conv$ denotes a unit conversion factor that converts m units to GJ/m. As shown in Table 2, the R^2 values for the previously proposed unit energy formulas are 0.9962 and 0.9952. For the previously proposed function, regression analysis was performed based on five data points, and the function was expressed as an exponential function [14,21]. In this study, various analyses were performed based on the five given data points in the same way as the method used in the existing literature. However, when various functions (linear, log, exponential, and polynomial) were used to calculate the unit energy, the 4th-order polynomial function had the highest R^2 value of 1.0000 for the unit fabrication and disposal energy. However, because of the uniqueness of the polynomial function, when calculating the unit energy using pipes from locations other than those proposed by Filion et al., the energy decreased as the pipe diameter increased [21]. In the case of individual pipes forming a constant pipe, as the diameter increases, the energy required to manufacture the pipe increases. However, in the case of the 4th-order polynomial presented in Table 2, when data other than the given data are input, the energy required to manufacture the pipe decreases. Consequently, considering the specific nature of the polynomial function, it was excluded from the unit energy formula. An exponential function that realized the highest R^2 value was chosen for the new unit energy formula. The proposed unit fabrication and disposal energy formula can be expressed as follows:

$$e_{N\_fab} = \left(4.2905 \times D^{1.9677}\right) \times \text{Conv} \tag{4}$$

$$e_{N\_dis} = \left(0.3035 \times D^{1.9927}\right) \times \text{Conv} \tag{5}$$

where e_{N_fab} denotes the new fabrication unit of energy, e_{N_dis} denotes the new disposal unit of energy, D denotes the pipe diameter (m), and Conv denotes a unit conversion factor that converts m units to GJ/m.

2.4.2. Energy Calculation for Design of Water Distribution System

As mentioned in Section 2.4.1, for the optimal design of WDSs employing LCEA, it is imperative to calculate the energy generated during the fabrication, maintenance, and disposal stages. In this study, the energy for each stage, as suggested by Lee, was computed using the newly proposed unit energy calculation formula based on the diameter [41]. The energy calculation formula for each stage, which integrates the newly proposed unit energy calculation formula, is presented in Table 3.

Table 3. Energy calculation formula and description for each stage [41].

Stage	Formula	Description
Fabrication	$E_{fab} = \sum_{i=1}^{Tot} L_i \times 4.2905 \times D^{1.9677}$	This stage includes various processes such as raw material extraction, material processing and production, and pipe fabrication. The energy consumed in these processes can be defined as the fabrication energy.
Maintenance	$E_{main} = \left(E_{reh} + E_{rep} + E_{rpi}\right) - E_{rec}$ $E_{reh} = \sum_{i=1}^{Tot} L_i(N_{reh} \times 0.65 \times 4.2905 \times D^{1.9677})$ $E_{rep} = \sum_{i=1}^{Tot} L_i(N_{rep} \times 4.2905 \times D^{1.9677})$ $E_{rpi} = \sum_{i=1}^{Tot} L_i \times N(t) \times (2 \times L_b \times 4.2905 \times D^{1.9677})$ $E_{rec} = \frac{1}{pump_e} \sum_{i=1}^{Tot} \left(E^p_{p,i} - E^a_{p,i}\right)$	The maintenance process of a WDS comprises the rehabilitation, repair, and replacement of pipes, owing to aging or pipe failure. Maintenance energy defines all the energy consumed during the maintenance stage. Lee et al. (2015) defined recycling energy as the benefit obtained from the improved flow rate resulting from the rehabilitation of pipes during the maintenance process [41].
Disposal	$E_{dis} = \sum_{i=1}^{Tot} L_i \times 0.3035 \times D^{1.9927} \times (1 + N_{reh})$	The disposal stage involves the disposal of pipes that constitute the WDS. Disposal energy is the energy consumed purely for the disposal of pipes, which includes the energy required for the disposal of pipes during replacement as well as that consumed during the disposal of pipes in the maintenance stage.

In this table, E_{fab} denotes the fabrication energy consumption (GJ), Tot denotes the total number of pipes, L_i denotes the length of the ith pipe (m), D denotes the diameter of the pipe, E_{main} denotes the maintenance energy consumption (GJ), E_{reh} denotes the rehabilitation energy (GJ), E_{rep} denotes the replacement energy (GJ), E_{rpi} denotes the repair energy (GJ), E_{rec} denotes the recycled energy (GJ), N_{reh} denotes the number of pipe rehabilitations, N_{rep} denotes the number of pipe replacements, *N(t)* denotes the pipe failure probability function, L_b denotes the typical breakage length, $pump_e$ denotes the efficiency of the pump, $E^{p}_{p,i}$ denotes the pump energy required when the pipe is rough before rehabilitation, and $E^{a}_{p,i}$ denotes the pump energy required when the pipe is smooth after rehabilitation.

2.5. Development of New Resilience Index Using Insufficient Pressure

A new resilience index was proposed to evaluate and compare the performance of the proposed design using the improved LCEA. Using the new resilience index, we evaluated the performance of the WDS design through scenarios such as system failure caused by natural disasters. The proposed new resilience index is an index that can indicate the performance of the system based on the pressure per node within the WDS, based on the resilience proposed by Cimellaro et al. (2016) [23]. The proposed new resilience index is as follows.

$$R = R_1 \times R_2 \times R_3 \times R_P \tag{6}$$

$$R_1 = \sum_{T_L=0}^{LC} \frac{1-\frac{\sum_i n^i_{p,e}}{n_{Tot}}}{T_L} \text{for i} = 1,2,\cdots,\text{n}, R_2 = \sum_{T_L=0}^{LC} \frac{F_{(t)}}{T_L}, F_{(t)} = \begin{cases} \frac{h(t)}{h_{Reserve}} h \le h_{reserve} \\ 1\ h > h_{reserve} \end{cases},$$
$$R_3 = \sum_{T_L=0}^{LC} \frac{\frac{Q(t)}{Q^*}}{T_L}, R_P = \sum_{T_L=0}^{LC} \frac{1-\frac{\sum_j^{Nan}\left(h_{min}-h_{jun}\right)}{\sum_j^{Ntn} h_{min}}}{T_L} \tag{7}$$

where T_L is the control time (life cycle), $n^i_{p,e}$ is the number of users receiving insufficient pressure, n_{Tot} is the number of users in the WDS, n is the number of nodes affected by the outage, *h(t)* is the tank water level at time *t*, $h_{Reserve}$ is the reserve capacity of the tank, *Q(t)* is the water quality at time *t*, Q^* is the water quality factor, Ntn is the total number of nodes, h_{min} is the minimum required water pressure, Nan is the number of abnormal nodes, and h_{jun} is the pressure of node *j*. Additionally, R_1 is the number of households affected by a water outage in the event of a system failure due to a natural disaster, etc., which is proportional to the system serviceability index proposed by Todini (2000) [22]. In the case of R_1, if it is assumed that no failure occurs in the WDS when a natural disaster occurs, it can be represented as 1, and if a failure occurs at all nodes in the WDS, it can be represented as 0 [22]. R_2 is a proposed index based on the water level of the tank in the network, R_3 is a factor for water quality, and R_4 is a newly proposed resilience index calculated through all nodes and abnormal nodes within the WDS. In the case of R_2, if there is a tank in the WDS and the reserve capacity in the tank is higher than the tank water level at time t, it can be represented as 1, and in the opposite case, the value of the index is calculated according to the ratio of the tank water level at time t to the reserve capacity in the tank. R_3 is the ratio of the water quality concentration to the bad concentration at time t due to the occurrence of a natural disaster. Therefore, if it does not meet the water quality standard, it will have a value between 0 and 1 depending on the result. R_4 is a coefficient that quantifies the pump energy required to restore a disaster situation according to the reduced pressure at each node in the WDS due to the occurrence of a natural disaster. R_4 has a value less than 1 when the pressure at each node decreases due to a natural disaster and has a value of 0 when the pressure at all nodes is 0%. R_1~R_4 are coefficients for indexing the resilience of the system to a disaster situation when a natural disaster occurs and a failure occurs in the WDS, and they always have a value less than 1 for an abnormal situation where a problem occurs within the WDS. If no failure occurs in the WDS, the resilience index has a maximum value of 1, and if the WDS does not function as intended, it has a minimum value of 0.

3. Results

3.1. Study Area

The Goyang network located in South Korea was selected as the target area for the energy optimization design using the improved LCEA model. The layout of the Goyang network is shown in Figure 4.

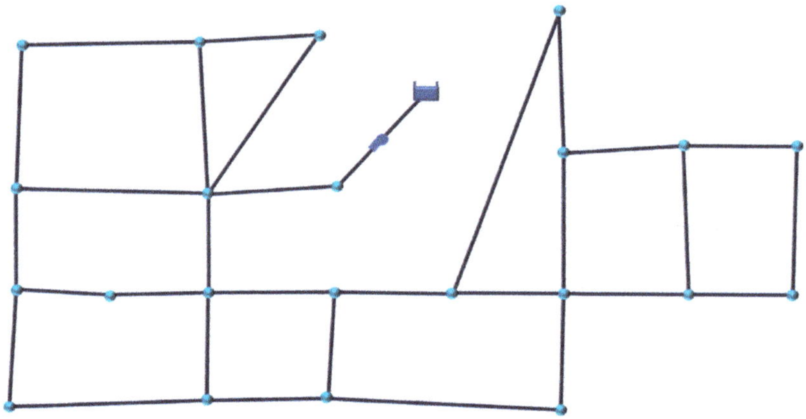

Figure 4. Layout of Goyang network.

The Goyang network comprises 1 reservoir, 25 nodes, and 30 pipes. The Hazen–Williams coefficient for calculating the head loss of the Goyang network was set to 100. The minimum water pressure required for each node in the Goyang network was 15 m. Table 4 lists the pipe diameter and cost per pipe diameter used for the optimal design of the Goyang network.

Table 4. Cost per unit length based on diameter.

Diameter (mm)	Cost (USD/m)	Diameter (mm)	Cost (USD/m)	Diameter (mm)	Cost (USD/m)	Diameter (mm)	Cost (USD/m)
80	37.890	125	40.563	200	47.624	300	62.109
100	38.933	150	42.554	250	54.125	350	71.524

3.2. Optimal WDS Design Using Improved LCEA Model

In this study, the proposed LCEA model was used to analyze the difference between energy- and cost-based optimal designs. There are a total of six parameters (CG, CGSR, DR_1, DR_2, CF, and AF) that need to be set for the MHVCA. Among the parameters of the MHVCA, candidate glasses (CG) are the storage space within the algorithm. The parameters of the MHVCA were selected through sensitivity analysis, and Table 5 shows the parameters of the MHVCA set to apply the MHVCA to the Goyang network.

Table 5. Values for each parameter of MHVCA.

Parameter	Value
CG	190
CGSR	0
DR_1	0.1
DR_2	0.7
CF	30
AG	45

The optimal design was performed using the MHVCA with parameters set according to Table 5. The optimal design was performed according to cost- and energy-based optimal designs, and the results showing the best value after 30 uses of each optimal design were analyzed. Table 6 shows the results of the optimal design of the Goyang network used 30 times using a cost-based design and energy-based design. Table 6 shows the results of the optimal design of the Goyang network used 30 times using a cost-based design and energy-based design.

Table 6. Comparison of cost and energy for each optimal design.

Analysis by Cost	Cost-Based Design	Energy-Based Design
Mean cost (USD)	177,064.779	177,492.181
Best cost (USD)	177,010.359	177,010.359
Worst cost (USD)	177,020.938	177,141.106
Standard deviation	21.172	130.392
Analysis by energy	**Cost-based design**	**Energy-based design**
Mean energy (GJ)	805.549	578.210
Best energy (GJ)	757.747	478.317
Worst energy (GJ)	788.369	535.783
Standard deviation	9.480	21.034

According to Table 6, in terms of cost, the best cost of the both cost-based design and energy-based design was the same at 177,010.359, which was the optimal value. However, the mean cost of the cost-based design was 177,020.938, and that of the energy-based design was 177,141.106, which was higher than that in the energy-based design. In terms of energy, the mean energy of the energy-based design was about 252.586 lower than that of the cost-based design, and the best energy was about 279.429 lower. The standard deviation was higher in the energy-based design than in the cost-based design, but based on the comparison of the best energy, worst energy, and mean energy, it can be seen that the energy-based design shows better results in terms of energy. Figure 5 shows the convergence curves according to the cost-based and energy-based optimal designs in terms of cost, and the convergence curves are expressed as the average of the results of 30 uses.

According to Figure 5, in the case of the cost-based design, it can be seen that it converges quickly and converges after about 2500 iterations. However, it can be seen that the energy-based design converges after about 6000 iterations. Figure 5 shows that the purpose of the optimal design is to minimize cost. Therefore, it can be seen that the cost-based design converges to a lower value than the energy-based design. Figure 6 shows the convergence curves according to the cost-based and energy-based optimal designs in terms of energy, and the convergence curves are expressed as the average of the results of 30 uses.

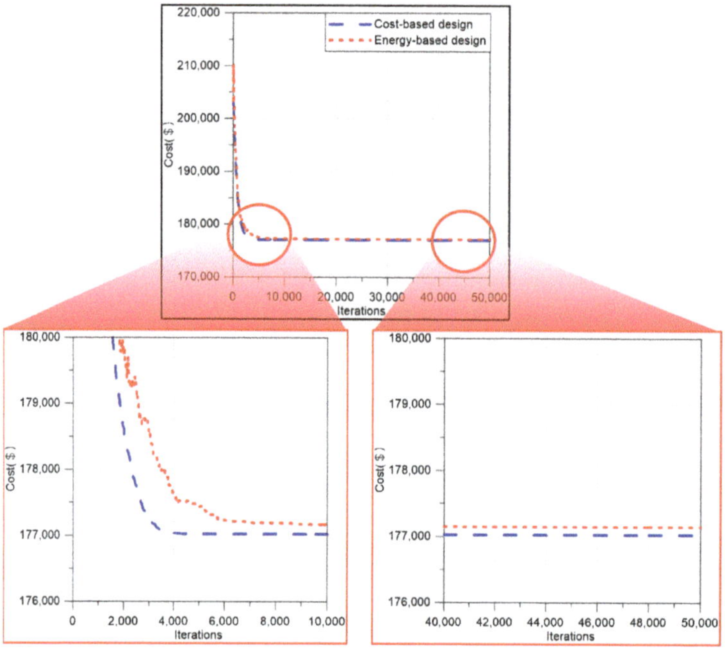

Figure 5. Convergence curves for optimal design in terms of cost.

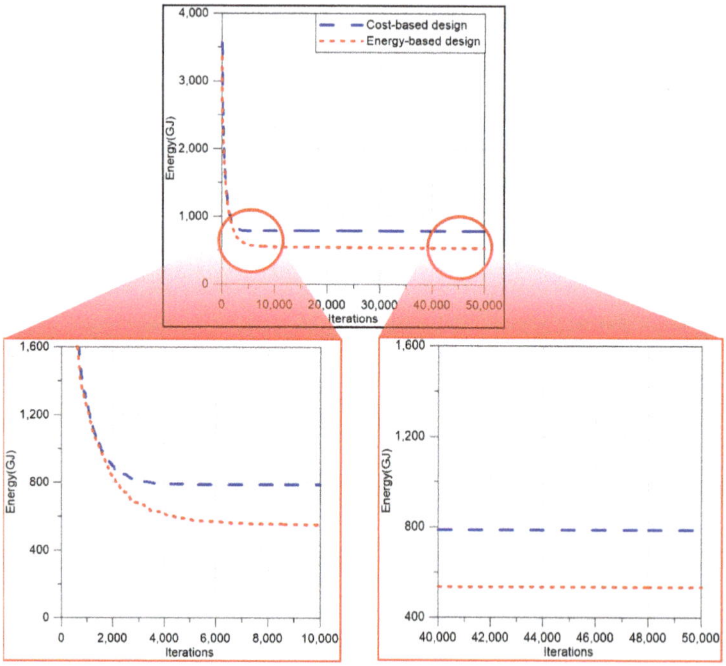

Figure 6. Convergence curves for optimal design in terms of energy.

According to Figure 6, in the case of the cost-based design, it can be seen that it converges quickly and then converges after about 4000 iterations. However, in the case of the energy-based design, it can be seen that it converges after about 8000 iterations, but it can be seen that it converges to a lower value than the cost-based design. Figure 5 shows that the purpose of the optimal design is to minimize energy. Therefore, it can be seen that the energy-based design converges to a lower value than the cost-based design. Table 7 shows the pipe diameters by the location of the design plan that showed the best results among the optimal design results of the cost-based design and energy-based design that were used to minimize energy.

Table 7. Pipe diameters by location of design plan.

Location	Pipe Diameter (mm) Cost-Based Optimal Design	Pipe Diameter (mm) Energy-Based Optimal Design	Location	Pipe Diameter (mm) Cost-Based Optimal Design	Pipe Diameter (mm) Energy-Based Optimal Design
1	47.624	47.624	16	37.89	37.89
2	40.563	42.554	17	37.89	37.89
3	40.563	38.933	18	37.89	37.89
4	38.933	38.933	19	37.89	37.89
5	37.89	37.89	20	37.89	37.89
6	37.89	37.89	21	37.89	37.89
7	37.89	37.89	22	37.89	37.89
8	37.89	37.89	23	37.89	37.89
9	37.89	37.89	24	37.89	37.89
10	37.89	37.89	25	37.89	37.89
11	37.89	37.89	26	37.89	37.89
12	37.89	37.89	27	37.89	37.89
13	37.89	37.89	28	37.89	37.89
14	37.89	37.89	29	37.89	37.89
15	37.89	37.89	30	37.89	37.89

According to Table 7, we can see that the design plans for the cost-based design and energy-design are different in locations 2 and 3. In the case of the cost-based design, the same pipe was used in locations 2 and 3 to minimize cost, and in the case of the energy-based design, different pipes were used in locations 2 and 3 to minimize energy.

Table 8 presents the results of the cost- and energy-based optimal designs using the MHVCA based on the WDS in the study area.

Table 8. Cost- and energy-based optimal design results using MHVCA.

	Cost-Based Optimal Design	Energy-Based Optimal Design
Total energy expenditure * (GJ)	805.549	494.769
Total energy expenditure per year (GJ/year)	33.565	20.615
Life cycle (year)	24	24
E_{fab} (GJ)	173.730	174.560
E_{reh} (GJ)	418.806	426.196
E_{rpi} (GJ)	24.623	0.000
E_{rep} (GJ)	173.730	174.560
E_{dis} (GJ)	23.195	23.310
E_{rec} (GJ)	8.535	303.858
Cost per year (USD/year)	7375.432	7377.704
Cost (USD)	177,010.359	177,064.903

Note(s): * Total energy expenditure is sum of all consumed energies ($E_{all} = E_{fab} + (E_{reh} + E_{rep} + E_{rpi} - E_{rec}) + E_{dis}$).

According to Table 8, the energy-based optimal design achieves approximately 39% more energy savings compared to the cost-based design, with a total energy expenditure difference of about 310.779 GJ. On an annual basis, the energy-based design leads to savings of approximately 24.949 GJ compared to the cost-based design throughout the life cycle. While the cost-based design exhibits lower fabrication, rehabilitation, repair, and disposal energies, the energy-based design excels in repair energy efficiency. Focused on minimizing energy consumption, the energy-based design results in lower repair energies during the maintenance stage. Moreover, the energy-based design outperforms in recycled energy during maintenance, with around 303.858 GJ compared to the cost-based design's approximately 8.535 GJ. The E_{fab} of the energy-based optimal design is 0.48% higher than that of the cost-based optimal design, and in the case of E_{dis}, it is about 0.49% higher. Unlike the cost-based optimal design, the energy-based optimal design is a design method for reducing CO_2 emissions. The energy-based optimal design considers the amount of carbon generated during maintenance stages such as rehabilitation and repair during the life cycle. Therefore, the energy-based optimal design derives an optimal design by selecting a pipe with a relatively large diameter. This is because pipes with a large diameter decrease less in aging over time than pipes with a small diameter. However, the energy-based optimal design has a lower E_{rep} than the cost-based optimal design. Based on the results of E_{rep}, it can be seen that the energy-based optimal design can use WDSs without replacement because it uses pipes with a relatively large diameter. In addition, the E_{rec} of the energy-based optimal design is about 35 times higher than that of the cost-based optimal design. It can be seen that the optimal design of the energy-based optimal design recovers more energy than the optimal design of the cost-based optimal design, and in terms of energy, the energy-based optimal design is better than the cost-based optimal design. E_{rpi} is lower in the energy-based optimal design than in the cost-based optimal design. The probability of pipe failure is calculated according to the diameter. The energy-based optimal design uses larger pipes than the cost-based optimal design, and this shows that the energy-based optimal design has a lower probability of pipe failure in areas where failure mainly occurs. E_{rep} is the same as E_{fab} for each design method. Erep is the energy generated when replacing a pipe. Therefore, if all individual pipes in the WDS reach the pipe replacement point, all pipes are replaced. Since the life cycle is longer than the pipe replacement point, all pipes are replaced, producing the same result as E_{fab}. Figure 7 shows the results of the cost- and energy-based optimal designs based on the consumed energy.

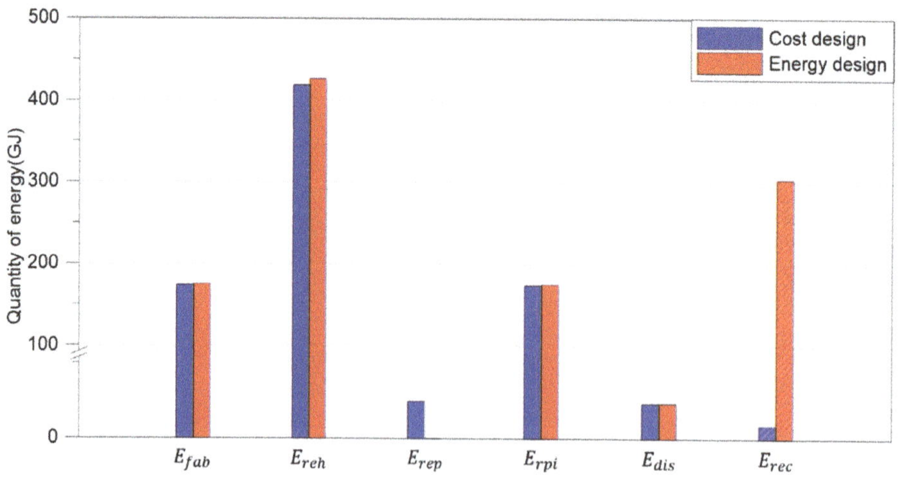

Figure 7. Energy consumption by optimal design methods.

It is evident that the energy-based design results in higher values for E_{fab}, E_{reh}, E_{rpi}, and E_{dis} compared to the cost-based design. However, the energy-based design exhibits a considerably lower E_{rep} value than that of the cost-based design. The E_{rec} plot (represented by the diagonal lines) in Figure 6 indicates that higher recycled energy values lead to improvements in energy efficiency gains. The energy-based design exhibits considerably higher E_{rec} values than those of the cost-based design. Moreover, Figure 7 shows that the energy-based design performs optimization based on various energy consumption values and, overall, is more effective in achieving energy savings compared to the cost-based design.

In this study, the proposed LCEA model was used to analyze the results of energy- and cost-based optimal designs for the Goyang network. The analysis revealed that the energy-based optimal design showed lower total energy expenditure and total energy expenditure per year than the cost-based optimal design. To conduct additional analysis, the benefits of reducing energy consumption were converted into costs. To convert energy savings into costs, all fuels used during the Goyang network's life cycle were assumed to be petroleum-based. A method of converting energy into kWh and subsequently into costs was employed, as suggested by the Korea Energy Economics Institute (KEEI), and the settlement unit prices for each fuel type provided by the Korean Statistical Information Service (KOSIS) were used [42,43]. According to the KEEI, 1 GJ is equivalent to 277.8236 kWh, and according to the KOSIS, the cost per 1 kWh is KRW 299.78. Additionally, recent exchange rates were used to convert KRW to USD. By applying this conversion method, the energy consumption results shown in Table 5 were analyzed based on the energy- and cost-based optimal designs. When implementing the energy-based optimal design for the Goyang network, approximately USD 20,526.22 was saved compared to the cost-based optimal design, with an annual saving of USD 855.26. However, the direct application of the results, obtained by converting energy into costs using the current exchange rate, was not entirely accurate, owing to potential fluctuations in the exchange rate.

Based on the energy-based and cost-based optimal designs, the new resilience and pipe failure scenarios proposed were established and applied to compare the performance of each design. In the failure simulation of the WDS, various failure scenarios, such as single-pipe failure and multiple-pipe failure, can be established [24]. According to Beker and Kansal, it was mentioned that when natural disasters such as earthquakes and landslides occur, multiple pipe failures rarely occur [44]. Jung et al. (2014) simulated single-pipe failure conditions after optimization, and Pagano et al. (2019) evaluated the impact of single-pipe failure in the water pipe network [45,46]. Therefore, in this study, resilience according to failure was calculated using failure scenarios for all pipes of the energy-based and cost-based optimal designs for the Goyang pipe network. Based on each design plan, PDA was performed using EPANET 2.2. To simulate PDA using EPANET 2.2, the minimum pressure was set to 0, required pressure was set to 15, which is the minimum required water pressure of the Goyang pipe network, and the pressure component was set to 0.5 [47,48].

Since the Goyang network used in this study is a network with no reservoir within the network, R_2 was assumed to be 1. In the case of water quality, R_3 was assumed to be 1, assuming that all factors subject to water quality testing are maintained above the standards set by the law [23]. Additionally, in the case of the population affected by a water outage, it was assumed that the same population lives at all nodes. The pipe failure scenario was constructed as a single-pipe failure scenario for the pipes existing in the WDS from the first pipe to the final pipe. Table 9 is a table presenting the resilience calculated according to the scenario based on each optimal design.

According to Table 9, the energy-based optimal design yielded high or equal resilience results in all but three scenarios. Upon calculating the average of resilience indices, it was determined to be approximately 0.8377 for the cost-based optimal design and around 0.8512 for the energy-based optimal design. The energy-based design showed a maximum increase of approximately 9.2176% in the resilience index compared to the cost-based design and an average increase of approximately 1.95%. Both the overall and average results indicate that the energy-based optimal design outperforms the cost-based optimal design in terms of disaster recovery. This suggests that the energy-based optimal design possesses a superior ability to restore the system to its original state in the event of a disaster compared to the cost-based optimal design.

Table 9. Comparison of resilience between cost-based and energy-based optimal designs by scenario.

Pipe Failure Scenario	Cost-Based Optimal Design (A)	Energy-Based Optimal Design (B)	Difference (%)	Pipe Failure Scenario	Cost-Based Optimal Design (A)	Energy-Based Optimal Design (B)	Difference (%)
1	0.0000	0.0000	0	16	1.0000	1.0000	0.0000
2	0.1490	0.1548	3.7628	17	1.0000	1.0000	0.0000
3	0.4975	0.5001	0.5263	18	1.0000	1.0000	0.0000
4	0.4807	0.4831	0.4893	19	0.9544	1.0000	4.5599
5	0.8055	0.8526	5.5220	20	0.9079	0.9543	4.8716
6	0.7996	0.8050	0.6776	21	1.0000	1.0000	0.0000
7	0.8941	0.8953	0.1262	22	0.8062	0.8083	0.2669
8	0.9041	0.9048	0.0852	23	0.8054	0.8116	0.7576
9	0.9078	1.0000	9.2176	24	0.9023	0.9063	0.4498
10	0.9543	1.0000	4.5744	25	0.9541	1.0000	4.5888
11	1.0000	1.0000	0.0000	26	0.8096	0.8118	0.2779
12	1.0000	1.0000	0.0000	27	0.8479	0.8496	0.1972
13	0.9077	0.9545	4.9062	28	1.0000	1.0000	0.0000
14	1.0000	1.0000	0.0000	29	0.8936	0.8947	0.1170
15	1.0000	1.0000	0.0000	30	0.9494	0.9500	0.0639

4. Discussion

In this study, sensitivity analysis was performed on each parameter of the MHVCA used. A total of six parameters were subjected to sensitivity analysis. Sensitivity analysis was conducted by setting initial values for CG, CGSR, DR_1, DR_2, CF, and AF based on previous studies and then applying various values to analyze the results and select values. Sensitivity analysis was performed 10 times in total with the goal of cost minimization, and the maximum cost, minimum cost, mean cost, and standard deviation were compared and analyzed to set the values. Tables 10–15 show the results of sensitivity analysis and selected values for each parameter.

Table 10. CG sensitivity analysis results and parameter value selection.

	100	110	120	130	140	150
Mean Cost (USD)	177,071.823	177,015.813	177,021.255	177,064.903	177,015.636	177,031.778
Best Cost (USD)	177,010.359	177,010.359	177,010.359	177,010.359	177,010.359	177,010.359
Worst Cost (USD)	177,294.363	177,064.903	177,064.903	177,026.697	177,063.124	177,064.903
Standard Deviation	78.212	16.363	21.793	24.957	15.830	26.253
Selection						
	160	**170**	**180**	**190**	**200**	-
Mean Cost (USD)	177,026.521	177,026.685	177,031.600	177,015.427	177,026.158	-
Best Cost (USD)	177,010.359	177,010.359	177,010.359	177,010.359	177,010.359	-
Worst Cost (USD)	177,064.903	177,064.779	177,064.903	177,061.038	177,064.903	-
Standard Deviation	24.692	24.938	26.033	15.204	24.149	-
Selection				O		-

Table 11. CGSR sensitivity analysis results and parameter value selection.

	0	0.1	0.2	0.3	0.4	0.5
Mean Cost (USD)	177,015.427	177,029.236	177,784.975	179,523.054	183,325.455	184,490.049
Best Cost (USD)	177,010.359	177,014.772	177,583.053	178,884.215	181,980.392	183,364.440
Worst Cost (USD)	177,061.038	177,064.903	178,028.545	180,156.294	184,345.112	185,891.822
Standard Deviation	15.204	22.111	134.937	396.245	707.003	877.799
Selection	O					
	0.6	**0.7**	**0.8**	**0.9**	**1.0**	**-**
Mean Cost (USD)	184,521.035	192,010.403	191,279.585	194,634.407	198,455.637	-
Best Cost (USD)	182,186.073	188,093.943	182,095.278	190,812.490	195,333.524	-
Worst Cost (USD)	187,807.926	194,201.503	194,767.095	197,291.409	201,015.790	-
Standard Deviation	1532.913	1933.878	3502.999	1826.033	1872.217	-
Selection						-

Table 12. DR_1 sensitivity analysis results and parameter value selection.

	0	0.1	0.2	0.3	0.4	0.5
Mean Cost (USD)	177,119.522	177,015.427	177,015.813	177,015.813	177,015.801	177,015.813
Best Cost (USD)	177,061.038	177,010.359	177,010.359	177,010.359	177,010.359	177,010.359
Worst Cost (USD)	177,216.720	177,061.038	177,064.903	177,064.903	177,064.779	177,064.903
Standard Deviation	57.427	15.204	16.363	16.363	16.326	16.363
Selection		O				
	0.6	**0.7**	**0.8**	**0.9**	**1.0**	**-**
Mean Cost (USD)	177,015.813	177,015.813	177,015.813	177,033.023	177,015.801	-
Best Cost (USD)	177,010.359	177,010.359	177,010.359	177,010.359	177,010.359	-
Worst Cost (USD)	177,064.903	177,064.903	177,064.903	177,236.998	177,064.779	-
Standard Deviation	16.363	16.363	16.363	67.992	16.326	-
Selection						-

Table 13. DR_2 sensitivity analysis results and parameter value selection.

	0	0.1	0.2	0.3	0.4	0.5
Mean Cost (USD)	177,199.296	177,039.121	177,043.797	177,042.471	177,036.864	177,031.753
Best Cost (USD)	177,014.772	177,010.359	177,010.359	177,010.359	177,010.359	177,010.359
Worst Cost (USD)	177,828.255	177,085.639	177,072.511	177,064.779	177,064.779	177,064.779
Standard Deviation	222.011	29.482	27.392	26.242	26.523	26.222
Selection						
	0.6	**0.7**	**0.8**	**0.9**	**1.0**	**-**
Mean Cost (USD)	177,020.881	177,015.427	177,026.697	177,015.801	177,021.255	-
Best Cost (USD)	177,010.359	177,010.359	177,010.359	177,010.359	177,010.359	-
Worst Cost (USD)	177,064.903	177,061.038	177,064.903	177,064.779	177,064.903	-
Standard Deviation	21.062	15.204	24.957	16.326	21.793	-
Selection		O				-

According to Tables 10–15, since the CGSR, DR_1, and DR_2 of the parameters of the MHVCA are probability parameters, sensitivity analysis was conducted in units of 0.1, and CG, CF, and AF were analyzed by creating about 11 to 13 cases depending on the range of parameters generally applied. When proceeding with the optimal design of the Goyang pipeline using the MHVCA, it can be seen that the best result was shown when CG = 190, CGSR = 0, DR_1 = 0.1, DR_2 = 0.7, CF = 30, and AF = 45. The MHVCA set with the values of the corresponding parameters was mostly the same as the MHVCA applying other parameters in terms of the minimum cost, but the maximum cost was lower, and it could be confirmed that this led to the lowest standard deviation.

Table 14. CF sensitivity analysis results and parameter value selection.

	0	10	20	30	40	50
Mean Cost (USD)	177,053.963	177,026.311	177,026.520	177,015.427	177,026.697	177,021.243
Best Cost (USD)	177,010.359	177,010.359	177,010.359	177,010.359	177,010.359	177,010.359
Worst Cost (USD)	177,286.754	177,064.779	177,064.779	177,061.038	177,064.903	177,064.779
Standard Deviation	81.170	24.386	24.689	15.204	24.957	21.768
Selection				O		
	60	**70**	**80**	**90**	**100**	**-**
Mean Cost (USD)	177,026.311	177,021.243	177,021.225	177,015.813	177,015.813	-
Best Cost (USD)	177,010.359	177,010.359	177,010.359	177,010.359	177,010.359	-
Worst Cost (USD)	177,064.779	177,064.779	177,068.339	177,064.903	177,064.903	-
Standard Deviation	24.386	21.768	21.793	16.363	16.363	-
Selection						-

Table 15. AF sensitivity analysis results and parameter value selection.

	0	15	30	45	60	75	90
Mean Cost (USD)	177,010.359	177,010.359	177,010.359	177,010.359	177,010.359	177,010.359	177,010.359
Best Cost (USD)	177,015.801	177,021.078	177,020.881	177,015.427	177,020.869	177,026.323	177,020.869
Worst Cost (USD)	177,064.779	177,064.779	177,064.903	177,061.038	177,064.779	177,064.903	177,064.779
Standard Deviation	16.326	21.440	21.062	15.204	21.036	24.406	21.036
Selection				O			
	105	**120**	**135**	**150**	**165**	**180**	**-**
Mean Cost (USD)	177,010.359	177,010.359	177,010.359	177,010.359	177,010.359	177,010.359	-
Best Cost (USD)	177,042.110	177,031.778	177,015.801	177,026.336	177,025.937	177,026.311	-
Worst Cost (USD)	177,064.903	177,064.903	177,064.779	177,064.903	177,064.779	177,064.779	-
Standard Deviation	25.958	26.253	16.326	24.425	23.815	24.386	-
Selection							-

5. Conclusions

In this study, an improved LCEA model was proposed to design an optimal WDS based on cost and energy, and performance was compared through the developed new resilience index. Using the specifications of the WDS under consideration and the regulations of the target watershed, a criterion was established to determine the Hazen–Williams coefficient, which serves as a basis for pipe renewal and replacement. The energy-based design has several advantages in that it aims to minimize energy consumption at all stages, including fabrication, maintenance, and disposal, due to the strength of the improved LCEA model. The energy-based design showed better performance than the cost-based design, especially in terms of energy recycling, which increased by approximately 36 times. Recycled energy, pointing to the energy benefits gained from the pipe rehabilitation and replacement process, had a positive impact on energy consumption. Additionally, the energy-based design demonstrated zero energy consumption for pipe repairs. As a result, the energy-based optimal design can consider a wide range of energy consumption aspects to achieve a more efficient design over the entire WDS life cycle.

To compare the performance of the proposed WDS design using the improved LCEA, an analysis was performed based on a new resilience index. As a result of analyzing the resilience index according to the scenario, it was confirmed that in most pipe failure scenarios, the energy-based optimal design showed better performance than the cost-based optimal design when pipe failure occurred. Based on the resilience index analysis results, it was found that the energy-based optimal design applied to a single-pipe failure scenario can respond to disaster more effectively than the cost-based optimal design when a pipe failure occurs.

The improved LCEA in this study relies on assumptions from prior research (e.g., energy consumption based on pipe diameter). To align better with real-world conditions, future studies should involve laboratory-scale experiments using the proposed equations. Further and extensive research can facilitate the practical application of the life cycle-based optimal design to actual drainage systems, allowing for the development of diverse optimal design techniques for various networks and objectives beyond energy.

Author Contributions: Y.m.R. and E.H.L. carried out the literature survey and drafted the manuscript. Y.m.R. worked on the subsequent draft of the manuscript. Y.m.R. performed the simulations. E.H.L. conceived the original idea of the proposed method. All authors have read and agreed to the published version of the manuscript.

Funding: This work was supported by Korea Environment Industry & Technology Institute (KEITI) through Technology development project to optimize planning, operation, and maintenance of urban flood control facilities Project, funded by Korea Ministry of Environment (MOE) (RS-2024-00398012).

Data Availability Statement: The data presented in this study are available on request from the corresponding author.

Conflicts of Interest: The authors declare no conflicts of interest.

References

1. Bolognesi, A.; Bragalli, C.; Marchi, A.; Artina, S. Genetic heritage evolution by stochastic transmission in the optimal design of water distribution networks. *Adv. Eng. Softw.* **2010**, *41*, 792–801. [CrossRef]
2. Diao, K.; Jung, D.; Farmani, R.; Fu, G.; Butler, D.; Lansey, K. Modular interdependency analysis for water distribution systems. *Water Res.* **2021**, *201*, 117320. [CrossRef] [PubMed]
3. Alperovits, E.; Shamir, U. Design of optimal water distribution systems. *Water Resour. Res.* **1977**, *13*, 885–900. [CrossRef]
4. Geem, Z.W. Optimal cost design of water distribution networks using harmony search. *Eng. Optim.* **2006**, *38*, 259–277. [CrossRef]
5. Shakhsi-Niaei, M.; Esfandarani, M.S. Multi-objective deterministic and robust models for selecting optimal pipe materials in water distribution system planning under cost, health, and environmental perspectives. *J. Clean. Prod.* **2019**, *207*, 951–960. [CrossRef]
6. Lansey, K.E.; Mays, L.W. Optimization model for water distribution system design. *J. Hydraul. Eng.* **1989**, *115*, 1401–1418. [CrossRef]
7. Reca, J.; Martínez, J. Genetic algorithms for the design of looped irrigation water distribution networks. *Water Resour. Res.* **2006**, *42*, W05416. [CrossRef]
8. Jung, D.; Kang, D.; Kim, J.H. Development of a hybrid harmony search for water distribution system design. *KSCE J. Civ. Eng.* **2018**, *22*, 1506–1514. [CrossRef]
9. Lee, E.H. Application of self-adaptive vision-correction algorithm for water-distribution problem. *KSCE J. Civ. Eng.* **2021**, *25*, 1106–1115. [CrossRef]
10. Ryu, Y.M.; Lee, E.H. Application of Modified Hybrid Vision Correction Algorithm for Water Distribution Systems in Civil Engineering. *KSCE J. Civ. Eng.* **2023**, *27*, 3617–3631. [CrossRef]
11. Lee, E.H.; Lee, H.M.; Yoo, D.G.; Kim, J.H. Application of a meta-heuristic optimization algorithm motivated by a vision correction procedure for civil engineering problems. *KSCE J. Civ. Eng.* **2018**, *22*, 2623–2636. [CrossRef]
12. Saldarriaga, J.; Páez, D.; Salcedo, C.; Cuero, P.; López, L.L.; León, N.; Celeita, D. A direct approach for the near-optimal design of water distribution networks based on power use. *Water* **2020**, *12*, 1037. [CrossRef]
13. Sangroula, U.; Han, K.H.; Koo, K.M.; Gnawali, K.; Yum, K.T. Optimization of water distribution networks using genetic algorithm based SOP–WDN program. *Water* **2022**, *14*, 851. [CrossRef]
14. Lee, S.; Yoo, D.G.; Jung, D.; Kim, J.H. Optimal life cycle design of water pipe system using genetic algorithm. *J. Korea Acad. -Ind. Coop. Soc.* **2015**, *16*, 4216–4227.
15. Risch, E.; Gutierrez, O.; Roux, P.; Boutin, C.; Corominas, L. Life cycle assessment of urban wastewater systems: Quantifying the relative contribution of sewer systems. *Water Res.* **2015**, *77*, 35–48. [CrossRef]
16. Shamir, U.; Howard, C.D. An analytic approach to scheduling pipe replacement. *J.-Am. Water Work. Assoc.* **1979**, *71*, 248–258. [CrossRef]
17. Sharp, W.W.; Walski, T.M. Predicting internal roughness in water mains. *J.-Am. Water Work. Assoc.* **1988**, *80*, 34–40. [CrossRef]
18. Mononobe, N. Hydraulics. *Iwanami* **1960**, 155–158.
19. Male, J.W.; Walski, T.M.; Slutsky, A.H. Analyzing water main replacement policies. *J. Water Resour. Plan. Manag.* **1990**, *116*, 362–374. [CrossRef]
20. Kleiner, Y.; Adams, B.J.; Rogers, J.S. Long-term planning methodology for water distribution system rehabilitation. *Water Resour. Res.* **1988**, *34*, 2039–2051. [CrossRef]
21. Filion, Y.R.; MacLean, H.L.; Karney, B.W. Life-cycle energy analysis of a water distribution system. *J. Infrastruct. Syst.* **2004**, *10*, 120–130. [CrossRef]

22. Todini, E. Looped water distribution networks design using a resilience index based heuristic approach. *Urban Water* **2000**, *2*, 115–122. [CrossRef]
23. Cimellaro, G.P.; Tinebra, A.; Renschler, C.; Fragiadakis, M. New resilience index for urban water distribution networks. *J. Struct. Eng.* **2016**, *142*, C4015014. [CrossRef]
24. Choi, Y.H.; Lee, H.M.; Yoo, D.G.; Kim, J.H. Optimal design of water supply system using multi-objective harmony search algorithm. *J. Korean Soc. Water Wastewater* **2015**, *29*, 293–303. [CrossRef]
25. Mandl, C.E. A survey of mathematical optimization models and algorithms for designing and extending irrigation and wastewater networks. *Water Resour. Res.* **1981**, *17*, 769–775. [CrossRef]
26. Monbaliu, J.; Jo, J.; Fraisse, C.W.; Vadas, R.G. Computer aided design of pipe networks. *Tech. Rev. -Mitsubishi Heavy Ind.* **1990**, *2*, 789–797.
27. Gessler, J.; Walski, T.M. *Water Distribution System Optimization (No. WES/TR/EL-85-11)*; Army Engineer Waterways Experiment Station Vicksburg MS Environmental Lab: Vicksburg, MA, USA, 1985.
28. Su, Y.C.; Mays, L.W.; Duan, N.; Lansey, K.E. Reliability-based optimization model for water distribution systems. *J. Hydraul. Eng.* **1987**, *113*, 1539–1556. [CrossRef]
29. Vasan, A.; Simonovic, S.P. Optimization of water distribution network design using differential evolution. *J. Water Resour. Plan. Manag.* **2010**, *136*, 279–287. [CrossRef]
30. Geem, Z.W.; Kim, J.H.; Loganathan, G.V. A new heuristic optimization algorithm: Harmony search. *Simulation* **2001**, *76*, 60–68. [CrossRef]
31. Maier, H.R.; Simpson, A.R.; Zecchin, A.C.; Foong, W.K.; Phang, K.Y.; Seah, H.Y.; Tan, C.L. Ant colony optimization for design of water distribution systems. *J. Water Resour. Plan. Manag.* **2003**, *129*, 200–209. [CrossRef]
32. Reca, J.; Martínez, J.; Gil, C.; Baños, R. Application of several meta-heuristic techniques to the optimization of real looped water distribution networks. *Water Resour. Manag.* **2008**, *22*, 1367–1379. [CrossRef]
33. Baek, C.W. Development of Optimal Decision-Making System for Rehabilitation of Water Distribution Systems Using ReHS. Master's Thesis, Korea University, Seoul, Republic of Korea, 2002.
34. Kim, J.H.; Kim, T.G.; Kim, J.H.; Yoon, Y.N. A study on the pipe network system design using non-linear programming. *J. Korean Water Resour. Assoc.* **1994**, *27*, 59–67.
35. Menon, K.K.; Narulkar, S.M. Application of heuristic-based algorithm in water distribution network design. *J. Water Resour. Pollut. Stud.* **2016**, *1*, 1–14.
36. U.S. Environmental Protection Agency (US EPA). *EPANET 2.0 User's Manual*; US EPA: Washington, DC, USA, 2000.
37. Jinesh Babu, K.S.; Mohan, S. Extended period simulation for pressure-deficient water distribution network. *J. Comput. Civ. Eng.* **2012**, *26*, 498–505. [CrossRef]
38. Mahmoud, H.A.; Savić, D.; Kapelan, Z. New pressure-driven approach for modeling water distribution networks. *J. Water Resour. Plan. Manag.* **2017**, *143*, 04017031. [CrossRef]
39. Hamed, M.M.; Elsayad, M.A.; Mahfouz, S.Y.; Khadr, W.M.H. Graphical user interface for water distribution network pressure-driven analysis using artificial elements. *Sustain. Water Resour. Manag.* **2022**, *8*, 89. [CrossRef]
40. Khadr, W.M.H.; Hamed, M.M.; Nashwan, M.S. Pressure Driven analysis of water distribution systems for preventing siphonic flow. *J. Hydro-Environ. Res.* **2022**, *44*, 102–109. [CrossRef]
41. Lee, S.Y. Research on Life Cycle Energy Analysis of Water Distribution System. Master's Thesis, Korea University, Seoul, Republic of Korea, 2014.
42. Korea Energy Economics Institute. Available online: https://www.keei.re.kr/main.nsf/index.html?open&p=/web_keei/change.nsf/UnitConverF&s=?OpenForm (accessed on 25 August 2023).
43. Korean Statistical Information Service. Available online: https://kosis.kr/statHtml-/statHtml.do?orgId=388&tblId=TX_38804_A010 (accessed on 25 August 2023).
44. Beker, B.A.; Kansal, M.L. Fuzzy logic-based integrated performance evaluation of a water distribution network. *AQUA—Water Infrastruct. Ecosyst. Soc.* **2022**, *71*, 490–506. [CrossRef]
45. Jung, D.; Kang, D.; Kim, J.H.; Lansey, K. Robustness-based design of water distribution systems. *J. Water Resour. Plan. Manag.* **2014**, *140*, 04014033. [CrossRef]
46. Pagano, A.; Sweetapple, C.; Farmani, R.; Giordano, R.; Butler, D. Water distribution networks resilience analysis: A comparison between graph theory-based approaches and global resilience analysis. *Water Resour. Manag.* **2019**, *33*, 2925–2940. [CrossRef]
47. Rossman, L.; Woo, H.; Tryby, M.; Shang, F.; Janke, R.; Haxton, T. *EPANET 2.2 User Manual*; Report EPA/600/r-20/133; Environmental Protection Agency: Washington, DC, USA, 2020.
48. Yoon, J.S.; Yoo, D.G.; Lee, H.M.; Kim, J.H. Optimal leakage detection model of water distribution systems using semi-pressure driven analysis and harmony search. *J. Korean Soc. Hazard Mitig.* **2012**, *12*, 23–32. [CrossRef]

water

MDPI

Article

Improvement of Removal Rates for Iron and Manganese in Groundwater Using Dual-Media Filters Filled with Manganese-Oxide-Coated Sand and Ceramic in Nepal

Ankit Man Shrestha [1], Shinobu Kazama [2], Benyapa Sawangjang [1] and Satoshi Takizawa [1,*]

[1] Department of Urban Engineering, Graduate School of Engineering, The University of Tokyo, 7-3-1 Hongo, Bunkyo-ku, Tokyo 113-8654, Japan; ankit@g.ecc.u-tokyo.ac.jp (A.M.S.); benyapa@env.t.u-tokyo.ac.jp (B.S.)

[2] Department of Socio-Cultural Environmental Studies, Graduate School of Frontier Sciences, The University of Tokyo, 5-1-5 Kashiwanoha, Kashiwa, Chiba 277-8563, Japan; kazama@edu.k.u-tokyo.ac.jp

* Correspondence: takizawa@env.t.u-tokyo.ac.jp

Abstract: Iron and manganese in groundwater impair the quality of drinking water; however, the rates of iron and manganese removal with conventional aeration and rapid sand filtration (RSF) processes vary extensively. Five full-scale aeration–RSF processes in Nepal also showed varying efficiencies of iron and manganese removal; while the iron concentration was below the national standard (0.30 mg/L) in 31 out of the 37 treated waters, the manganese concentration was higher than the standard (0.20 mg/L) in all of the treated waters. Re-aeration and stirring of the treated water did not oxidize soluble manganese, and this caused the poor removal rates for manganese. Bench-scale dual-media filters comprising anthracite on top of sand/ceramic layers with dosages of poly aluminum chloride and chlorine worked well by removing coagulated iron in the anthracite layer and then removing manganese in the sand/ceramic layers. A manganese-oxide-coated ceramic filter provided the highest manganese removal from 1.10 mg/L to <0.01 mg/L, followed by manganese-oxide-coated sand and quartz sand. Increasing the pH from 7.5 to 9.0 stabilized the manganese removal. Therefore, we propose a re-design of the present treatment processes and the selection of suitable filter media for better removal of iron and manganese.

Keywords: dual-media filtration; ceramic media; manganese sand; quartz sand; treatment efficiency

1. Introduction

Iron and manganese occur naturally in soils, rocks, and minerals. In aquifers, groundwater dissolves iron and manganese, giving rise to elevated concentrations thereof [1]. The iron and manganese concentrations in groundwater depend on several factors, such as the iron and manganese contents in rocks, the depth of the aquifer, hydraulic residence time, temperature, the porosity of the soil, rainfall, humidity, and the acidity of the soil and rocks [2].

Because of their common chemical characteristics, iron and manganese coexist in groundwater [3]. Iron and manganese in water cause an unpleasant taste, color, and increased turbidity. Iron and manganese stain laundry and plumbing fixtures at levels of >0.30 mg/L and >0.02 mg/L, respectively [4]. Dissolved iron can promote the growth of iron bacteria in groundwater and piped water [5]. Even at a concentration of 0.02 mg/L, manganese may be deposited in pipes, and it may slough off as a black precipitate [6]. Long-term exposure to high concentrations of manganese in groundwater has been linked to chronic poisoning, which poses a severe threat to human health [7]. A high concentration of manganese in water may lead to learning disabilities, deficits in intellectual function, compulsive behavior, and attention disorders in children [8].

Iron concentrations in groundwater are higher than 0.30 mg/L in many countries, including Bangladesh [3,9], China [7], Finland [10], India [3], Indonesia [11], Italy [12],

Nigeria [13], Slovakia [7] and Vietnam [3]. More than 50 countries supply drinking water containing manganese at concentration of >0.40 mg/L [14]; in South Asia, people's awareness of the occurrence and potential health effects of manganese in drinking water is low despite its frequent detection at elevated levels [8]. Manganese concentrations above 0.20 mg/L in drinking water have also been reported in Afghanistan [8], Bangladesh [9], China [7], Egypt [3], Finland [10], India [3], Indonesia [11], Italy [12], Korea [15], Myanmar [8], Nigeria [13], and Sri Lanka [8].

Considerable laboratory studies have been conducted to remove iron and manganese from drinking water; these include studies of membrane filtration [16], electrocoagulation [17,18], coagulation/flocculation [18], ion exchange [19], flotation [20], biological [21] and chemical oxidation (chlorine, hypochlorite [22], ozone [23], or potassium permanganate [22]), and filtration and adsorption [24]. However, application of the above methods in full-scale water treatment plants (WTPs) is limited due to the following limitations: electrocoagulation is effective only for colloidal contaminants and consumes energy [25], membrane fouling causes high operational costs and low permeate flux [26], and chemical coagulation requires large dosages of coagulants and produces sludge containing toxic elements. In addition, coagulation–flocculation followed by gravity sedimentation may be time-consuming [26]; ion exchange has challenges related to waste disposal, as resins can be fouled due to the precipitation of iron and manganese with dissolved oxygen [25]. Floatation requires high capital, maintenance, and operation costs [26]; the manganese concentration in effluents from biological oxidation filters fluctuates under complex operating conditions [25], and chemical oxidation has limitations such as difficulty in the storage and transport of oxidants, corrosion, the formation of solid manganese compounds that may interfere with system operations, and the formation of by-products [27]. The chemical oxidation of iron and manganese can be combined with aeration; it was reported that the oxidation of iron using aeration and chlorination used less chlorine than a system using chlorination only, which indicated that adopting aeration before chlorination reduced the amount of chlorine required for the oxidation of iron [28]. Dissolved oxygen and free chlorine readily oxidize iron in a pH range of 6.0–8.5, but manganese oxidation is very slow, while manganese removal through sorption and surface-catalyzed chlorine oxidation is effective [29].

Adsorption processes are widely used for metal removal due to the simplicity of operation, high removal efficiency, ease of recovery, and cost-effectiveness. Furthermore, adsorbents can be regenerated through desorption processes for multiple uses [24]. Iron and manganese oxides are well known as effective adsorbents for heavy metal removal due to their microporous structure and high surface area. In general, manganese oxides (MnO_2) have a higher adsorption capacity for manganese than iron oxides do [30]. To remove iron and manganese through adsorption, it is necessary to oxidize dissolved iron and manganese. In addition, metal oxides are used extensively as heterogeneous catalysts for oxidation reactions. Manganese has stable high-oxidation states, so it is more thermally and chemically stable than iron [31]. Manganese oxide is a catalyst for manganese oxidation and adsorbs the oxidized manganese in a filter bed packed with MnO_2-coated sand (manganese sand) [32]. Manganese ore has better manganese removal efficiency than that of quartz sand; however, manganese sand is the most stable and has the highest efficiency of manganese removal [33]. The adsorption process and the rate of manganese removal in water depend on several factors, including the pH, adsorbent dosage, adsorbent particle size, contact time, temperature, and manganese concentration [34].

At a full-scale WTP in Hingna, India, a cascade aerator and rapid sand filter (RSF) partially removed iron (0.08–1.51 mg/L) and manganese (0.23–1.70 mg/L) in raw water, reaching levels of 0.02–0.80 mg/L and 0.08–0.70 mg/L, respectively [22]. At a WTP in Mazandaran, Iran, tray aerators and RSF removed iron (0.93 ± 0.91 mg/L) and manganese (0.24 ± 0.10 mg/L) by 57.0% and 48.7%, respectively, whereas with additional pre-chlorine injection, the removal efficiencies moderately increased to 77.7% and 62.5%, respectively [35]. At the Koskuuslahde WTP in Ilmajoki, Finland, manganese (0.90 mg/L) in raw water was effectively removed through the addition of potassium permanganate to an aeration–RSF process; however, without potassium permanganate, the manganese reduction was only 5% on average, which improved to >91% with potassium permanganate [10]. At 26 WTPs in Korea, an average manganese removal rate of 93.5% (0.05–0.97 mg/L) was achieved with MnO_2-coated media [15]. Experimental results with a full-scale test filter at a WTP in Belgium showed that filters containing MnO_2-coated anthracite (MOCA) had a very high (>90%) manganese removal rate compared with filters without an MOCA layer (~10%) [36]. Aeration followed by RSF is a common method for drinking water treatment to remove iron and manganese [37], but due to a lack of comprehensive understanding regarding the efficiencies of iron and manganese removal using aeration–RSF, the effective removal of iron and manganese has not been achieved in different WTPs. Information gathered from over 100 WTPs in the Netherlands, Belgium, Germany, Jordan, and Serbia showed that manganese removal through aeration–RSF is complex and simultaneously influenced by several parameters of water quality, such as the pH, oxygen in the filtrate, and iron loading per filter run, as well as process parameters, such as the filtration rate, filter bed depth, and contact time [38]. Recently, researchers in the Netherlands reported that manganese removal rates varied extensively from merely 9% to 100% depending on the operational periods of dual-media RSFs using anthracite and sand, whereas the iron and ammonia removal rates were stable at 97–100% and 80–100%, respectively [39].

Groundwater in the southern part of Nepal, i.e., the Terai region, is abundant and, thus, a reliable and sustainable source of water [40]. In most of the Terai region, the depth of wells is approximately 150 m, with an average well yield of 25 L/s [41]. However, the quality of the groundwater as a source of drinking water is a major concern [42]. In three districts of the eastern Terai region, 107/175 samples and 88/175 samples of groundwater had iron and manganese concentrations exceeding the National Drinking Water Quality Standards of Nepal (NDWQS), which are 0.30 mg/L and 0.20 mg/L, respectively [43]. The existing groundwater treatment systems installed for iron and manganese removal are aeration–RSF systems because of their low investment cost and simple operation [37]. Although there are some methods that have higher removal rates for iron and manganese than those of the aeration–RSF processes mentioned above, these processes are not employed in Nepal and other developing countries due to their high investment costs, requirements for chemicals, and difficulty of operation and maintenance. While iron oxidation is fast in simple aeration systems [44], manganese oxidation and removal in aeration–RSF treatment plants are slow and influenced by several parameters, including the water quality and process design [38]. Therefore, this study aimed to evaluate the removal efficiencies for iron and manganese in the existing aeration–RSF plants in the Terai region and to evaluate the removal efficiency of MnO_2-coated media as an appropriate method for the removal of iron and manganese in the study area. This study compared the removal efficiencies for iron and manganese using dual-media filters filled with anthracite on the top and three different filter media (quartz sand, manganese sand, and MnO_2-coated ceramics (ferrolite) with pre-chlorination and coagulation under various conditions.

2. Materials and Methods

2.1. Study Area

The study area was the urban municipalities in the Sunsari district, one of 14 districts in the Koshi province in southern Terai, Nepal [45]. The district has a total area of 1275 km^2 and is located in the southeastern part of Nepal (Figure 1) [46]. This district is the second most populous district in the Koshi province [45], with a total population of 926,962 and an annual population growth rate of 1.86% as per the census in 2021 [47]. The study area is characterized by a subtropical monsoon climate, and the average monthly temperature ranges from 9 to 40 °C. The mean annual precipitation is 1794 mm, and 85% of the precipitation occurs in the rainy season from June to September [46]. The district consists of 12 municipalities, 6 urban municipalities (2 sub-metropolitan cities and 4 municipalities), and 6 rural municipalities [48] based on the constitutional provision of Nepal [49].

Most of the aquifers in the study area are recharged by rivers and precipitation at higher altitudes [50].

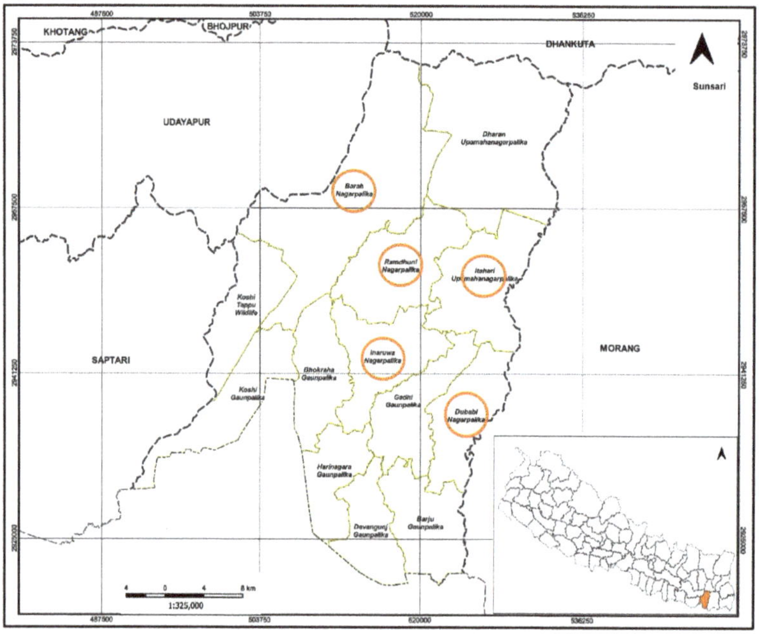

Figure 1. Urban municipalities in the Sunsari district selected for this study. The red area on the index map indicates the Sunsari district in Nepal (upamahanagarpalika = sub-metropolitan city, nagarpalika = municipality, gaupalika = rural municipality).

2.2. Water Supply Projects in the Study Area

The water supply projects (WSPs) in the study area are operated by water users and sanitation associations (WUSAs); one of the main water service providers is managed by the communities, which are responsible for the planning, construction, operation, and management of water supply systems with support from the central/provincial/municipal government [51]. Members of WUSAs are mostly elected by the community, and they do not require any qualifications or prior experience with water supply systems [52]. In the study area, there are 119 WSPs operated by WUSAs [53]. Out of the 119 WSPs, 14 (out of 23 WSPs) in Baraha municipality and all (81) WSPs in the sub-metropolitan city of Dharan (Table A1) use surface water as the main water source; most of these projects depend on the tributaries of the Sardhukhola River for their water supply [54]. Only 22 WSPs in 5 urban municipalities, namely, Baraha, Duhabi, Inaruwa, Itahari, and Ramdhuni, abstract groundwater from deep tube wells as their main source of water; of these, 12 WSPs have water treatment plants [53]. One WSP with a groundwater treatment plant was selected from each of the five urban municipalities (Figure 1), totaling five WSPs, namely, Duhabi, Inaruwa, Jhumka, Mahendranagar, and Pakali, with a total of sixteen tube wells (Table A1). The depth of the wells in the study area ranges from 80 to 125 m.

2.3. Iron and Manganese Removal Processes in the Study Area

Among the five WSPs, Duhabi, Inaruwa, and Pakali have two treatment plants that treat raw water from different groundwater sources (Figure 2a), whereas the other two WSPs, Jhumka and Mahendranagar, have only one treatment plant that treats raw water from one or more groundwater sources in time-shifted modes (Figure 2b). Different numbers of tube wells are connected to the water treatment plants in different WSPs.

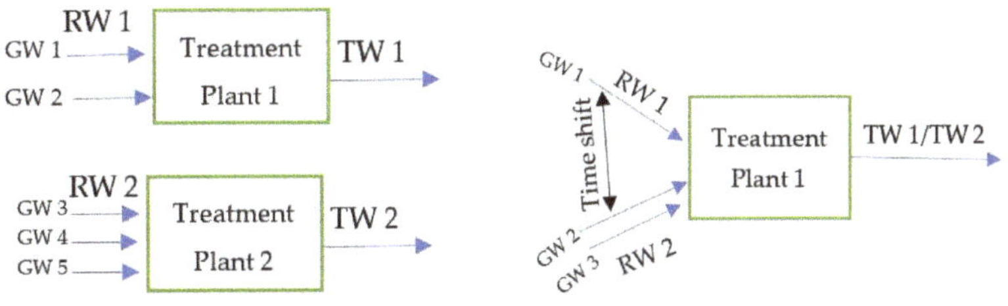

(**a**) Duhabi, Inaruwa, and Pakali (**b**) Jhumka and Mahendranagar

WSP	Duhabi	Inaruwa	Pakali	Jhumka	Mahendranagar
TP#/RW 1	TP1/GW 1, 2	TP1/GW 1, 2	TP1/GW 1	TP1/GW 1	TP1/GW 1
TP#/RW 2	TP2/GW 3, 4, 5	TP2/GW 3, 4	TP2/GW 2	TP1/GW 2	TP1/GW 2, 3

RW: raw water; TW: treated water; GW: groundwater; TP: treatment plant

Figure 2. Treatment processes of water supply projects: (**a**) two treatment plants treat raw water from different groundwater sources; (**b**) one treatment plant treats raw water from different groundwater sources in a time-shifted mode.

These water treatment plants have aeration–RSF processes for iron and manganese removal (Figure 3a). Both TPs in Duhabi and the TP in Mahendranagar have been operated since 2017; the RSFs are backwashed twice per week in both WSPs. Both TPs in Pakali and TP2 in Inaruwa have been operated since 2021; the RSFs are backwashed daily. The TP in Jhumka and TP1 in Inaruwa have been operated since 2006 and 2000, respectively; the filter media (including MnO_2-coated pebbles in the aeration unit of Inaruwa) are replaced every 5 years, and the filter media and pebbles were replaced in 2020; RSFs are backwashed daily in these TPs.

(**a**) (**b**) (**c**)

Figure 3. Groundwater treatment system: (**a**) two sets of aeration–RSF units; (**b**) pall rings in aeration units; (**c**) MnO_2-coated pebbles in aeration units.

2.3.1. Aeration Unit

Air is supplied to the raw water in the aeration tanks with air compressors. To enhance the aeration and oxidation of iron and manganese, two types of aeration media—pall rings (Figure 3b) or MnO_2-coated pebbles (Figure 3c)—are used in the different water supply projects (Table 1).

Table 1. Aeration and filtration media used in the groundwater treatment plants of the WSPs.

WSP	Duhabi TP1/TP2	Inaruwa		Jhumka TP1	Mahendranagar TP1	Pakali TP1/TP2
		TP1	TP2			
Aeration Media	Not in operation	MnO_2-coated pebbles	Pall rings	Pall rings	MnO_2-coated pebbles	Pall rings
Filtration Media	Manganese sand + Quartz sand	Quartz sand	Quartz sand	Quartz sand	Quartz sand	Quartz sand

Note: TP: Treatment plant.

2.3.2. Filtration Unit

In all WSPs, the filters are filled with quartz sand (effective size (d_{10}) = 0.45–0.60 mm, uniformity coefficient d_{60}/d_{10} = 1.3–1.7) at a thickness of 600–750 mm. At the Duhabi WSP, a dual-medium filter with manganese sand (250 mm) on top of a quartz sand layer (400 mm) has been installed.

2.4. Field Survey

Field surveys were conducted three times during the following periods:

1. 23 February–13 March 2023 (dry season);
2. 8 August–26 September 2023 (rainy season);
3. 27 February–15 March 2024 (dry season).

In the first field survey, one sample was collected from each of the WSPs in the study area; the WSPs with high iron and/or manganese contents in the treated water were selected to evaluate the treatment efficiency for iron and manganese. In the second field survey, triplicate samples of raw water, i.e., groundwater from deep tube wells, and treated water from each TP of the five selected WSPs were collected for analyses in the field on the same day. No rainfall was recorded from December of 2022 to 15 March 2023 in the study area, while an average precipitation of 1384.8 mm was recorded at three rainfall stations near the study area in the rainy season (June–September 2023) [55]. In the third field survey, the iron and manganese removal efficiencies were compared using bench-top experimental units filled with three different filter media with pre-chlorination and coagulation under different operating conditions.

2.5. Iron and Manganese Oxidation through the Re-Aeration and Stirring of Treated Water

Stirring water in the presence of suspended reactant particles can promote the oxidation of iron and manganese by increasing the frequency of collision, leading to oxidation reactions [56]. Therefore, to examine the effects of stirring for the oxidation of iron and manganese, bench-top experiments were performed using the devices shown in Figure A1. In the stirring experiments, treated water containing iron and manganese was stirred in a 500 mL vessel at 1500 rpm for up to 60 min; the pH, dissolved oxygen, ferrous iron, and soluble manganese were measured every 15 min.

The efficiency of iron and manganese removal may be affected by low concentrations of dissolved oxygen [38] and short detention times; thus, saturated dissolved oxygen in water may facilitate the better oxidation of iron and manganese with an extended aeration time. In the aeration experiments, 1 L of treated water containing iron and manganese was aerated with an air pump for up to 90 min; the pH, dissolved oxygen, ferrous iron, and soluble manganese were measured every 15 min.

2.6. Filtration Experiment

To conduct bench-scale filtration experiments, raw water was transported from the Jhumka and Inaruwa WSPs to the laboratory at the Urban Water Supply and Sanitation Project, Regional Project Management Office, Itahari, Sunsari, which is 8 and 15 km away from Jhumka and Inaruwa, respectively. In the bench-scale filtration experiments for iron and manganese removal, we compared three different filter media: quartz sand, manganese sand, and MnO_2-coated ceramics (ferrolite, Tohkemy Corporation, Osaka, Japan). The three filtration columns were made of acrylic pipes with a diameter of 30 mm and a height of 1 m (Figure 4a). The bottom sections of the columns were filled with gravel (diameter 2–12 mm) to a thickness of up to 150 mm (Figure 4b). Above the gravel layers, three different filter media, namely, quartz sand (effective size = 0.6 mm, density = 2.63–2.64 g/cm^3, uniformity coefficient d_{60}/d_{10} = 1.4–1.5), manganese sand (effective size = 0.6 mm, density = 2.58–2.65 g/cm^3, uniformity coefficient d_{60}/d_{10} = 1.4–1.5, MnO_2 content >0.5 mg/gm), and ferrolite (effective size = 0.3 mm, apparent density = 0.8–1.0 gm/cm^3, MnO_2 content >3.4 mg/gm), were filled to a thickness of 400 mm in Column 1, Column 2, and Column 3, respectively. Anthracite (effective size = 1.2 mm) was filled to a thickness of 200 mm above the quartz sand and manganese sand in Columns 1 and 2, respectively, whereas fine anthracite (effective size = 0.7 mm) was filled to a thickness of 200 mm in Column 3. The remaining 250 mm of space was used for bed expansion during the backwashing of the filter media after each filtration experiment.

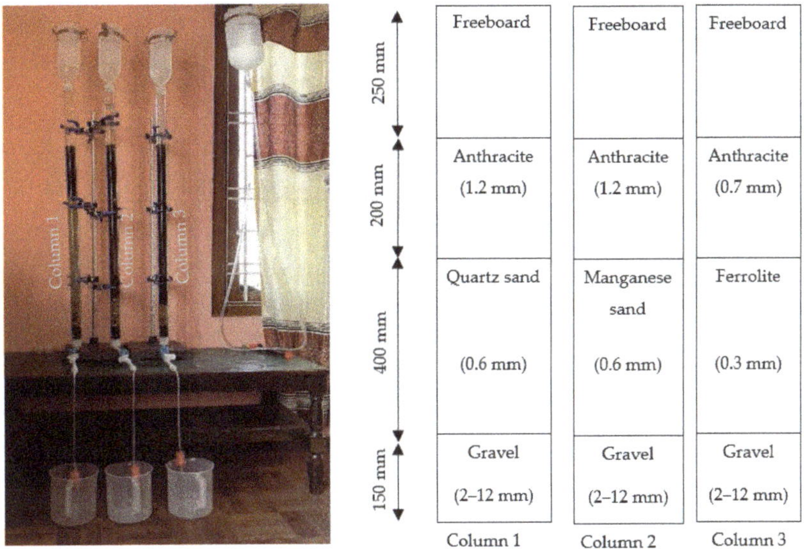

(a) (b)

Figure 4. Filtration experiments using different filter media: (**a**) assembly of the filtration equipment; (**b**) thickness of different filter media (numbers in () show the effective sizes of the media).

Dual-media filtration was performed to retain oxidized iron flocs in the anthracite layer and to maintain the high reactivity of the quartz sand, manganese sand, or ferrolite layer for manganese removal. A comparatively fine size of anthracite (0.7 mm) was used in Column 3; this depended on the small size of ferrolite (0.3 mm) used in the lower layer, which was expected to trap fine particles and improve the filtration efficiency in comparison with Columns 1 and 2.

To activate the filter media, they were soaked in water containing 200 mg/L of chlorine for 1 h; this was prepared using sodium hypochlorite (NaClO). After washing the media with tap water, the filter media were put into the filtration columns. Before conducting the filtration experiments, the filter media were cleaned by filtering clean water containing chlorine at 2.0–3.0 mg/L for 3 h. Then, the filter media were backwashed with clean water with about 30% of bed expansion at average flow rates of 0.388 L/h, 0.380 L/h, and 0.200 L/h for Columns 1, 2, and 3, respectively. After backwashing, raw water containing iron and manganese was filtered through the filtration columns at a filtration rate, i.e., linear velocity, of 5 m/h.

Chlorine was injected into raw water for the chemical and catalytic oxidation (MnO_2 as a catalyst) of iron and manganese. The chlorine dosage in the raw water before filtration was based on the stoichiometric dose to oxidize iron and manganese, namely, 0.62 mg/L of chlorine per mg/L of iron and 1.29 mg/L of chlorine per mg/L of manganese plus chlorine consumption by the filter media, which was measured by filtering water containing 2–3 mg/L of chlorine added to the 0.50 mg/L of residual chlorine in the treated water. Every 30 min, raw water and treated water were collected for water quality analyses. Sodium hydroxide was added in the experiments to study the effect of pH on manganese, as it was reported that a high pH enhances manganese removal in MnO_2-coated filter media [30] and the chlorine oxidation of manganese [28].

Poly aluminum chloride (PACl) solution was added to the raw water to coagulate suspended iron oxide if the total iron concentration was above 0.30 mg/L in the treated water. To determine the effects of the PACl dose, the raw water was filtered through a filter paper (pore size: 5 μm) after being mixed with PACl at increasing dosages of 5 mg/L and 10 mg/L until the total iron in the filtered water was below 0.10 mg/L.

2.7. Water Quality Analysis

The water quality was analyzed through 10 physicochemical parameters. Temperature, electrical conductivity, total dissolved solids (TDSs), and pH were measured with LAQUAtwin (HORIBA Advanced Techno Co., Ltd., Kyoto, Japan); turbidity was measured with a Eutech™ TN-100 Turbidimeter (Gul Circle, Singapore); dissolved oxygen (DO) was measured with a D-210PD-S LAQUA (HORIBA Advanced Techno Co., Ltd., Kyoto, Japan); oxidation–reduction potential (ORP) was measured with portable HMP6 m (HACH LANGE®, Loveland, CO, USA) using a Ag/AgCl reference electrode; then, the measured ORP values were converted into Eh, i.e., the ORP of the standard hydrogen electrode (SHE), by adding C (electrode potential across the reference electrode relative to the SHE) using the formula $C = -0.828T + 229$, where T is the water temperature in °C [57].

Soluble manganese, chlorine, and ammonia were analyzed using a portable colorimeter (DR 900, HACH LANGE®, Loveland, CO, USA) and powdered reagents according to the USEPA periodate oxidation method, USEPA DPD method, and salicylate method, respectively. In the first and third field surveys, ferrous and total iron were analyzed using HACH DR 900 (HACH LANGE®, Loveland, CO, USA) and powdered reagents according to the phenanthroline method and USEPA Ferro Ver® method, respectively, whereas in the second field survey, both ferrous and total iron were analyzed with an HI 83399 (Hannah Instruments, Cluj-Napoca, Romania) and powdered reagents according to the phenanthroline method. The results of the water quality analysis were compared with those of NDWQS [58] to ascertain the treatment efficiency.

2.8. Data Analysis

A paired t-test was performed using R (version 4.3) to compare the ferrous iron, total iron, and soluble manganese in groundwater in the dry and rainy seasons. The results of the statistical tests were considered significant at $p < 0.05$.

3. Results

3.1. Physiochemical Water Quality of Raw and Treated Water

Table A2 shows the physiochemical water quality parameters collected in the first field survey (23 February–13 March 2023, dry season). The treated water of Duhabi was not measured, as the aeration unit was not in operation during the survey period. The groundwater temperature was 24–29 °C, and the treated water temperature was 25–28 °C. The groundwater pH was 6.66–7.85, which was within the range of the pH standard (6.50–8.50). The groundwater pH was comparatively high (>7.32) in Duhabi but comparatively low in Mahendranagar (<6.80); at low pH, the physiochemical removal of manganese may be difficult in aeration–RSF treatment plants [38]. After treatment, the pH increased in all WSPs, except in Inaruwa, which was possibly because of the decarbonation through aeration.

The groundwater in Inaruwa contained a high ferrous iron concentration (>3.28 mg/L), and the ferrous iron in the raw water was oxidized and removed below the detection limit (BDL, <0.02 mg/L); the oxidation of ferrous iron follows a stoichiometric equation ($4Fe^{2+} + O_2(aq) + 10H_2O \rightarrow 4Fe(OH)_3 + 8H^+$) [59], thus increasing the proton concentration [60] and decreasing the pH in Inaruwa [61].

The turbidity of all groundwater was lower than the standard of 5 NTU, except in Pakali (11.7 NTU), which was possibly because the sample was taken just after starting the water pump; however, it was reduced below the standard of 5 NTU after the treatment. The TDSs in the groundwater of Duhabi were high (215–244 mg/L), but they were low in Inaruwa (111–137 mg/L); however, the TDSs in all groundwater was lower than the standard of 1000 mg/L. TDSs in groundwater may vary naturally due to the dissolution of minerals contained in aquifers' rocks and soils [60]. The TDSs in the treated water (128–203 mg/L) remained almost unchanged with respect to those in raw water.

The ORP (Eh) in the groundwater was in a range of −71 to 80 mV (Eh: 0.14–0.29 V); half (8/16) of the groundwater had negative ORP values (Eh < 0.20 V), which indicated a reduction in the environment of groundwater that prevented the oxidation of ferrous iron and soluble manganese [59]. In Inaruwa, all groundwater had negative ORP values of −71 to −11 mV (Eh: 0.14–0.20 V), and the ferrous iron concentration was high (2.08–4.32 mg/L), whereas in Jhumka, Mahendranagar, and Pakali, groundwater with positive ORP values (Eh: > 0.21 V) had low ferrous iron concentrations. In all of the treated water, Eh increased after aeration; in Inaruwa, Jhumka, and Mahendranagar, it increased to 0.37–0.41 V, 0.41–0.43 V, and 0.41 V, respectively, while in Pakali, Eh slightly increased to 0.25–0.30 V. A high ORP above 100 mV (Eh: 0.33 V) is characterized by the presence of oxygen in water [62], so the aeration systems in Inaruwa, Jhumka, and Mahendranagar had high aeration efficiency, while that in Pakali had low aeration efficiency.

The ferrous iron concentrations in groundwater were in a range from BDL (<0.02 mg/L) to 4.22 mg/L, and the total iron concentration was 0.05–4.54 mg/L. In Duhabi, all groundwater had a total iron concentration below the standard of 0.30 mg/L, except for one groundwater sample (0.44 mg/L). In Inaruwa, Jhumka, and Mahendranagar, all groundwater had a total iron concentration higher than 0.30 mg/L, while in Pakali, one groundwater sample had a total iron concentration of 0.24 mg/L, and another had a concentration of 1.3 mg/L.

In all treated water, the ferrous iron concentration was <0.30 mg/L, except for one treated water sample in Jhumka (0.87 mg/L). Although aeration did not work in Duhabi and the DO was low (1.48–2.39 mg/L), the ferrous iron concentration decreased from <0.15 mg/L in the raw water to <0.03 mg/L in the treated water. Due to the presence of ferric iron, the total iron concentration in the treated water of Inaruwa was above 0.30 mg/L; the high ferric iron concentration in the treated water indicated poor filtration efficiency. The soluble manganese concentration in the groundwater was 0.80–1.40 mg/L; all of the groundwater failed to meet the national standard for manganese (0.20 mg/L). The soluble manganese concentration was lowered to 0.40 mg/L in the treated water in Jhumka, while all other treated water had high soluble manganese concentration (0.50–1.10 mg/L), indicating inefficiency in the removal of soluble manganese.

Figure 5a shows the physiochemical water quality parameters (temperature, turbidity, and TDSs) collected in the second field survey (8 August–26 September 2023, rainy season). The temperature of the raw water was 26.4–30.1 °C, and it was 26.5–32.0 °C for the treated water. Compared with that in the dry season, the water temperature was slightly higher due to the higher ambient temperature in the rainy season [55]. The turbidity of all raw water (0.24–4.75 NTU) was within the range of the national standard of 5 NTU. In the first survey, one of the groundwater samples in Pakali had high turbidity, as it was collected just after starting the pump; however, in the second survey, the sample collected during the continuous operation of the pump had low turbidity (3.79–4.52 NTU). The turbidity in all of the treated water decreased to 0.03–2.00 NTU. The TDSs in the raw water varied from 95.5 to 225.0 mg/L; similarly to the dry season, TDSs were high in Duhabi (198–200 mg/L) and low in Inaruwa (95.5–155 mg/L). However, the TDSs were within the range of the national standard of 1000 mg/L. The TDSs in the treated water (109–216 mg/L) remained almost unchanged.

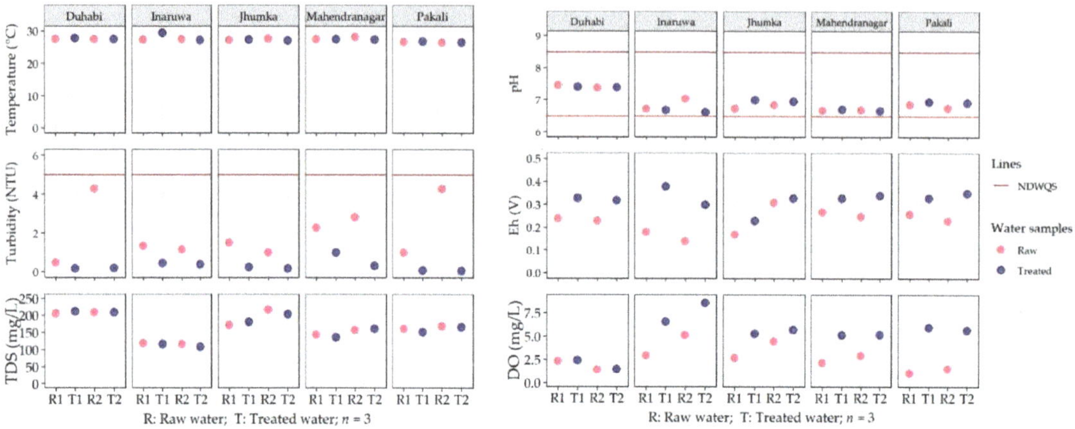

(a) (b)

Figure 5. Average water quality parameters in raw and treated water: (**a**) temperature, turbidity, and TDSs; (**b**) pH, Eh, and DO (sampling date: 8 August–26 September 2023).

Figure 5b shows the physiochemical water quality parameters (pH, Eh, and DO) in the rainy season. The pH in raw water (pH 6.57–7.58) was within the range of the national standard of 6.50–8.50. Similarly to the dry season, the pH in the groundwater was comparatively high (>7.27) in Duhabi and low (<6.79) in Mahendranagar, making it difficult to remove soluble manganese in aeration–RSF treatment plants [38]. In most of the treated water, the pH increased after the treatment; however, in some of the treated water, the pH was decreased. As mentioned before, decarbonation through aeration increases pH [61], whereas the oxidation of ferrous iron decreases pH. The Eh values of the raw water at different WSPs varied from 0.14 to 0.38 V and increased in the treated water. Similarly to the dry season, raw water with low Eh values (0.11–0.91 V) in Inaruwa had high ferrous iron concentrations (3.22–4.31 mg/L), whereas in Jhumka, Mahendranagar, and Pakali, groundwater with higher Eh values had lower ferrous iron concentrations.

The dissolved oxygen (DO) in the raw water was 0.12–5.50 mg/L. The Eh value in the RW 2 of Inaruwa was the lowest, while the DO was the highest among all of the raw water samples; the high DO values may have been due to the recharging of the aquifer with water containing high DO values in the rainy season, whereas the lower Eh values may have been due to the presence of reducing species in groundwater; many studies have reported that reducing species can coexist with DO at disequilibrium due to slow redox equilibrium processes [63]. After treatment, the DO values increased in all of the treated water (>4.33 mg/L), except for that of Duhabi due to the breakdown of the aeration system, with a maximum DO value of 8.83 mg/L in Inaruwa.

The total iron concentration of the raw water was 0.03–4.49 mg/L (Figure 6a). Ferrous iron comprised most (83.35%) of the total iron, while ferric iron was a minor component (16.65%). The total iron concentration was above the national standard of 0.30 mg/L in 7 out of the 10 raw water samples; however, in Duhabi, the total iron concentration in all raw water was below the standard of 0.30 mg/L, which was possibly due to unique geology in the southernmost location in the Sunsari district (Figure 1).

The soluble manganese concentration in the raw water at different WSPs varied from 1.40 to 1.77 mg/L (Figure 6b), which was higher than the national standard of 0.20 mg/L. The efficiency of soluble manganese removal was low (8.28–27.56%); as a result, the soluble manganese concentration in the treated water was 1.07–1.37 mg/L. Due to the high concentration of soluble manganese in treated water, a black precipitate was found in piped water supplied to the community (Figure A4). As per the field visit and information from the WUSA members, the black precipitate of manganese was mostly found at the ends of the water supply networks.

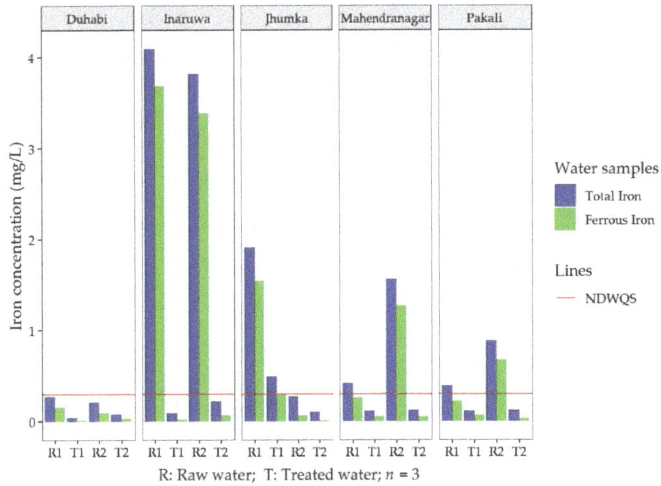

(**a**)

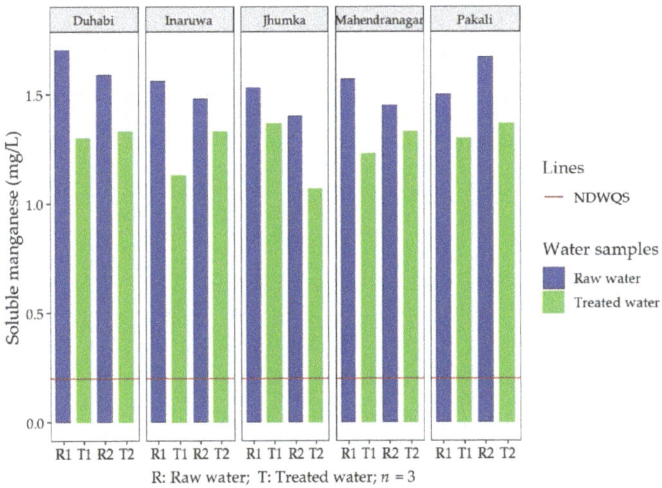

(**b**)

Figure 6. Water quality in raw and treated water: (**a**) average ferrous iron and total iron concentrations; (**b**) average soluble manganese concentrations (n = 3 for all samples; sampling date: 8 August–26 September 2023).

The soluble manganese removal rates in Jhumka [TW 2 (23.6%) > TW 1 (10.5%)] and Mahendranagar [TW 1 (21.7%) > TW 2 (8.3%)] were higher for raw water that had low iron concentrations (Jhumka [RW 2 (0.27 mg/L) < RW 1 (1.91 mg/L)] and Mahendranagar [RW 1 (0.42 mg/L) < RW 2 (1.56 mg/L)]). Similarly, in Inaruwa, because both raw water samples had high iron concentrations (>3.80 mg/L), the iron removal was efficient, but the manganese removal rate was low. The manganese removal in TP-1 (27.6%) was higher than that in TP-2 (10.1%), which was possibly due to the MnO_2-coated pebbles in the aeration unit of TP-1 [64]. In Duhabi, even though the aeration system was not in operation due to dysfunction, the manganese removal rates were 16.4–23.5% because of the low iron concentration in the raw water (<0.27 mg/L), the higher pH of the raw water (pH 7.26–7.56) in comparison with the other WSPs (pH 6.65–7.12), and the manganese sand layer on top of the quartz sand [10,34].

Figure 7 shows a comparison of iron and manganese in the rainy and dry seasons. The mean ferrous iron concentration in the rainy season (1.25 mg/L) was lower than that in the dry season (1.42 mg/L) (Figure 7); however, the difference of 0.17 mg/L was not significant (paired *t*-test: $p > 0.05$) due to the large variations in the ferrous iron concentrations in the groundwater. Similarly, the average total iron concentration in the groundwater in the dry season was higher than that in the wet season by 0.35 mg/L, but the difference was not significant (paired *t*-test: $p > 0.05$). In the rainy season, the increased infiltration of rainwater could dilute the iron concentration [65]; however, these effects were not obvious in the study area.

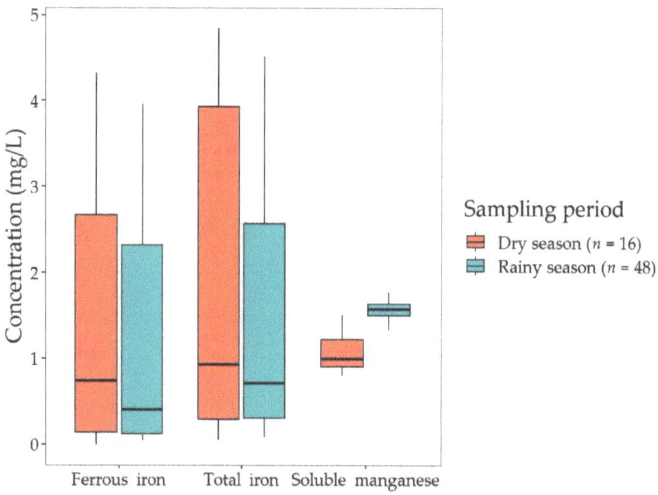

Figure 7. Variations in iron and manganese in the dry and rainy seasons (dry season: 23 February–13 March 2023; rainy season: 8 August–26 September 2023).

Unlike the iron concentration, the average concentration of soluble manganese in the rainy season (1.55 mg/L) was significantly higher than that in the dry season (1.07 mg/L), by 0.48 mg/L (Figure 7) (paired *t*-test: $p < 0.05$). The manganese concentrations in groundwater are dependent on aquatic chemistry, dissolution from rocks and manganese-bearing minerals [65], and soil [66]. As most of the aquifers in the study area are recharged by rivers and precipitation [50], the recharging of groundwater by rivers could influence the manganese concentration [67]. Variations in the Eh values in the groundwater (0.14–0.28 V) could be a cause of the larger variations in ferrous/total iron than in soluble manganese in the groundwater (Figure 7).

3.2. Iron and Manganese Oxidation through Stirring and Re-Aeration

The ferrous iron (0.05–0.32 mg/L) remaining in the treated water from the full-scale aeration–RSF plants was oxidized to BDL within 60 min of stirring; however, the soluble manganese (1.10–1.20 mg/L) slightly decreased by less than 8.33% even after 60 min of stirring (Figure A2), indicating the slow oxidation of manganese compared with ferrous iron. For the treated water of Duhabi, a stirring experiment was not performed due to the low DO value (<2.48 mg/L). As a result of aeration to saturate DO, the ferrous iron (0.05–0.44 mg/L) in treated water was oxidized below the BDL within 90 min; however, the soluble manganese (1.10–1.40 mg/L) decreased by less than 27.27% even after 90 min, remaining at concentrations above the national standard of 0.20 mg/L (Figure A3). These results show the inefficiency of aeration and the requirement of oxidizing chemicals and catalysts for the oxidation of soluble manganese.

3.3. Bench-Top Filtration Experiments on Raw Water

Raw water from two WSPs, i.e., Jhumka and Inaruwa, was filtered separately using dual-media filters filled with three different filter media: quartz sand, manganese sand, and ferrolite beneath anthracite. Table 2 shows the raw water quality measured after transportation from the WSPs to the laboratory. The pH of the raw water was weakly alkaline, with a slightly higher pH in Jhumka (7.60) than in Inaruwa (7.45), which favored the adsorption of iron and manganese [10]. The turbidity of the raw water from Inaruwa was very high (30.28 NTU); this indicated increased turbidity compared with that measured at the sampling point (2.01 NTU) due to the oxidation of a high concentration of ferrous iron (3.54 mg/L) during transportation from the sampling point to the experimental laboratory. High turbidity in raw water interferes with manganese adsorption on filter media [68]; thus, a PACl solution was added to the raw water from Inaruwa to remove the turbidity in the anthracite layer of the filtration columns.

Table 2. Raw water quality after transportation from the Jhumka and Inaruwa WSPs.

Water Quality Parameters	Jhumka	Inaruwa
pH	7.60	7.45
Temperature (°C)	24.0	25.3
Turbidity (NTU)	3.57	30.28
Electrical conductivity (μS/cm)	414	246
Ammonia (mg/L)	BDL (<0.02)	0.06
Ferrous iron (mg/L)	0.03	0.42
Total iron (mg/L)	0.82	3.92
Manganese (mg/L)	1.10	1.10

Among the three filtration columns, Column 3 (with ferrolite on the lower layer) consumed the maximum amounts of chlorine by filtering clean water containing chlorine through the filter media. The filter media used to filter the raw water from Jhumka were cleaned with 2 mg/L of chlorine, which resulted in 1.12 mg/L of chlorine consumption by Column 3 even after 3 h of cleaning; then, the filter media used to filter the raw water from Inaruwa were cleaned with 3 mg/L of chlorine for up to 5 h, which resulted in chlorine consumption by Column 3 of up to 0.42 mg/L. Based on these results, the chlorine dosage to be injected into the raw water from Jhumka was calculated to be approximately 3.05 mg/L as follows: the stoichiometric dose needed to oxidize iron and manganese ($0.62 \times 0.03 + 1.29 \times 1.10 = 1.43$ mg/L) + chlorine consumed by filter media (1.12 mg/L) + minimum residual chlorine (0.50 mg/L). The chlorine dosage for the raw water from Inaruwa was calculated to be approximately 2.59 mg/L as follows: the stoichiometric dose needed to oxidize iron and manganese ($0.62 \times 0.42 + 1.29 \times 1.10 = 1.67$ mg/L) + chlorine consumed by filter media (0.42 mg/L) + minimum residual chlorine (0.50 mg/L).

Figure 8a shows the total iron concentrations in the raw water and treated water from Jhumka in the bench-scale filtration experiments, which were operated for 180 min with a pre-dose of 3.10–3.18 mg/L chlorine and without PACl. The total iron concentration in the raw water (0.82 mg/L) was reduced below the national standard of 0.30 mg/L through filtration with any of the three filter media. However, the total iron concentration in the treated water decreased when using quartz sand (0.09–0.19 mg/L), manganese sand (0.04–0.12 mg/L), and ferrolite (0.01–0.05 mg/L). Figure 8b shows the soluble manganese concentrations in the raw and treated water from Jhumka in the bench-scale filtration experiments, which were operated for 180 min with a chlorine pre-dose of 3.10–3.18 mg/L and without PACl. After filtration, the soluble manganese concentration in raw water (1.10 mg/L) decreased below 0.20 mg/L when using any of the three filter media. However, 0.06 mg/L of soluble manganese remained in the water filtered with quartz sand until 120 min, but the concentration was reduced to below the BDL (soluble manganese 0.01 mg/L) after a filtration time of 150 min. The soluble manganese in the raw water was reduced to the BDL after 90 min of filtration with manganese sand, whereas the soluble manganese in the water filtered with ferrolite was below the BDL before 30 min of filtration.

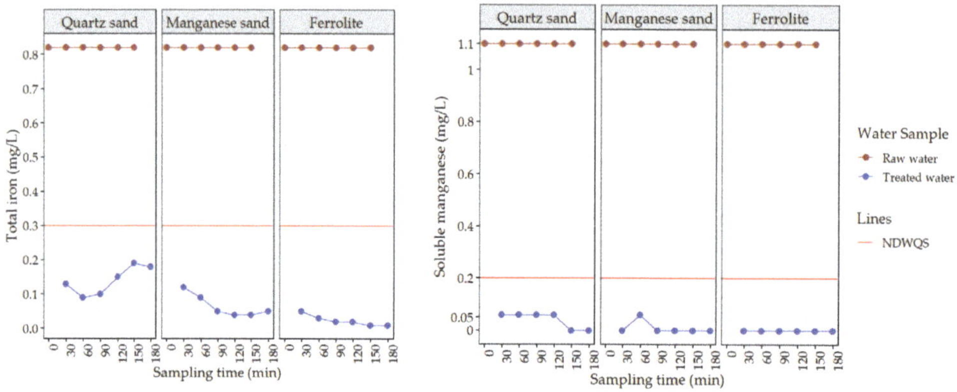

(**a**) (**b**)

Figure 8. (**a**) Total iron and (**b**) soluble manganese in raw water (pH = 7.60) and filtered water from the Jhumka WSP.

To determine the effects of the PACl dosage, paper filtration using a paper filter (pore size 5 μm) was conducted at various PACl doses of 0, 5, and 10 mg/L. Since 0.42 mg/L of ferrous iron in 3.92 mg/L of total iron was present in the raw water, which could not be filtered using paper filtration, chlorine was dosed at 1 mg/L to oxidize the ferrous iron before filtration; then, filtering the raw water through the filter paper without a PACl dosage reduced the 3.92 mg/L of total iron to 2.74 mg/L. At a PACl dosage of 5 mg/L, the total iron was decreased to 0.24 mg/L through paper filtration. Then, at a PACl dosage of 10 mg/L, the total iron concentration in the filtered water was decreased to 0.10 mg/L. Based on these paper filtration experiments, the bench-scale filtration experiments for the raw water of Inaruwa were conducted at PACl dosages of 0, 5, and 10 mg/L.

Figure 9a shows the total iron concentrations in the raw water and treated water from Inaruwa in the bench-scale filtration experiments with different PACl dosages of 0, 5, and 10 mg/L. After filtration, without a PACl dose, the total iron concentration (3.92 mg/L) decreased to 0.38 mg/L in 90 min using ferrolite, which was above the national standard of 0.30 mg/L. The total iron concentrations in the filtered water decreased from those at 30 min to 1.37–2.54 mg/L, 0.93–1.66 mg/L, and 0.38–0.84 mg/L when using quartz sand, manganese sand, and ferrolite, respectively, after 60–90 min of filtration due to the clogging and retention of iron-oxide particles in the filter layers; however, all of the filtered water contained total iron concentrations above the standard of 0.3 mg/L without the addition of the coagulant (PACl).

After backwashing the filter media, PACl at a dosage of 5 mg/L was injected into the raw water before the bench-scale filtration experiments; then, the total iron concentrations in the filtered water were reduced to 0.10–0.18 mg/L, 0.08–0.09 mg/L, and 0.04–0.07 mg/L when using quartz sand, manganese sand, and ferrolite, respectively, after 90 min. A further increase in the PACl dose to 10 mg/L slightly decreased the total iron concentrations in the filtered water to 0.05–0.10 mg/L, 0.04–0.05 mg/L, and 0.02–0.03 mg/L when using quartz sand, manganese sand, and ferrolite, respectively, after 90 min.

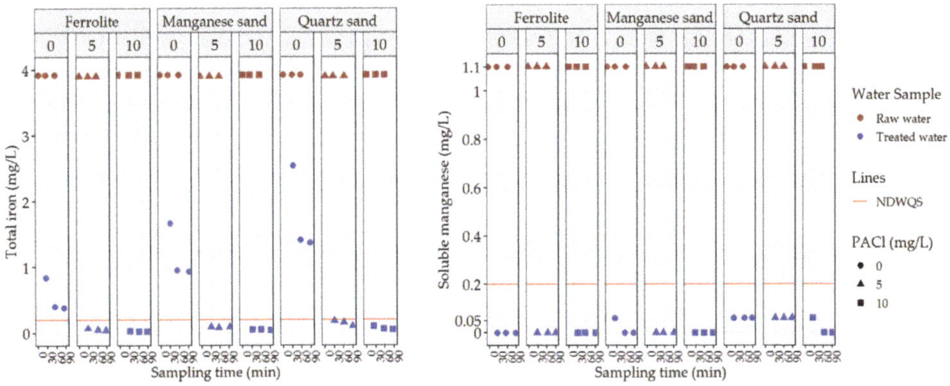

(**a**) (**b**)

Figure 9. Effects of PACl on the removal of (**a**) total iron and (**b**) soluble manganese in raw water from Inaruwa (pH = 7.40).

Figure 9b shows the soluble manganese concentrations in the raw water and treated water from Inaruwa in the bench-scale filtration experiments and in the filtered water over the course of 90 min of filtration with PACl doses of 0, 5, and 10 mg/L. Even without the PACl doses, the soluble manganese concentrations in the raw water (1.10 mg/L) decreased below the standard of 0.2 mg/L for all three of the filter media; ferrolite decreased the soluble manganese concentration to the BDL within 30 min of filtration, whereas it took 60 min of filtration to bring the soluble manganese concentration to the BDL using manganese sand. However, soluble manganese was present at a concentration of 0.06 mg/L when using no PACl and a 5 mg/L dose in the filtered water when using the quartz sand filter, whereas, with a PACl dose of 10 mg/L, the manganese concentration decreased to the BDL after 60 min of filtration.

Figure 10a shows the soluble manganese concentrations in the raw water and filtered water from Jhumka at various pH levels with chlorine doses of 3.10–3.18 mg/L and without PACl. The concentration of soluble manganese in the raw water was 1.00 mg/L, which was reduced to the BDL at a pH of 7.70–9.01 when using all three media. Figure 10b shows the soluble manganese concentration in the raw water from Inaruwa and in the filtrate at various pH levels with a chlorine dose of 2.60–2.78 mg/L and 10 mg/L of PACl in the bench-scale filtration experiments. The concentration of soluble manganese in the raw water was 1.10 mg/L, which was reduced to the BDL using manganese sand and ferrolite at a pH of 7.50–9.03. For quartz sand, a soluble manganese concentration of 0.06 mg/L was present at a pH below 8.26; however, when the pH was increased to 8.26–9.03, the soluble manganese concentration was reduced to the BDL. These results show that pH levels in the range of 7.5–9.1 had little effect on the removal of soluble manganese.

Figure A5 shows the chlorine consumed in the bench-scale filtration experiments. Free chlorine at 3.10–3.18 mg/L and 2.60–2.78 mg/L was injected into the raw water from Jhumka and Inaruwa, respectively. Figure A5a shows the chlorine consumption of the three filter media in the filtration of raw water from Jhumka, namely, 0.64–0.46 mg/L, 1.18–0.89 mg/L, and 2.67–2.12 mg/L for quartz sand, manganese sand, and ferrolite, respectively. By increasing the pH in the raw water (Figure A5b), the consumption of chlorine by the filter media decreased from 0.78 mg/L to 0.43 mg/L, from 1.21 mg/L to 0.43 mg/L, and from 2.09 mg/L to 0.92 mg/L for quartz sand, manganese sand, and ferrolite, respectively. Similarly, the consumption of chlorine by the filter media in the filtration of raw water from Inaruwa was as follows: 0.78–0.60 mg/L, 1.05–0.80 mg/L, and 1.60–1.07 mg/L for quartz sand, manganese sand, and ferrolite, respectively (Figure A5c). With a PACl dosage of 10 mg/L, increasing the pH of the raw water decreased the chlorine consumption from 0.92 mg/L to 0.53 mg/L, from 1.10 mg/L to 0.62 mg/L, and from 1.38 mg/L to

0.87 mg/L for quartz sand, manganese sand, and ferrolite, respectively (Figure A5d). The higher chlorine consumption of ferrolite and manganese sand indicated the higher reactivity of these filter media than of quartz sand. The chlorine consumption decreased with increased filtration time and pH in the raw water. The pH in the raw water from Jhumka (7.70) was increased to 8.23–9.01; after filtration, the pH in the treated water decreased to 8.12–8.95, 8.10–8.89, and 8.01–8.75 for quartz sand, manganese sand, and ferrolite, respectively (Figure A6a). The pH in the raw water from Inaruwa (7.50) was increased to a pH of 8.01–9.03; after filtration, the pH in the treated water was 7.95–8.79, 7.83–8.49, and 7.72–8.35 for quartz sand, manganese sand, and ferrolite, respectively (Figure A6b). For the filtered water, at 30 min, the increased pH was due to the addition of NaClO (pH = 13), as the pH of the raw water at 0 min was measured without the addition of NaClO. However, when the pH in the raw water was increased to 9 after the addition of sodium hydroxide and NaClO, there was a decrease in the pH of the filtered water. At a high pH (>6), the quantity of hypochlorite ions increased through the dissociation of hypochlorous acid ($HOCl \rightarrow OCl^- + H^+$ [69]) according to the following equation: $NaClO + H_2O \rightarrow HOCl + NaOH$. Thus, H^+ ions were released; this may have been the reason for the decrease in the pH of the filtered water. The pH reduction in the treated water from Inaruwa was greater than that in the water from Jhumka because of the dose of PACl for coagulation [70].

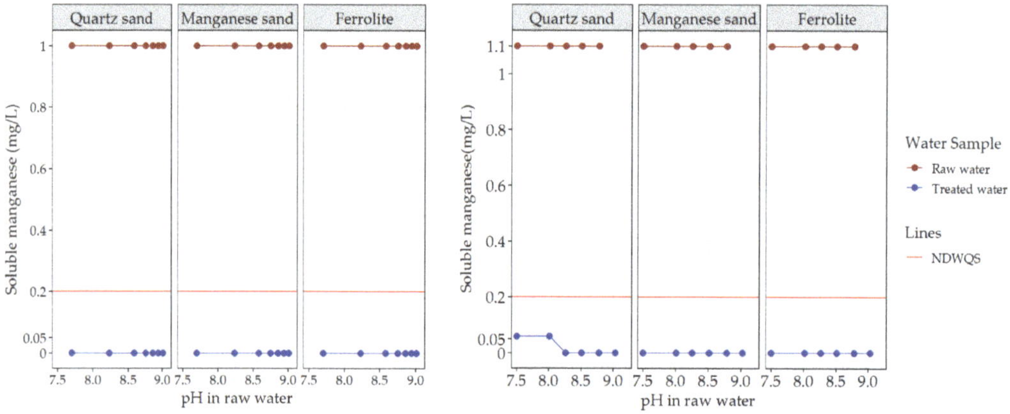

(a) (b)

Figure 10. Effects of pH on the removal of soluble manganese from raw water from (**a**) Jhumka and (**b**) Inaruwa after the filtration experiment.

4. Discussion

4.1. Variations in Iron and Manganese in the Groundwater in the Study Area

The water quality parameters, such as the pH, TDSs, and iron concentration, of the groundwater were similar among some WSPs, indicating that the groundwater might have been extracted from aquifers with similar geological conditions. However, the iron concentrations in the groundwater varied extensively at each location of the WSPs, indicating the effects of the reduction–oxidation (redox) conditions in the local aquifers on the iron concentration. Although the iron concentrations in the groundwater varied extensively in both the dry and rainy seasons, the mean iron concentration decreased from the dry season to the rainy season. However, the soluble manganese concentration in the groundwater in the rainy season significantly increased on that in the dry season, indicating increased leaching from manganese-bearing minerals and rocks [66,71]. These results suggest that it is important to monitor groundwater quality throughout the year or at least to set the design and operational conditions of iron and manganese removal processes in the dry and rainy seasons.

4.2. Iron and Manganese Removal through Aeration–RSF in the WSPs

The WSPs installed and operated aeration–RSF processes for the removal of iron and manganese, with some WSPs employing MnO_2-coated pebbles and manganese sand. Although ferrous iron was oxidized and removed to concentrations below the national standard of 0.30 mg/L through aeration–RSP processes, small amounts of ferrous iron remained in the treated water. With additional stirring and re-aeration, the remaining ferrous iron in the treated water was removed to the BDL within 60 min, which indicated the importance of effective and extended aeration for oxidation for the removal of ferrous iron in raw water. The oxidation of manganese in the existing aeration–RSF processes was very slow, even in those using MnO_2-coated pebbles and manganese sand; additional stirring and re-aeration of treated waters could not oxidize soluble manganese. The Eh−pH diagrams of iron and manganese (Figure A7) [71] show the most stable forms of iron and manganese based on the Eh and pH of water [72]; because the pH and Eh values of water treated through aeration–RSF were 6.68–7.37 and 0.25–0.43 V, respectively, in the dry season and 6.62–7.42 and 0.23–0.38 V, respectively, in the rainy season, the most stable form of iron was found to be Fe_2O_3, whereas the most stable form of manganese was found to be Mn[2^+]. Therefore, ferrous iron is readily oxidized through aeration, whereas soluble manganese is not [64]; thus, to oxidize soluble manganese to MnO_2, a high pH (>10) and/or high Eh (>0.40 V) for oxidation through the addition of alkaline solution and/or oxidizing chemicals is required.

Because the iron loading per filter run is inversely correlated with manganese removal efficiency in aeration–RSF [38], for the treatment plants at Jhumka and Mahendrangar, the raw water with high iron concentrations may have resulted in comparatively low manganese removal. The iron concentration was high in the raw water from Jhumka (1.91 mg/L) and Mahendranagar (1.56 mg/L) and in both raw water samples from Inaruwa (3.81 and 4.09 mg/L); therefore, the installation of additional pre-treatment or dual-media filters for iron removal may improve the manganese removal efficiency in those WSPs.

The low pH in the raw water from Mahendrangar may be one reason for the low soluble manganese removal rate, as the physiochemical removal of manganese in aeration–RSF is poor at a low pH [38]; with the lowest value overall, a soluble manganese removal rate of only 8.28% was found in the process of treating raw water with a low pH of 6.70 in Mahendranagar. High iron concentrations (1.56 mg/L) and poor backwashing of the filter media (twice a week) might also have caused low soluble manganese removal rates. The MnO_2-coated pebbles in the aeration unit of TP1 in Inaruwa showed comparatively better manganese removal efficiency than that of TP2 in Inaruwa, which had pall rings in its aeration unit (without MnO_2-coated pebbles). MnO_2-coated pebbles were also present in the aeration unit of Mahendranagar, but the manganese removal in Inaruwa and Mahendranagar was not sufficient to decrease the soluble manganese below the national standard of 0.20 mg/L. Large sizes of the MnO_2-coated media (pebbles) and improper placement of coated media (in the aeration unit) are possible reasons for the incomplete removal of soluble manganese. A high specific surface area is an important property of adsorbents, as they must provide a large number of active sites for adsorption [73]; however, because the raw water had a high iron concentration, the iron attached to the surfaces of MnO_2-coated pebbles in the aeration unit might have covered the adsorption sites for manganese [68]. In the Inaruwa WSP, the treated water had a high total iron concentration comprising mostly ferric iron, which indicated the poor performance of the RSF unit despite the daily backwashing of the RSF. Although the oxidation of iron can be increased by providing effective oxidation, the removal of ferric iron in colloidal form requires efficient filtration; proper filter media, proper backwashing, regular monitoring of media (condition and level), and addition/replacement are required. Therefore, the design and operation of the aeration–RSF systems in WSPs should be closely examined and revised if necessary.

In the Duhabi WSP, even though the aeration system was not in operation, a manganese removal efficiency of 16.4–23.5% was achieved; this was mostly due to the adsorption of soluble manganese to the filter media. The low iron concentration, high pH of the raw water, and additional manganese sand layer, along with the quartz sand, may have contributed to the removal of manganese [10,34]. Operation of the aeration system or setting up another oxidant system would further improve the manganese removal rate in Duhabi.

4.3. Improvement of Manganese Removal Using Dual-Media Filters Filled with Different Filter Media

Dual-media filtration was effective in the two-step removal of iron and manganese; iron was removed in the top anthracite layer even without the injection of a coagulant (PACl); this was followed by the removal of manganese in the sand or ceramic layer. The removal of iron flocs in the anthracite layer reduced the clogging of the sand or ceramic layer below the anthracite layer. Coagulant injection is recommended only if the iron concentration in the raw water is high, which interferes with the auto-catalytic oxidation and removal of soluble manganese. Among the three filter media tested, ferrolite was the most efficient in the removal of soluble manganese, while manganese sand and quartz sand were also able to reduce the concentration of soluble manganese to below the national guideline of 0.2 mg/L through the addition of chlorine. In addition, the filter media used in this study can be regenerated using chorine or permanganate solutions when they lost adsorption capacities of manganese. Thus, it is expected that we can use these filter media for several years. The dual-media filtration process could be used as a measure for controlling high iron and manganese concentrations in Water Safety Plans proposed by the World Health Organization [74].

When the pH of the raw water is >7.50, the oxidation of soluble manganese using chlorine with all three media was effective at a linear velocity of 5 m/h, while further increasing the pH contributed to a reduction in the required chlorine dose and the stabilization of manganese removal. The full-scale treatment plants of WSPs are operated at a very high linear filtration velocity of above 20 m/h, which is another reason for the low efficiency of soluble manganese removal. Thus, increasing the pH might allow us to increase the filtration velocity by increasing the adsorption capacity of MnO_2-coated filter media [30] and making chlorine oxidation more effective [28].

Although the dual-media filters used in this study showed high iron and manganese removal efficiencies, the initial and operational costs may increase from the conventional RSFs. While chlorine and PACl are easily available and affordable because these chemicals are used in the water treatment plants sourced from surface waters, high-quality products of MnO_2-coated filter media, both sand and ceramics, are not available locally; thus, the initial construction cost might increase from those of conventional RSFs. Therefore, considering the importance of iron and manganese removal from groundwater, national and regional governments should subsidize construction of dual-media filters. In the design and operation of full-scale plants, it is important to prevent by-pass flow that might cause leakages of iron and manganese. Furthermore, chlorine dosages for oxidation of iron and manganese might produce disinfection by-products (DBPs). Therefore, to prevent DBP formation, chlorine doses should be carefully controlled to a level minimally adequate for oxidation of iron and manganese.

5. Conclusions

The iron concentration in the groundwater of five WSPs in the Sunsari district, Nepal, varied extensively, while the variations in manganese concentrations were small among these WSPs. These results show the susceptibility of iron leaching to local geological conditions—especially the redox potential—in comparison with manganese, which indicates that the design and operation of aeration–RSF processes should be modified even in the same areas based on the maximum concentrations of iron and manganese. While the iron concentration decreased in the rainy season compared with the dry season, the manganese concentration increased, which indicated that different processes influenced the iron and manganese concentrations in groundwater. Therefore, we should monitor iron and manganese concentrations throughout the year for the design and operation of aeration–RSF processes.

Although ferrous iron was removed through aeration–RSF processes in the WSPs, soluble manganese was not oxidized even with an extended duration of stirring and aeration. Manganese-oxide-coated pebbles and sand were not effective in the oxidation and removal of soluble manganese; this was possibly due to the large surface areas (pebbles) and clogging of sand filters by iron oxide flocs. The chlorine oxidation of manganese followed by dual-medium filtration using anthracite and sand or ceramics was found to be efficient in removing soluble manganese by retaining iron oxide flocs in the top anthracite layer followed by the auto-catalytic oxidation of soluble manganese in the lower filter layer, which was filled with manganese sand or MnO_2-coated ceramics (ferrolite). Among the three filter media used in this study, ferrolite was the most efficient, as it showed a stable manganese concentration in the treated water, followed by manganese sand and quartz sand. A chlorine dose is indispensable for the oxidation of manganese; maintaining 0.5 mg/L of residual chlorine in the effluent from filters was sufficient to oxidize soluble manganese.

Although sand and manganese sand are used as filter media in the WSPs in Nepal, the results of this study suggest the importance of designing RSF units with dual-medium filters, selecting filter media with a high capacity for manganese adsorption, and dosing chlorine to remove manganese from groundwater. Because iron concentrations are higher than manganese concentrations and iron is readily oxidized through aeration, it is suggested to oxidize ferrous iron completely in an aeration unit, followed by a chlorine dose for the oxidation of soluble manganese to reduce the chlorine dosage. Dual-media filtration using anthracite and sand/ceramics is an efficient method for removing iron and manganese in small filter units. Therefore, we propose the standardization of the design and operation of processes for removing iron and manganese from groundwater in developing countries, including Nepal.

Author Contributions: Conceptualization, S.T. and S.K.; methodology, S.T. and B.S.; validation, B.S., S.K. and S.T.; formal analysis, A.M.S.; investigation, A.M.S.; data curation, A.M.S.; writing—original draft preparation, A.M.S.; writing—review and editing, B.S., S.K. and S.T.; visualization, A.M.S.; supervision, B.S., S.K. and S.T.; project administration, S.T.; funding acquisition, S.T. All authors have read and agreed to the published version of the manuscript.

Funding: This research was supported by the Japan International Cooperation Agency (JICA) through a collaborative program with the University of Tokyo and a scholarship provided to Ankit Man Shrestha for his graduate studies. This study was also supported by a Grant-in-Aid for Scientific Research (No. 22H01621) provided by the Japan Society for the Promotion of Sciences (JSPS).

Data Availability Statement: The data presented in this study are available upon request.

Acknowledgments: The authors would like to express their gratitude to Duhabi, Inaruwa, Jhumka, Mahendranagar, and Pakali Water Supply and Sanitation Users Committees, the Urban Water Supply and Sanitation Sector Project, the Regional Project Management Office, Sunsari and Tohkemy Corporation, Osaka, Japan for their support in conducting this research.

Conflicts of Interest: The authors declare no conflicts of interest.

Abbreviations

DBPs	Disinfection by-products
PACl	Poly aluminum chloride
MnO_2	Manganese oxide
WTP	Water treatment plant
Redox	Reduction–oxidation
RSF	Rapid sand filter
MOCA	Manganese-oxide-coated anthracite
NDWQS	National Drinking Water Quality Standards, Nepal

WUSA	Water Users and Sanitation Association
RW	Raw water
TW	Treated water
GW	Groundwater
TP	Treatment plant
TDSs	Total dissolved solids
DO	Dissolved oxygen
ORP	Oxidation reduction potential
SHE	Standard hydrogen electrode
BDL	Below detection limit
NaClO	Sodium hypochlorite

Appendix A

Table A1. Selected water supply projects in the study.

Name of Urban Municipality	No. of WSPs	No. of WSPs with DTB as Main Source	No. of WSPs with GWTP	Selected WSPs (No. of DTBs)	Name of Selected WSPs	Remarks
Dharan	81	0	0	0		WSP based on surface water
Duhabi	1	1	1	1 (5)	Duhabi	One WSP from each of five urban municipalities is selected
Inaruwa	1	1	1	1 (4)	Inaruwa	
Ramdhuni	8	7	3	1 (2)	Jhumka	
Baraha	23	9	5	1 (3)	Mahendranagar	
Itahari	5	4	2	1 (2)	Pakali	
Total	119	22	12	5 (16)		

Notes: WSP: water supply project, DTB: deep tubewell, GWTP: groundwater treatment plant.

Table A2. Physicochemical water quality of the water samples (first field survey: 23 February–13 March 2023, dry season).

Water Supply Project	Water Source	Temperature (°C)	pH	Turbidity (NTU)	TDS (mg/L)	Eh (V)	Ferrous Iron (mg/L)	Total Iron (mg/L)	Soluble Manganese (mg/L)
Duhabi	GW 1	29.0	7.32	0.14	230	0.27	0.03	0.05	1.0
	GW 2	29.0	7.44	0.33	215	0.18	0.04	0.39	0.8
	GW 3	28.0	7.34	0.32	244	0.15	0.27	0.44	1.3
	GW 4	28.0	7.85	0.27	238	0.27	BDL	0.09	1.0
	GW 5	28.0	7.44	0.34	238	0.24	0.05	0.21	1.1
Inaruwa	GW 1	26.0	7.16	1.36	124	0.17	2.08	3.90	1.2
	GW 2	27.0	6.66	0.31	136	0.20	4.32	4.68	1.4
	TW1	26.0	6.68	0.56	128	0.41	BDL	0.52	1.0
	GW 3	27.4	7.02	0.76	111	0.16	3.00	4.02	0.8
	GW 4	27.0	7.01	2.54	137	0.14	3.24	4.84	0.9
	TW2	28.0	6.72	0.16	107	0.37	BDL	0.50	0.7
Jhumka	GW 1	26.8	6.85	3.07	228	0.18	4.22	4.54	1.1
	TW1	26.1	6.99	0.32	191	0.43	0.87	0.99	0.6
	GW 2	28.0	7.10	2.60	207	0.26	0.26	0.31	0.9
	TW2	26.0	7.11	0.10	200	0.41	BDL	0.04	0.4
Mahendranagar	GW 1	24.0	6.70	3.94	182	0.26	0.79	1.17	0.8
	GW 3	26.0	6.69	0.23	173	0.29	0.70	0.70	1.5
	TW1	25.0	7.03	0.37	185	0.41	0.03	0.16	1.1
	GW 2	25.0	6.80	0.33	212	0.21	2.56	2.72	1.0
Pakali	GW 1	24.0	6.95	0.72	183	0.23	0.17	0.24	0.9
	TW1	25.0	7.69	0.34	185	0.30	BDL	0.06	0.5
	GW 2	26.0	7.03	11.70	216	0.18	1.02	1.30	1.4
	TW2	26.0	7.37	0.71	203	0.25	BDL	0.07	0.6

Notes: GW: groundwater, TW: treated water, TDS: total dissolved solids, BDL (below detection limit): ferrous iron 0.02 mg/L.

(**a**) (**b**)

Figure A1. Oxidation experiments on treated water: (**a**) stirring in a sealed bottle; (**b**) aeration experiment.

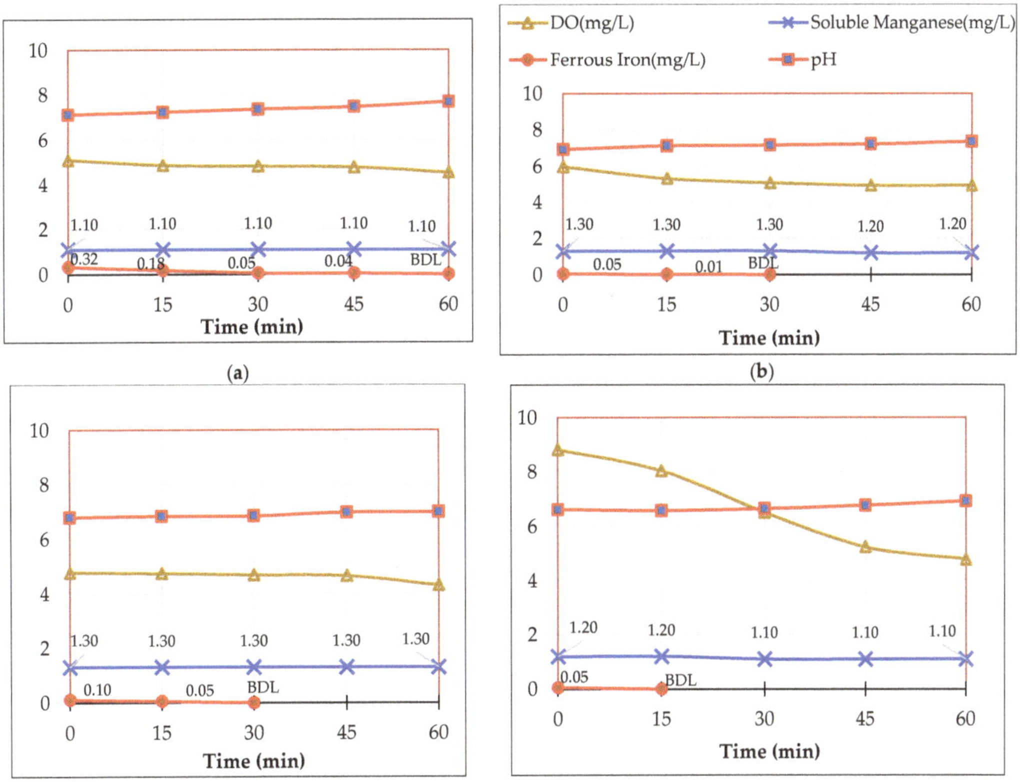

Figure A2. Parameters after stirring treated water from different water supply projects: (**a**) Jhumka; (**b**) Pakali; (**c**) Mahendranagar; (**d**) Inaruwa.

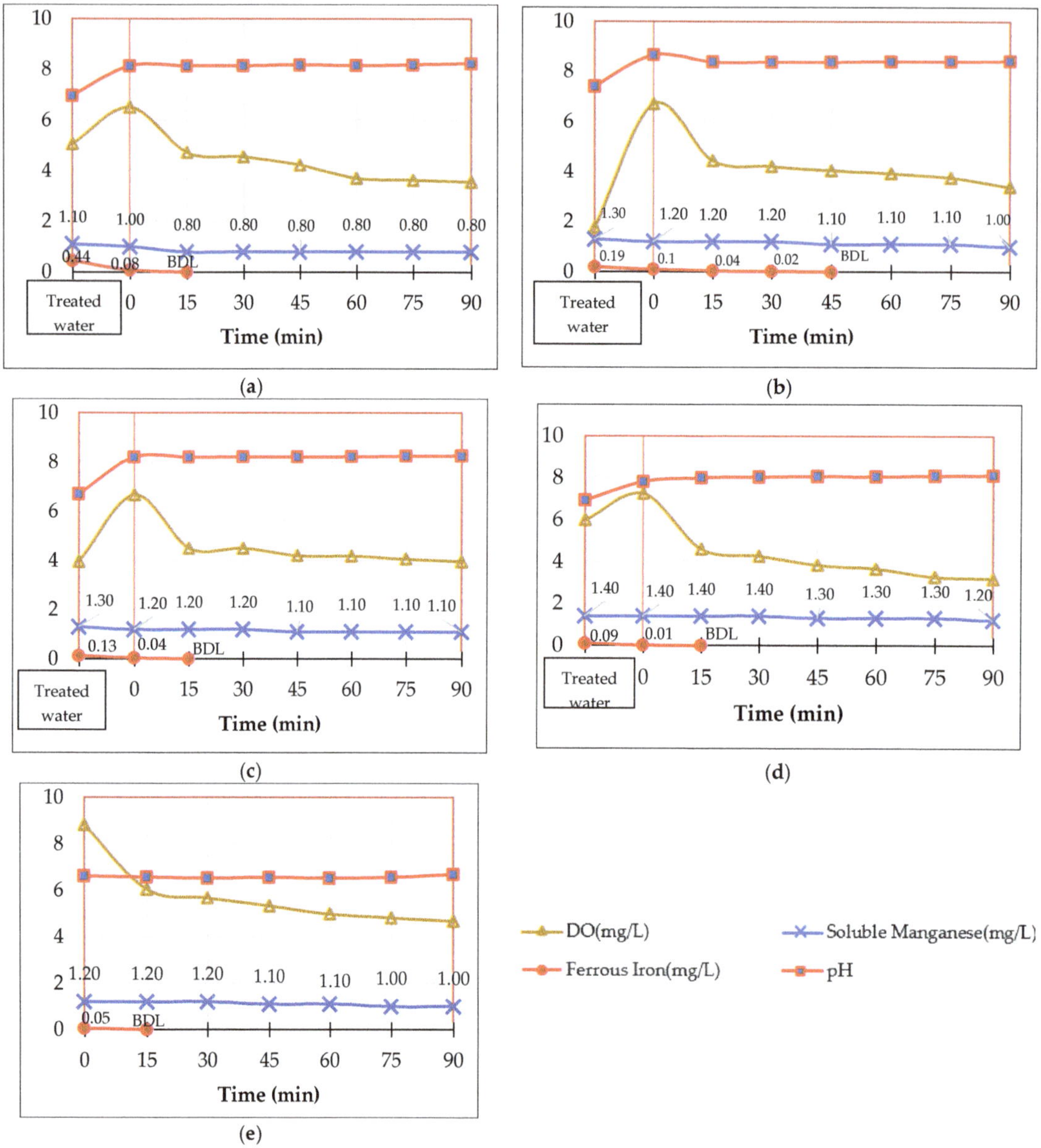

Figure A3. Parameters after the aeration of treated water from different water supply projects up to the saturation level of DO: (**a**) Jhumka; (**b**) Duhabi; (**c**) Mahendranagar; (**d**) Pakali; (**e**) Inaruwa (the DO in treated water from Inaruwa was at the saturation level).

(**a**) (**b**)

Figure A4. Black precipitate in treated water due to manganese at (**a**) household tap (in Jhumka) and (**b**) dead-end washout point (in Mahendranagar).

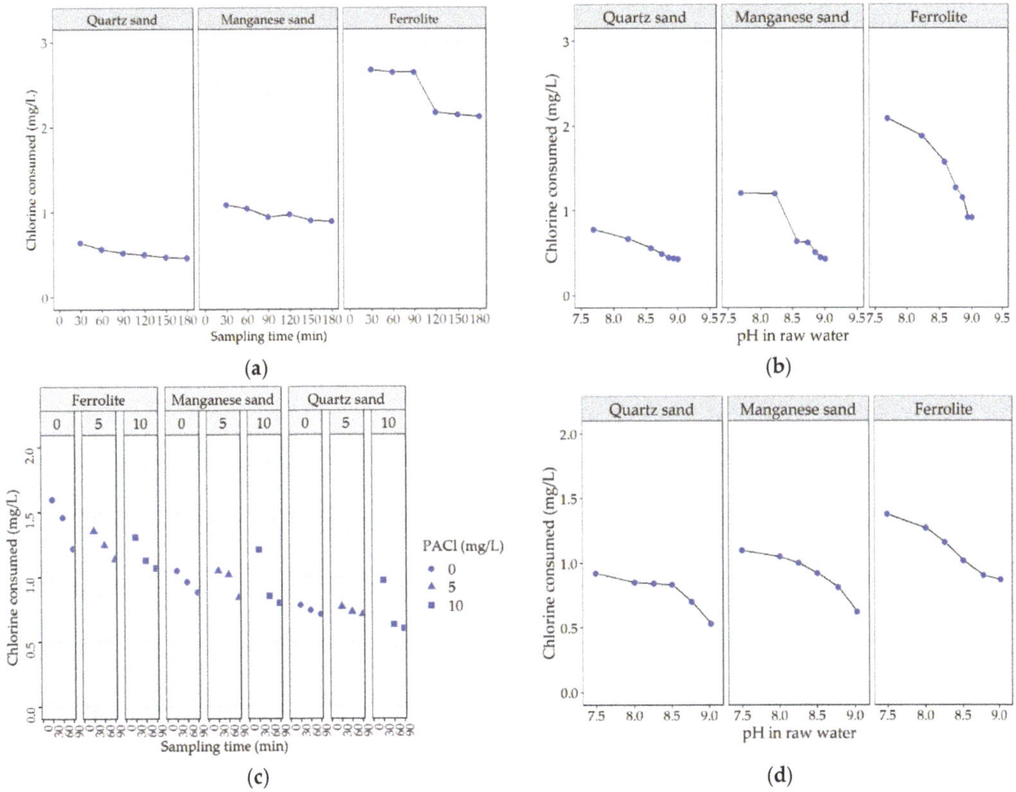

Figure A5. Chlorine consumption in the bench-scale filtration experiment for raw water from (**a**) Jhumka without pH adjustment (pH 7.60), (**b**) from Jhumka with pH adjustment (pH 7.70–9.01), (**c**) from Inaruwa without pH adjustment (pH 7.40), and (**d**) from Inaruwa with pH adjustment (pH 7.50–9.03).

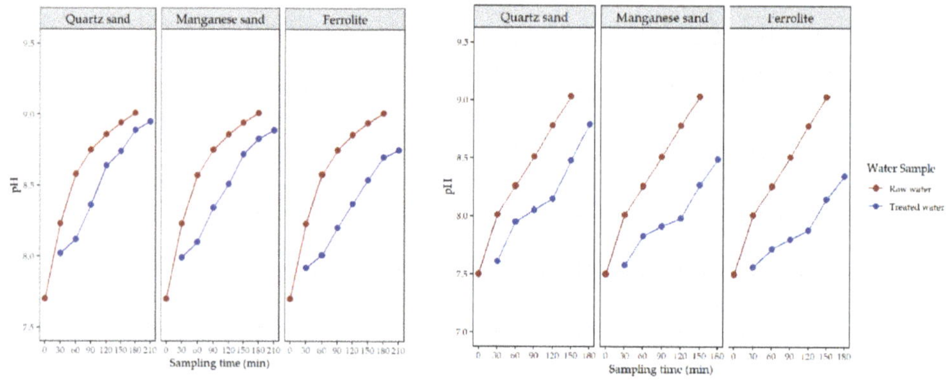

(a) (b)

Figure A6. pH in water samples after pH adjustment in the bench-scale filtration experiment for raw water from (**a**) Jhumka and (**b**) Inaruwa.

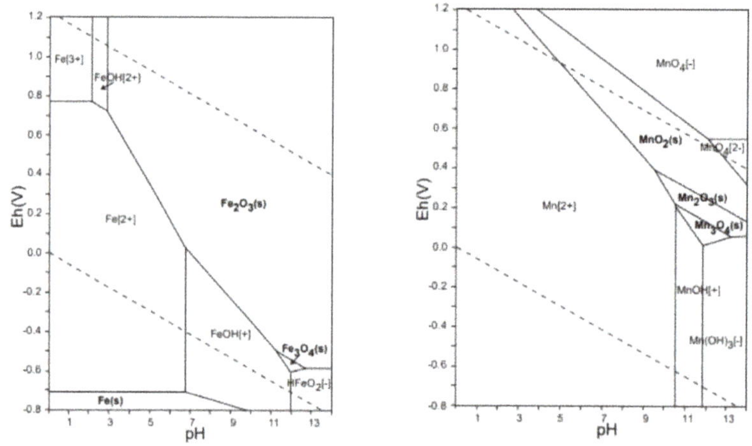

(a) (b)

Figure A7. Eh−pH diagrams of the system: (**a**) Fe-O-H; (**b**) Mn-O-H [71].

References

1. Zhang, X.; Hu, B.; Wang, P.; Chen, J.; Yang, L.; Xiao, K.; Zhang, X. Hydrogeochemical Evolution and Heavy Metal Contamination in Groundwater of a Reclaimed Land on Zhoushan Island. *Water* **2018**, *10*, 316. [CrossRef]
2. Sharma, G.K.; Jena, R.K.; Ray, P.; Yadav, K.K.; Moharana, P.C.; Cabral-Pinto, M.M.S.; Bordoloi, G. Evaluating the Geochemistry of Groundwater Contamination with Iron and Manganese and Probabilistic Human Health Risk Assessment in Endemic Areas of the World's Largest River Island, India. *Environ. Toxicol. Pharmacol.* **2021**, *87*, 103690. [CrossRef]
3. Islam, M.Z.; Mostafa, M.G. Iron, Manganese, and Lead Contamination in Groundwater of Bangladesh: A Review. *Water Pract. Technol.* **2024**, *19*, 745–760. [CrossRef]
4. WHO. *Fourth Edition Incorporating the First and Second Addenda Guidelines for Drinking-Water Quality*; World Health Organization: Geneva, Switzerland, 2022.
5. Krishnakumari, B.; Mohit Abhishek, V.; Puneeth, T.E.; Vignesh, S.; Mohamed, M.K. Irfan Removal of Iron and Manganese From Ground Water. *Int. J. Eng. Res.* **2018**, *V6*, 02. [CrossRef]
6. Amin, S.; Kazama, S.; Sawangjang, B.; Takizawa, S. Causes and Effects of Scale Deposition in Water Supply Pipelines in Surakarta City, Indonesia. *Water* **2024**, *16*, 2275. [CrossRef]
7. Wang, M.; Wang, M.; Yang, L.; Yang, T.; Li, J.; Chen, Y. Distribution Characteristics and Genesis of Iron and Manganese Ions in Groundwater of Eastern Sanjiang Plain, China. *Water* **2023**, *15*, 2068. [CrossRef]

8. Rahman, M.F.; Ali, M.A.; Chowdhury, A.I.A.; Ravenscroft, P. Manganese in Groundwater in South Asia Needs Attention. *ACS ES&T Water* **2023**, *3*, 1425–1428. [CrossRef]
9. Islam, M.S.; Mostafa, M.G. Occurrence, Source, and Mobilization of Iron, Manganese, and Arsenic Pollution in Shallow Aquifer. *Geofluids* **2023**, *2023*, 6628095. [CrossRef]
10. Piispanen, J.K.; Sallanko, J.T. Mn(II) Removal from Groundwater with Manganese Oxide-Coated Filter Media. *J. Environ. Sci. Health Tox. Hazard. Subst. Environ. Eng.* **2010**, *45*, 1732–1740. [CrossRef]
11. Rusydi, A.F.; Onodera, S.I.; Saito, M.; Ioka, S.; Maria, R.; Ridwansyah, I.; Delinom, R.M. Vulnerability of Groundwater to Iron and Manganese Contamination in the Coastal Alluvial Plain of a Developing Indonesian City. *SN Appl. Sci.* **2021**, *3*, 399. [CrossRef]
12. Palmucci, W.; Rusi, S.; Di Curzio, D. Mobilisation Processes Responsible for Iron and Manganese Contamination of Groundwater in Central Adriatic Italy. *Environ. Sci. Pollut. Res.* **2016**, *23*, 11790–11805. [CrossRef] [PubMed]
13. Moyosore, J.O.; Coker, A.O.; Sridhar, M.K.C.; Mumuni, A. Iron and Manganese Levels of Groundwater in Selected Areas in Ibadan and Feasible Engineering Solutions. *Eur. Sci. J.* **2014**, *10*, 11.
14. Frisbie, S.H.; Mitchell, E.J.; Sarkar, B. Urgent Need to Reevaluate the Latest World Health Organization Guidelines for Toxic Inorganic Substances in Drinking Water. *Environ. Health* **2015**, *14*, 63. [CrossRef]
15. Kim, J. Manganese Treatment to Reduce Black Water Occurrence in the Water Supply. *Environ. Eng. Res.* **2015**, *20*, 230–236. [CrossRef]
16. Tang, X.; Zhu, X.; Huang, K.; Wang, J.; Guo, Y.; Xie, B.; Li, G.; Liang, H. Can Ultrafiltration Singly Treat the Iron- and Manganese-Containing Groundwater? *J. Hazard. Mater.* **2021**, *409*, 124983. [CrossRef] [PubMed]
17. McBeath, S.T.; Hajimalayeri, A.; Jasim, S.Y.; Mohseni, M. Coupled Electrocoagulation and Oxidative Media Filtration for the Removal of Manganese and Arsenic from a Raw Ground Water Supply. *J. Water Process. Eng.* **2021**, *40*, 101983. [CrossRef]
18. Xu, L.; Cao, G.; Xu, X.; Liu, S.; Duan, Z.; He, C.; Wang, Y.; Huang, Q. Simultaneous Removal of Cadmium, Zinc and Manganese Using Electrocoagulation: Influence of Operating Parameters and Electrolyte Nature. *J. Environ. Manag.* **2017**, *204*, 394–403. [CrossRef]
19. Kononova, O.N.; Bryuzgina, G.L.; Apchitaeva, O.V.; Kononov, Y.S. Ion Exchange Recovery of Chromium (VI) and Manganese (II) from Aqueous Solutions. *Arab. J. Chem.* **2019**, *12*, 2713–2720. [CrossRef]
20. Wang, L.K.; Wang, M.-H.S. Evolutionary Progress in Science, Technology, Engineering, Arts and Mathematics (Steam). In *Series Innovative Process System Consisting of Dissolved Air Flotation, Recarbonation and Filtration, Using Lime and Sodium Aluminate for Removal of Hardness, Iron and Manganese from Groundwater*; Lenox Institute of Water Technology: Lenox, MA, USA, 2022; Volume 4. [CrossRef]
21. Hoyland, V.W.; Knocke, W.R.; Falkinham, J.O.; Pruden, A.; Singh, G. Effect of Drinking Water Treatment Process Parameters on Biological Removal of Manganese from Surface Water. *Water Res.* **2014**, *66*, 31–39. [CrossRef]
22. Khadse, G.K.; Patni, P.M.; Labhasetwar, P.K. Removal of Iron and Manganese from Drinking Water Supply. *Sustain. Water Resour. Manag.* **2015**, *1*, 157–165. [CrossRef]
23. Jasim, S.Y.; Mohseni, M. Ozone Application for Arsenic and Manganese Treatment at the City of White Rock, BC, Canada. *Ozone Sci. Eng.* **2019**, *41*, 322–331. [CrossRef]
24. Hua, M.; Zhang, S.; Pan, B.; Zhang, W.; Lv, L.; Zhang, Q. Heavy Metal Removal from Water/Wastewater by Nanosized Metal Oxides: A Review. *J. Hazard. Mater.* **2012**, *211–212*, 317–331. [CrossRef]
25. Jerroumi, S.; Amarine, M.; Gourich, B. Technological Trends in Manganese Removal from Groundwater: A Review. *J. Water Process. Eng.* **2023**, *56*, 104365. [CrossRef]
26. Patil, D.S.; Chavan, S.M.; Oubagaranadin, J.U.K. A Review of Technologies for Manganese Removal from Wastewaters. *J. Environ. Chem. Eng.* **2016**, *4*, 468–487. [CrossRef]
27. Kouzbour, S.; El Azher, N.; Gourich, B.; Gros, F.; Vial, C.; Stiriba, Y. Removal of Manganese (II) from Drinking Water by Aeration Process Using an Airlift Reactor. *J. Water Process. Eng.* **2017**, *16*, 233–239. [CrossRef]
28. El-Sheikh, M.A.; Guirguis, H.S.; Fathy, A. A Comparative Study of Methods Used for Fe and Mn Oxidation and Removal from Groundwater Assessing the Impact of Climate Change on the Eastern Nile View Project Iron and Manganese Removal from Groundwater View Project. *J. Eng. Appl. Sci.* **2016**, *63*, 277–292.
29. Tobiason, J.E.; Bazilio, A.; Goodwill, J.; Mai, X.; Nguyen, C. Manganese Removal from Drinking Water Sources. *Curr. Pollut. Rep.* **2016**, *2*, 168–177. [CrossRef]
30. Kan, C.C.; Aganon, M.C.; Futalan, C.M.; Dalida, M.L.P. Adsorption of Mn^{2+} from Aqueous Solution Using Fe and Mn Oxide-Coated Sand. *J. Environ. Sci.* **2013**, *25*, 1483–1491. [CrossRef] [PubMed]
31. Mabate, T.P.; Maqunga, N.P.; Ntshibongo, S.; Maumela, M.; Bingwa, N. Metal Oxides and Their Roles in Heterogeneous Catalysis: Special Emphasis on Synthesis Protocols, Intrinsic Properties, and Their Influence in Transfer Hydrogenation Reactions. *SN Appl. Sci.* **2023**, *5*, 196. [CrossRef]
32. Michel, M.M.; Reczek, L.; Papciak, D.; Wlodarczyk-Makula, M.; Siwiec, T.; Trach, Y. Mineral Materials Coated with and Consisting of MnOx—Characteristics and Application of Filter Media for Groundwater Treatment: A Review. *Materials* **2020**, *13*, 2232. [CrossRef]
33. Jez-Walkowiak, J.; Dymaczewski, Z.; Szuster-Janiaczyk, A.; Nowicka, A.B.; Szybowicz, M. Efficiency of Mn Removal of Different Filtration Materials for Groundwater Treatment Linking Chemical and Physical Properties. *Water* **2017**, *9*, 498. [CrossRef]

34. Rudi, N.N.; Muhamad, M.S.; Te Chuan, L.; Alipal, J.; Omar, S.; Hamidon, N.; Abdul Hamid, N.H.; Mohamed Sunar, N.; Ali, R.; Harun, H. Evolution of Adsorption Process for Manganese Removal in Water via Agricultural Waste Adsorbents. *Heliyon* **2020**, *6*, e05049. [CrossRef]
35. Kalvani, N.; Mesdaghinia, A.; Yaghmaeian, K.; Abolli, S.; Saadi, S.; Rashidi Mehrabadi, A.; Alimohammadi, M. Evaluation of Iron and Manganese Removal Effectiveness by Treatment Plant Modules Based on Water Pollution Index: A Comprehensive Approach. *J. Environ. Health Sci. Eng.* **2021**, *19*, 1005–1013. [CrossRef]
36. Bruins, J.H.; Petrusevski, B.; Slokar, Y.M.; Huysman, K.; Joris, K.; Kruithof, J.C.; Kennedy, M.D. Reduction of Ripening Time of Full-Scale Manganese Removal Filters with Manganese Oxide-Coated Media. *J. Water Supply Res. Technol. AQUA* **2015**, *64*, 434–441. [CrossRef]
37. Haukelidsaeter, S.; Boersma, A.S.; Piso, L.; Lenstra, W.K.; van Helmond, N.A.G.M.; Schoonenberg, F.; van der Pol, E.; Hurtarte, L.C.C.; van der Wielen, P.W.J.J.; Behrends, T.; et al. Efficient Chemical and Microbial Removal of Iron and Manganese in a Rapid Sand Filter and Impact of Regular Backwash. *Appl. Geochem.* **2024**, *162*, 105904. [CrossRef]
38. Bruins, J.H.; Vries, D.; Petrusevski, B.; Slokar, Y.M.; Kennedy, M.D. Assessment of Manganese Removal from over 100 Groundwater Treatment Plants. *J. Water Supply: Res. Technol. AQUA* **2014**, *63*, 268–280. [CrossRef]
39. Haukelidsaeter, S.; Boersma, A.S.; Kirwan, L.; Corbetta, A.; Gorres, I.D.; Lenstra, W.K.; Schoonenberg, F.K.; Borger, K.; Vos, L.; van der Wielen, P.W.J.J.; et al. Influence of Filter Age on Fe, Mn and NH4+ Removal in Dual Media Rapid Sand Filters Used for Drinking Water Production. *Water Res.* **2023**, *242*, 120184. [CrossRef]
40. Pathak, D. Water Availability and Hydrogeological Condition in the Siwalik Foothill of East Nepal. *Nepal J. Sci. Technol.* **2016**, *17*, 31–38. [CrossRef]
41. Pathak, D. Status of Groundwater Exploitation and Investigation in Terai and Inner Terai Region of Nepal. *Bull. Nepal Hydrogeol. Assoc.* **2018**, *3*, 77–83.
42. Shrestha, S.R.; Tripathi, G.N.; Laudari, D. Groundwater Resources of Nepal: An Overview. In *Groundwater of South Asia*; Springer: Berlin/Heidelberg, Germany, 2018; pp. 169–193. [CrossRef]
43. Mahato, S.; Mahato, A.; Karna, P.K.; Balmiki, N. Investigating Aquifer Contamination and Groundwater Quality in Eastern Terai Region of Nepal. *BMC Res. Notes* **2018**, *11*, 321. [CrossRef]
44. Liu, J.; Chen, Q.; Yang, Y.; Wei, H.; Laipan, M.; Zhu, R.; He, H.; Hochella, M.F. Coupled Redox Cycling of Fe and Mn in the Environment: The Complex Interplay of Solution Species with Fe- and Mn-(Oxyhydr)Oxide Crystallization and Transformation. *Earth Sci. Rev.* **2022**, *232*, 104105. [CrossRef]
45. National Statistics Office. National Population and Housing Census Provincial Report (Koshi Province). 2021. Available online: https://censusnepal.cbs.gov.np/results/files/result-folder/National%20Report_English.pdf (accessed on 24 August 2024).
46. Ghimire, D.; Dhakal, D.R.; Bhusal, G.; Khadka, A.K.; Shrestha, S.; Regmi, R.P. Post Monsoon Season Spatial Distribution of Groundwater Depth over the Low-Lands of Sunsari District of Eastern Nepal. *J. Nepal Phys. Soc.* **2022**, *8*, 32–38. [CrossRef]
47. National Statistics Office. National Population and Housing Census 2021 National Report. Available online: https://censusnepal.cbs.gov.np/results/files/result-folder/National%20Report_English.pdf (accessed on 24 August 2024).
48. Khadka, D.B.; Bhattarai, M. Delineation of Ground Water Potential Zoning Using GIS and Remote Sensing by AHP of Sunsari District (Koshi Basin) Area of Nepal. *J. Eng. Res. Rep.* **2023**, *24*, 42–61. [CrossRef]
49. Acharya, K.K. Local Governance Restructuring in Nepal: From Government to Governmentality. *Dhaulagiri J. Sociol. Anthropol.* **2018**, *12*, 37–49. [CrossRef]
50. Neupane, S.; Gajurel, A.P.; Shakya, N.; Lupker, M.; Hada, R. Assessment on Connection between Shallow and Deep Aquifers Using Isotope Analysis of Surface Water and Groundwater in Sunsari and Morang Districts. *J. Nepal Geol. Soc.* **2019**, *59*, 73–78. [CrossRef]
51. Shrestha, A.; Kazama, S.; Takizawa, S. Influence of Service Levels and COVID-19 on Water Supply Inequalities of Community-Managed Service Providers in Nepal. *Water* **2021**, *13*, 1349. [CrossRef]
52. Kumar Mishra, A.; Kumar Karna, A. Assessment of Management Transfer Process of Water Supply Systems in Sunsari District of Nepal. *Int. J. Res.-Granthaalayah* **2019**, *7*, 1–24. [CrossRef]
53. NWASH. Available online: https://nwash.gov.np/ (accessed on 25 November 2022).
54. Bhandari, M.R.; Aryal, K.; Bhanadari, M.R. Status of Water Sources and Its Impact on Livelihood Practices in Sardhukhola Watershed of Eastern Nepal Agriculture and Forestry Journal Status of Water Sources and Its Impact on Livelihood Practices in Sardhukhola Watershed of Eastern Nepal (Sunsari). *Agric. For. J.* **2020**, *4*, 19–31. [CrossRef]
55. Government of Nepal Ministry of Energy, Water Resource and Irrigation Department of Hydrology and Meteorology Climate Division. Preliminary Precipitation and Temperature Summary. 2023. Available online: https://www.dhm.gov.np/uploads/dhm/climateService/Preliminary_precipitation_and_temperature_summary_July_2023.pdf (accessed on 24 August 2024).
56. The Collision Theory. Available online: https://chem.libretexts.org/Bookshelves/Physical_and_Theoretical_Chemistry_Textbook_Maps/Supplemental_Modules_(Physical_and_Theoretical_Chemistry)/Kinetics/06:_Modeling_Reaction_Kinetics/6.01:_Collision_Theory/6.1.06:_The_Collision_Theory (accessed on 22 May 2024).
57. Hach Company Oxidation Reduction Potential (ORP) Direct Measurement Method Method 10228-2000 to 2000 MV ORP Electrode. 2021. Available online: https://images.hach.com/ (accessed on 24 August 2024).
58. Government of Nepal National Drinking Water Quality Standard 2023: Implementation Directive. Available online: https://dwssm.gov.np/storage/listies/March2023/khanepani-mapdanda-kitab-final.pdf (accessed on 8 May 2024).

59. Fu, X.; Niu, Z.; Lin, M.; Gao, Y.; Sun, W.; Yue, T. Strengthened Oxygen Oxidation of Ferrous Ions by a Homemade Venturi Jet Microbubble Generator towards Iron Removal in Hydrometallurgy. *Minerals* **2021**, *11*, 1342. [CrossRef]
60. Kent, R.; Landon, M.K. Trends in Concentrations of Nitrate and Total Dissolved Solids in Public Supply Wells of the Bunker Hill, Lytle, Rialto, and Colton Groundwater Subbasins, San Bernardino County, California: Influence of Legacy Land Use. *Sci. Total Environ.* **2013**, *452–453*, 125–136. [CrossRef]
61. Siwiec, T.; Michel, M.M.; Reczek, L.; Nowak, P. The Influence of Aeration on the Change in Corrosiveness and Aggressiveness of Groundwater. *Water Supply* **2016**, *16*, 445–452. [CrossRef]
62. Goncharuk, V.V.; Bagrii, V.A.; Mel'nik, L.A.; Chebotareva, R.D.; Bashtan, S.Y. The Use of Redox Potential in Water Treatment Processes. *J. Water Chem. Technol.* **2010**, *32*, 1–9. [CrossRef]
63. Wątor, K.; Dobrzyński, D.; Sugimori, K.; Kmiecik, E. Redox Potential Research in the Field of Balneochemistry: Case Study on Equilibrium Approach to Bioactive Elements in Therapeutic Waters. *Int. J. Biometeorol.* **2020**, *64*, 815–826. [CrossRef]
64. Vries, D.; Bertelkamp, C.; Schoonenberg Kegel, F.; Hofs, B.; Dusseldorp, J.; Bruins, J.H.; de Vet, W.; van den Akker, B. Iron and Manganese Removal: Recent Advances in Modelling Treatment Efficiency by Rapid Sand Filtration. *Water Res.* **2017**, *109*, 35–45. [CrossRef]
65. Islam, M.Z.; Mostafa, M.G. Seasonal Variation of Fe, Mn, and Pb in Groundwater of Northwestern Bangladesh. *J. Chem. Environ.* **2024**, *3*, 77–97. [CrossRef]
66. Kousa, A.; Komulainen, H.; Hatakka, T.; Backman, B.; Hartikainen, S. Variation in Groundwater Manganese in Finland. *Environ. Geochem. Health* **2021**, *43*, 1193–1211. [CrossRef] [PubMed]
67. Liu, W.; Qin, D.; Yang, Y.; Guo, G. Enrichment of Manganese at Low Background Level Groundwater Systems: A Study of Groundwater from Quaternary Porous Aquifers in Changping Region, Beijing, China. *Water* **2023**, *15*, 1537. [CrossRef]
68. Cheng, L.H.; Xiong, Z.Z.; Cai, S.; Li, D.W.; Xu, X.H. Aeration-Manganese Sand Filter-Ultrafiltration to Remove Iron and Manganese from Water: Oxidation Effect and Fouling Behavior of Manganese Sand Coated Film. *J. Water Process. Eng.* **2020**, *38*, 101621. [CrossRef]
69. Basic Chemistry of Chlorination. Available online: https://www.hydroinstruments.com (accessed on 30 June 2024).
70. Wei, N.; Zhang, Z.; Liu, D.; Wu, Y.; Wang, J.; Wang, Q. Coagulation Behavior of Polyaluminum Chloride: Effects of PH and Coagulant Dosage. *Chin. J. Chem. Eng.* **2015**, *23*, 1041–1046. [CrossRef]
71. Takeno, N. Atlas of Eh-PH Diagrams Intercomparison of Thermodynamic Databases. 2005. Available online: https://www.nrc.gov/docs/ML1808/ML18089A638.pdf (accessed on 24 August 2024).
72. Kobylin, P.; Mäenpää, L.; Roine, A.; Anttila, K. E—PH (Pourbaix) Diagrams Module. 2014. Available online: https://www.metso.com/globalassets/portfolio/hsc-chemistry/17-eh---ph---diagrams.pdf (accessed on 24 August 2024).
73. Iwuozor, K.O.; Ighalo, J.O.; Chizitere Emenike, E.; Adaobi Igwegbe, C.; Adeniyi, A.G. Do Adsorbent Pore Size and Specific Surface Area Affect the Kinetics of Methyl Orange Aqueous Phase Adsorption? *J. Chem. Lett* **2021**, *2*, 188–198.
74. Water Safety Portal. Available online: https://wsportal.org/ (accessed on 23 August 2024).

water

MDPI

Article

Assessment of Human Errors in the Operation of the Water Treatment Plant

Jakub Żywiec [1,*], Barbara Tchórzewska-Cieślak [1] and Kateryna Sokolan [2]

[1] Department of Water Supply and Sewerage Systems, Faculty of Civil, Environmental Engineering and Architecture, Rzeszow University of Technology, Al. Powstańców Warszawy 6, 35-959 Rzeszow, Poland; cbarbara@prz.edu.pl

[2] Department of Mechanical Engineering Technology, Faculty of Engineering, Transport and Architecture, Khmelnytsky National University, 11 Instytutska Str., 29016 Khmelnytsky, Ukraine; sokolan.kateryna@gmail.com

* Correspondence: j.zywiec@prz.edu.pl

Abstract: The water supply system (WSS) is an anthropotechnical system whose reliability depends on proper human activity. Research indicates that 75% of WSS failures are due to human errors. The water treatment plant (WTP) is a key element of the WSS. The water treatment process requires human control as the operator. His task is to maintain an appropriate level of reliability and safety for the system by controlling the technological objects. The aim of the work was to assess the reliability of the WTP operator. The paper presents a Human Reliability Assessment (HRA) of the operator of the WTP using the Fuzzy-Bayes CREAM method. The values of the Human Error Probability (HEP) for operators were determined, which are key to carrying out further analyses of the human impact on the reliability and operational safety of anthropotechnical critical infrastructure systems. The HEP value of the water treatment plant operator varies in the range of 0.0005–0.0746 (depending on the technological process). Identification of new threats related to the impact of the human factor on the WSS's functioning and taking them into account in reliability calculations will allow for a better representation of actual operating conditions.

Keywords: anthropotechnical system; human error; human reliability; water supply; water treatment

1. Introduction

Critical infrastructure systems must be characterized by high technical reliability and operational safety [1]. The management of such systems should be based on a strategy aimed at minimizing the risk of their failure [1]. Among the causes of failure, apart from technical or environmental reasons, the direct or indirect impact of the human factor is also important [2]. It can affect the system already at the design stage, at the stage of system construction, or finally during its operation, creating conditions for failure [2]. Statistics show that human activities contribute to 60–90% of failures in various technical systems [2,3].

Access to drinking water is one of the most important aspects of society's functioning. Water supply systems (WSS) are classified as critical infrastructure systems [1]. They are also an example of an anthropotechnical system, i.e., a system consisting of three elements: man-technical object-environment [4]. Data available in the literature indicate that the human factor contributed to 75% of failures in water supply systems [5,6]. The impact of the human factor on the functioning of the technical system may be internal, including authorized activities (of staff), or external in relation to acts of vandalism, terrorism, or its incorrect use (actions of third parties) [4]. A man in the anthropotechnical system may turn out to be the weakest link. For this reason, in the early 1960s, research on human reliability [7] was started in order to determine the probability of the correct performance of the task by the technical staff, in a given time interval, at any time of the system operation. From the very beginning, there was a multidisciplinary approach to the subject of Human Reliability Assessment (HRA), which included the sciences of reliability, ergonomics, and psychology [2,8]. Over the last 60 years, over 70 different HRA methods have been presented [2,9,10]. In order to classify them, a division into three generations has been proposed [10]. The first generation of HRA covers the years 1960–1990 and includes such methods as HEART (Human Error Assessment and Reduction Technique) [11] and TESEO (Tecnica Empirica Stima Errori Operatori) [12]. These methods focus on quantifying the probability of human error. The second generation of HRA methods covers the years 1990–2005. The methods of this generation take into account the influence of the human's cognitive functions, such as memory, attention, thinking, and perception, as well as the context of the situation on the human's reliability. Examples of methods of this generation are ATHEANA (A Technique for Human Event Analysis) [13] and CREAM (Cognitive Reliability and Error Analysis Methods) [14]. The third generation of HRA, developed after 2005, focuses on the use of simulation methods to assess human reliability or proposes modifications of earlier generation methods, e.g., CARA (Controller Action Reliability Assessment) [15] and Bayes-SLIM (Success Likelihood Index Method) [16]. Methods of assessing human reliability have been widely used in many fields, including critical infrastructure sectors such as the energy sector [17–19], space sector [20], transport sector [15,21–26], and health sector [3,27]. The paper [17] presents the use of the HCR (Human Cognitive Reliability) method to determine the human error probability for hydropower station operation. The work [18] presents the use of the FTA (Fault Tree Analysis) method to calculate the human reliability of maintenance groups in power transmission grids. The work [19] presents an HRA analysis using the classic CREAM method for oil well workers. The CREAM method was also used in [20] to analyze human errors in the crew fueling process during a space launch. In paper [15], a tool for analyzing the human reliability of air traffic controllers was presented. HRA analyses related to air transport were also presented in the works [21,22], where a modified version of the CREAM method (with fuzzy logic) was used. HRA analyses are also presented for sea transport. The paper [23] presents an analysis using the fuzzy HEART method for cargo operation in offshore tanker units. Another modification of the HEART method using the Dempster–Shafer evidence theory approach was presented in [24] to assess human reliability for the gas-freeing process on chemical tankers. A fuzzy and Bayesian network CREAM model for human reliability analysis was presented in [25] for assessing the reliability of a tanker crew. A different approach to the CREAM method with fuzzy-based clustering was presented in [26] for assessing the human reliability of professional drivers. A proposal for a new HRA method was presented in the paper [3] for formal pharmaceutical human reliability assessment. The analysis of HRA in the medical area is also presented in paper [27], which presents human reliability assessment for medical devices based on failure mode and effects analysis and fuzzy linguistic theory. Based on the literature review, the continuous development of new HRA assessment methods and modifications of previous generation methods and their applications in various systems is observed. All this research allows us to better investigate human reliability and contribute to the development of knowledge in the field of reliability research.

The impact of the human factor on the functioning of the WSS is not fully recognized, which is reflected in the small number of research on this subject. Żywiec et al. [4] presented the concept of WSS safety assessment, including the influence of the system operator. Wu et al. [5] reviewed 62 failure cases of WSS in terms of identifying human errors. Tang et al. [6] presented an approach to water safety assessment based on resilience mechanisms. No direct HRA studies for WSS operators had been found in the available literature; what was the inspiration to undertake the research problem.

The aim of the work was to assess the reliability of the operator of the water treatment plant, whose task is to manage and control the operation of technological facilities. The obtained results will allow us to determine the impact of the human factor on the functioning of WTP facilities. The paper presents the results of Human Reliability Analysis for WTP operators, which was carried out using the Fuzzy-Bayes CREAM [25] method. The research was carried out with a group of 38 operators who work at 12 water treatment plants located in south-eastern Poland. The presented results show the Human Error Probability values (HEP) for different scenarios, including the operator's work in the processes of water intake, water treatment, water storage, and water pumping.

According to the authors, HRA should be included as a permanent element of the WSS reliability assessment. Recognition of the Human Error Probability (HEP) values allows for further WSS reliability research, which will take into account the human impact on the functioning of the anthropotechnical critical infrastructure system. Taking into account the previously ignored human factor in the reliability analysis will allow for a better representation of the actual operating conditions of the WSS, as a complex system whose reliable and safe operation depends on the interactions between man, the technical system, and the environment.

2. Materials and Methods

2.1. CREAM Framework

The CREAM method was introduced by E. Hollnagel in 1998 [14]. It has been used in many fields, such as: chemical industry, air transport, and sea transport [19,20,25]. It is based on the model of cognitive processes, the Contextual Control Model (COCOM), which describes the relationships between cognitive functions in the human mind [14]. In simple terms, this model assumes that every human action results from a controlled process of using their competences, depending on the requirements of a given situation. Human control can fluently change levels from complete lack to full control. In the COCOM model, there are four control modes, which are described by the values of the Human Error Probability (HEP) [14]. They are presented in Table 1.

Table 1. Human control modes and their probability [14].

Control Mode	Description	HEP Value
Scrambled (SC)	This mode characterizes situations in which the operator devotes little attention to the planning of subsequent actions or does not plan them at all. He takes action in a random, unplanned way. Most often, it concerns an unknown situation, when the operator loses the ability to think logically, does not analyze the possible solutions and effects of his actions.	$10^{-1} < P < 1$
Situational (ST)	The operator takes further actions based on the current state of the system, ignoring the achievement of the main task objective. Most often, this applies to situations in which there are time constraints, or the operator is unable to properly interpret the current state of the system.	$1 \times 10^{-2} < P < 5 \times 10^{-1}$
Tactical (TT)	The operator takes subsequent actions according to procedures or plans known to him. In the event of an unknown situation, the next action is taken after analyzing the operating parameters of the system and the context of the situation. It is based on familiar patterns of action.	$1 \times 10^{-3} < P < 1 \times 10^{-1}$
Strategic (SR)	The operator takes actions in a thoughtful and planned manner, knowing their consequences. His actions are directed toward achieving the main goal of the task. The operator relies on his knowledge and experience.	$5 \times 10^{-6} < P < 1 \times 10^{-2}$

The essence of the method is to determine the HEP value based on the operator's control mode [14]. To do this, nine Common Performance Conditions (CPC) factors must be assessed. They can have a positive, negative, or neutral impact on operator reliability. The 9 CPC factors are: work organization, working conditions, quality of the human-machine interface, availability of procedures/plans, number of simultaneously performed tasks, amount of time to complete the task, time of day, qualifications and experience, and team cooperation. The HEP value is determined based on the operator control mode, which is determined by the assessment of the CPC factors, presented in the form of a matrix [14]:

$$\{\sum N, \sum O, \sum P\} \tag{1}$$

where:

∑N—number of factors having a negative impact on operator reliability,

∑O—number of factors having a neutral impact on operator reliability,

∑P—number of factors having a positive impact on operator reliability.

Then, based on the model presented in Figure 1, the operator control mode is determined [14]. There are 52 possible result combinations. The most favorable is {0, 2, 7} and means that there are 0 factors negatively affecting the operator's reliability, 2 factors with a neutral impact, while 7 of the factors have a positive impact; the most unfavorable result possible to obtain is the combination {9, 0, 0}, where all 9 assessed CPC factors have a negative impact on operator reliability [14].

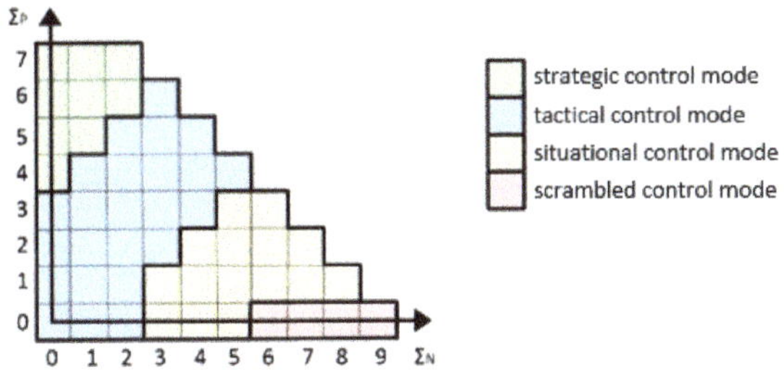

Figure 1. Control mode model [14].

2.2. *Use of Fuzzy Logic in CREAM*

The theory of fuzzy logic and fuzzy sets was introduced in 1965 by Zadeh [28]. The author introduced the concepts of fuzzy sets and fuzzy numbers, which can be used to mathematically model imprecise or uncertain quantities and have found wide practical applications in many fields of science [29,30]. Due to the fact that human action is complex and difficult to describe mathematically, fuzzy logic has often been used to modify HRA methods, including CREAM [22,25,27].

The fuzification process allows us to transform the input values into fuzzy numbers, which are described by the membership function, assigning values from 0 to 1 for all elements belonging to the set. With regard to the CREAM method, the works of Zhou et al. [25] and Jin et al. [22] present the use of fuzzy logic to modify the assessment of CPC factors. The proposal for assessing CPC factors for the WSS operator is presented in Table 2.

Table 2. Modification of CPC factors.

No.	CPC Name	CPC Level	Effect	Membership Function Parameters
CPC1	Adequacy of organization	adequate acceptable inadequate	positive neutral negative	(50, 90, 100, 100) (10, 50, 90) (0, 0, 10, 50)
CPC2	Working conditions	adequate acceptable inadequate	positive neutral negative	(50, 90, 100, 100) (10, 50, 90) (0, 0, 10, 50)
CPC3	Quality of the SCADA System	adequate acceptable inadequate	positive neutral negative	(50, 90, 100, 100) (10, 50, 90) (0, 0, 10, 50)
CPC 4	Availability of procedures and plans	adequate acceptable inadequate	positive neutral negative	(50, 90, 100, 100) (10, 50, 90) (0, 0, 10, 50)

Table 2. *Cont.*

No.	CPC Name	CPC Level	Effect	Membership Function Parameters
CPC 5	Number of simultaneous tasks	acceptable inadequate	neutral negative	(50, 90, 100, 100) (0, 0, 50, 90)
CPC 6	Available time	adequate acceptable inadequate	positive neutral negative	(50, 90, 100, 100) (10, 50, 90) (0, 0, 10, 50)
CPC 7	Time of day	acceptable inadequate	neutral negative	(50, 90, 100, 100) (0, 0, 50, 90)
CPC 8	Qualifications and training	adequate acceptable inadequate	positive neutral negative	(50, 90, 100, 100) (10, 50, 90) (0, 0, 10, 50)
CPC 9	Team collaboration quality	adequate acceptable inadequate	positive neutral negative	(50, 90, 100, 100) (10, 50, 90) (0, 0, 10, 50)

As a given factor may cause an increase or decrease in operator reliability or may have no impact on the level of operator reliability, a three-level rating scale was proposed for factors CPC1-CPC4, CPC6, CPC8, and CPC 9 as: appropriate, acceptable, inappropriate, and for factors CPC 5 and CPC7, a two-level scale as: acceptable and inappropriate [22,25]. Figure 2 shows a graphical interpretation of the proposed rating scale. The scale can be described using fuzzy numbers, described by membership functions according to Equations (2)–(4) for a three-level scale [22,25]:

$$\mu_{\text{INADEQUATE}}(x) = \begin{cases} 1, & x \leq 0 \\ 1, & 0 < x \leq 10 \\ (50 - x)/40, & 10 \leq x \leq 50 \\ 0, & x \geq 50 \end{cases} \tag{2}$$

$$\mu_{\text{ACCEPTABLE}}(x) = \begin{cases} 0, & x \leq 10 \\ (x - 10)/40, & 10 < x \leq 50 \\ (90 - x)/40, & 50 \leq x \leq 90 \\ 0, & x \geq 90 \end{cases} \tag{3}$$

$$\mu_{\text{ADEQUATE}}(x) = \begin{cases} 0, & x \leq 50 \\ (x - 50)/40, & 50 < x \leq 90 \\ 1, & 90 \leq x \leq 100 \\ 1, & x \geq 100 \end{cases} \tag{4}$$

and Equations (5) and (6) for the two-level scale [21,26]:

$$\mu_{\text{INADEQUATE}}(x) = \begin{cases} 1, & x \leq 0 \\ 1, & 0 < x \leq 50 \\ (90 - x)/40, & 50 \leq x \leq 90 \\ 0, & x \geq 90 \end{cases} \tag{5}$$

$$\mu_{\text{ACCEPTABLE}}(x) = \begin{cases} 0, & x \leq 50 \\ (x - 50)/40, & 50 < x \leq 90 \\ 1, & 90 \leq x \leq 100 \\ 1, & x \geq 100 \end{cases} \tag{6}$$

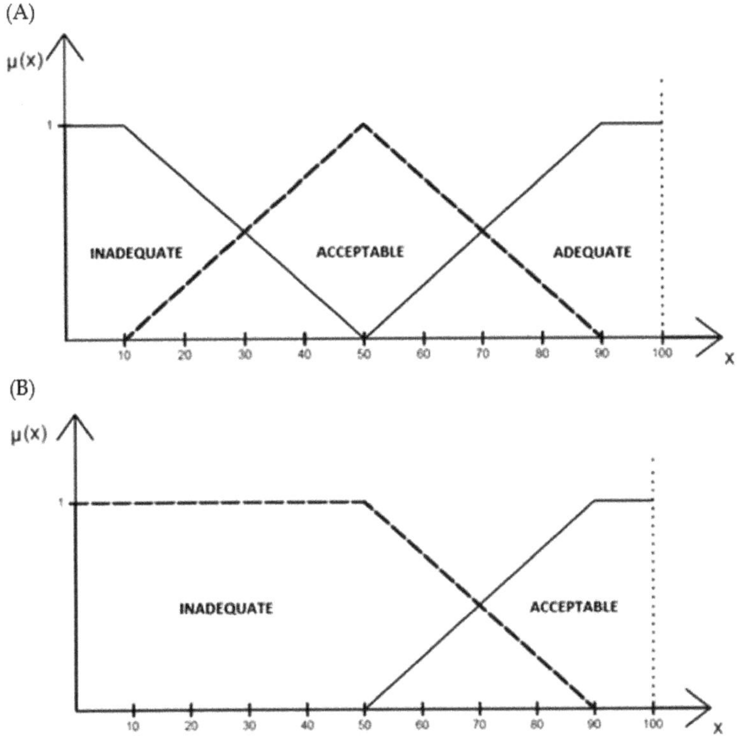

Figure 2. Membership function for the CPC factor rating scale: (**A**) three-level rating scale; (**B**) two-level rating scale (based on [22,25]).

Quantities describing control modes are also described by using fuzzy numbers. For this purpose, the HEP intervals were presented in the form of a decimal logarithm, as shown in Table 3. The values determined in this way were used for fuzzy modeling using triangular fuzzy numbers. The membership functions for individual control modes are described by Equations (7)–(10) [22,25]. Figure 3 shows a graphical interpretation of the adopted fuzzy model for control modes.

$$\mu_{SC}(x) = \begin{cases} 0, & x \le -1 \\ 2 \cdot x + 2, & -1 \le x \le -0.5 \\ -2 \cdot x, & -0.5 \le x \le 0 \\ 0, & x \ge 0 \end{cases} \tag{7}$$

$$\mu_{ST}(x) = \begin{cases} 0, & x \le -2 \\ \frac{x+2}{0.85}, & -2 \le x \le -1.15 \\ \frac{-0.3-x}{0.85}, & -1.15 \le x \le -0.3 \\ 0, & x \ge -0.3 \end{cases} \tag{8}$$

$$\mu_{TT}(x) = \begin{cases} 0, & x \le -3 \\ x+3, & -3 \le x \le -2 \\ -1-x, & -2 \le x \le -1 \\ 0, & x \ge -1 \end{cases} \tag{9}$$

$$\mu_{SR}(x) = \begin{cases} 0, & x \leq -5.3 \\ \frac{x+5.3}{1.65}, & -5.3 \leq x \leq -3.65 \\ \frac{-2-x}{1.65}, & -3.65 \leq x \leq -2 \\ 0, & x \geq -2 \end{cases} \quad (10)$$

Table 3. Dependence of control mode and HEP value.

Control Mode	Logarithmized HEP Value	Membership Function Parameters
Scrambled (SC)	0 < log10 (P) < −1	(−1; −0.5; 0)
Situational (ST)	−2 < log10 (P) < −0.3	(−2; −1.15; −0.3)
Tactical (TT)	−3 < log10 (P) < −1	(−3; −2; −1)
Strategic (SR)	−5.3 < log10 (P) < −2	(−5.3; −3.65; −2)

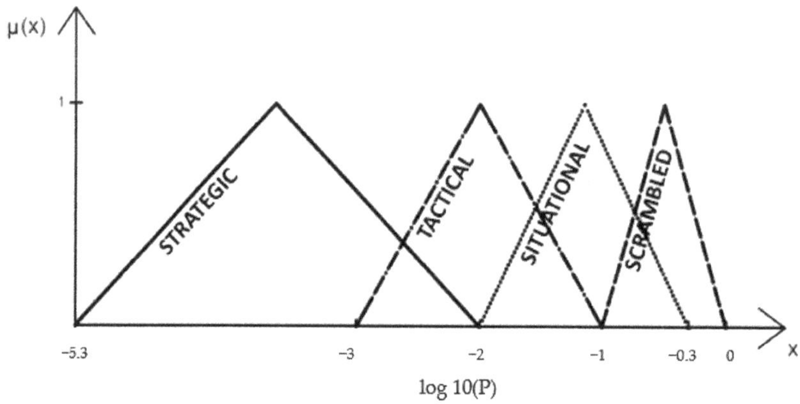

Figure 3. Membership function for control modes (based on [22,25]).

The opposite process of fuzzification is defuzzification. It involves transforming a number from a fuzzy form to a crisp form. A number of tools for defuzzification of fuzzy numbers have been presented in the literature [29–31]. One of the most popular defuzzification methods is the COG (Center of Gravity) method. In this method, the center of gravity of the figure obtained on the membership function graph is determined according to the equation [30]:

$$CV = \frac{\int x \cdot \mu(x) dx}{\int \mu(x) dx} \quad (11)$$

where:

- CV—crisp value–defuzzified value of fuzzy number,
- x—element of the real set (R), $x \in R$,
- μ(x)—membership function that assigns each element $x \in R$ its degree of membership in the fuzzy set.

2.3. Use of Bayesian Networks in CREAM

The Bayesian network is used to graphically present a probabilistic model for a group of random variables and the relationships between them [32]. It has many applications, including: in medicine, genetics, statistics, economics, and environmental engineering [33–35]. A Bayesian network is a pair (D, P), where D is a directed acyclic graph and P is a probability distribution. The network is composed of nodes representing random variables and directed arcs representing relationships between subsequent variables [32]. If the variable X_j has an influence on the variable X_i, there is an arc illustrating this relationship coming from the node X_j and entering the node X_i. Node X_i is then called the "parent" of node X_j. Each node in the network stores a probability distribution $P(X_i \mid X_{\pi(i)})$ where $X_{\pi(i)}$ is the set of nodes corresponding to π(i) parents of node i [32]. The conditional probability values stored at each node, form a table of conditional probabilities.

The probability distribution of all random variables in the network is described by the equation [32,36]:

$$P(X_1,\ldots,X_n) = \prod_{i=1}^{n} P\left(X_i \middle| X_{\pi(i)}\right) \tag{12}$$

Probability values for individual variables can be determined using the inference by enumeration method [32,36]. Let us assume that among the studied variables, we distinguish the variable Q, dependent on the variables E_1, E_2, ..., E_k and the remaining variables H_1, H_2, ..., H_r. Joint probability for variables Q and E_1, E_2, ..., E_n is described by the formula [32,36]:

$$P(Q, E_1, E_2, \ldots, E_k) = \sum_{h1\ldots hr} P(Q, H_1 H_2, \ldots, H_r, E_1, E_2, \ldots, E_k) \tag{13}$$

The works of Zhou et al. [25] and Jin et al. [22] present the possibility of using Bayesian networks in the CREAM method. The essence of the CREAM method is to determine the operator's control mode during the work and then the value of Human Error Probability. This value varies depending on the given circumstances, which are described by CPC factors. The relationship between CPC factors and control mode can be represented using a Bayesian network. Due to the fact that 7 CPC factors have 3 impact effects on operator reliability, and the remaining 2 factors have 2 impact effects, so for the entire set there are $3^7 \times 2^2$ = 8748 different combinations of CPC factor evaluation. For each combination, as shown in Figure 1, an appropriate control mode is assigned. By using a Bayesian network, it is possible to determine the probability distribution of the control mode for each combination. To reduce the complexity of calculations, CPC factors have been divided into three categories: organizational factors, environmental factors, and work-related factors [22,25]. Figure 4 shows a Bayesian network that can be used to determine the probability distribution for control modes. The first layer in the network is the fuzzified survey score of CPC factors, which are input data for the presented network. The second layer aims to simplify the calculations by dividing the CPC factors into groups:

- organizational factors–include training, experience, and skills of operators, organization, planning, supervision of work processes, enterprise administration, etc.,
- environmental factors–related to the work environment, e.g., lighting, noise, vibration, temperature, air humidity, ergonomics of the workplace,
- work-related factors–refer to the work process, e.g., length of shift, number of tasks to be performed simultaneously, characteristics of the task, and the stressful nature of the task.

The last layer allows you to determine the probability distribution for the control mode, as the output values for the analyzed network.

The relationships between individual nodes in the Bayesian network presented in Figure 4 are as follows: $\pi(X_{13}) = (X_{10}, X_{11}, X_{12})$, $\pi(X_{12}) = (X_4, X_5, X_6)$, $\pi(X_{11}) = (X_2, X_3, X_7)$, $\pi(X_{10}) = (X_1, X_8, X_9)$, $\pi(X_1) = (\varnothing)$, $\pi(X_2) = (\varnothing)$, $\pi(X_3) = (\varnothing)$, $\pi(X_4) = (\varnothing)$, $\pi(X_5) = (\varnothing)$, $\pi(X_6) = (\varnothing)$, $\pi(X_7) = (\varnothing)$, $\pi(X_8) = (\varnothing)$, $\pi(X_9) = (\varnothing)$. The probability value at each node is described by Equations (14)–(17).

$$P(X_{10}, X_1, X_8, X_9) = \sum_{X1}\sum_{X8}\sum_{X9} P(X_{10}|X_1, X_8, X_9) \cdot P(X_1) \cdot P(X_8) \cdot P(X_9) \tag{14}$$

$$P(X_{11}, X_2, X_3, X_7) = \sum_{X2}\sum_{X3}\sum_{X7} P(X_{11}|X_2, X_3, X_7) \cdot P(X_2) \cdot P(X_3) \cdot P(X_7) \tag{15}$$

$$P(X_{12}, X_4, X_5, X_6) = \sum_{X4}\sum_{X5}\sum_{X6} P(X_{12}|X_4, X_5, X_6) \cdot P(X_4) \cdot P(X_5) \cdot P(X_6) \tag{16}$$

$$P(X_{13}, X_{10}, X_{11}, X_{12}) = \sum_{X10}\sum_{X11}\sum_{X12} P(X_{13}|X_{10}, X_{11}, X_{12}) \cdot P(X_{10}) \cdot P(X_{11}) \cdot P(X_{12}) \tag{17}$$

The dependencies presented are based on probability theory and may be difficult to calculate due to their complexity. There are computer programs designed to perform such calculations for Bayesian networks. In this paper, the GeNIe 3.0 program from BayesFusion, LLC (Pittsburgh, PA, USA) [37] was used to perform calculations for the Bayesian network presented in Figure 4.

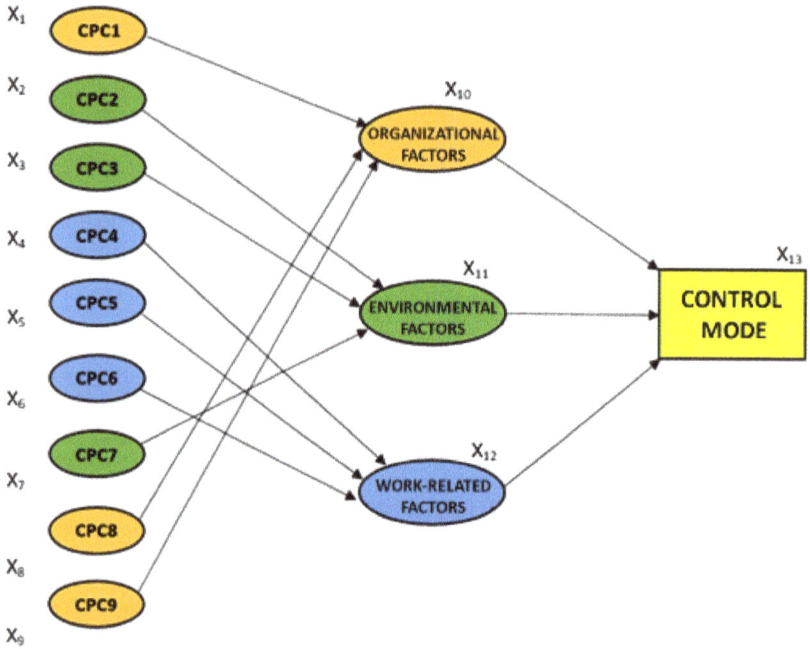

Figure 4. Bayesian network proposed for the determination of the control mode probability distribution.

The probability distribution for the operator control mode obtained in the final node should be defuzzified using the COG method to obtain a crisp value (CV) from the fuzzy form, in accordance with Equation (11). Based on Figure 3, the value of the Human Error Probability (HEP) should be calculated from the equation [22,25]:

$$\text{HEP} = 10^{\text{CV}} \tag{18}$$

2.4. Research Object

The research object was a group of water treatment plant operators, including 38 operators working in 12 water treatment plants located in south-eastern Poland. For the research, persons working as operators were qualified, whose task was to control the operation of technological objects and supervise the operation of the water treatment plants using available means (SCADA system, manual or local control of technological objects). HRA was carried out for 6 scenarios covering the operator's work in individual subsystems, i.e., the water intake subsystem, the water treatment subsystem (separately for technological processes of coagulation, filtration, and disinfection), the water storage subsystem, and the water pumping subsystem. Operators were subjected to surveys in which they assessed individual CPC factors on a scale of 0–100. A survey form is presented in Table 4.

Table 4. Survey questionnaire.

	CPC Factor	Score *Scale: 0 Inadequate, 50 Acceptable, 100 Adequate*
CPC 1	Adequacy of organization	(answer)
CPC 2	Working conditions	(answer)
CPC 3	Quality of the SCADA interface	(answer)
CPC 4	Availability of procedures and plans	(answer)
CPC 5	Number of simultaneous tasks	(answer)
CPC 6	Available time	(answer)
CPC 7	Time of day	(answer)
CPC 8	Qualifications and training	(answer)
CPC 9	Crew collaboration quality	(answer)

Tables S1 and S2 in the Supplementary Materials present the collective results of the survey carried out within the group of tested operators for the 6 analyzed scenarios.

To summarize the methodological approach, Figure 5 presents the individual stages of operator reliability analysis using the Fuzzy-Bayes CREAM method.

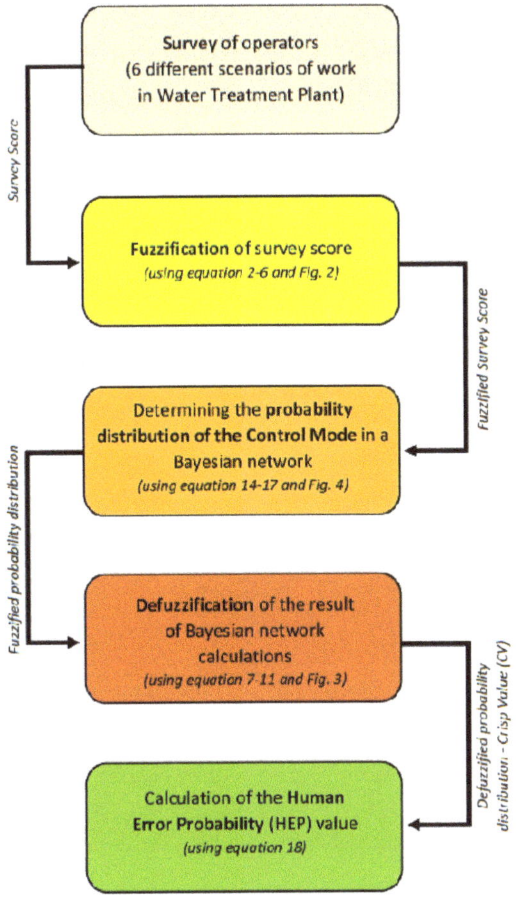

Figure 5. Stages of HRA analysis using the Fuzzy-Bayes CREAM method.

3. Results

In order to illustrate the reasoning, the sequence of calculations is presented with the example of one operator for the water filtration process. Table 5 presents the survey score for CPC factors assessment for operator O1 in the water filtration process and their form after the fuzzification process in accordance with Equations (2)–(4) and Figure 2. The fuzzy values were then entered into BayesFusion GeNIe 3.0 software, and Bayesian network calculations were performed using Equations (14)–(17). The results obtained with the program present the probability distribution for the control mode in which the tested operator works. The results are presented in Figure 6. The operators survey scores of O1–O19 are shown in Table S1, the operators survey scores of O20–O38 are shown in Table S2.

The next step is to defuzzify the obtained results using the COG method to obtain the Crisp Value (CV) according to Equation (11), and then the HEP value from Equation (18).

$$CV = \frac{\int x\cdot\mu_T(x)dx}{\int \mu_T(x)dx} == \frac{\int_{-5.3}^{-5.266} x\cdot\frac{x+5.3}{1.65}dx + \int_{-5.266}^{-2.980} x\cdot 0.020407dx +}{\int_{-5.3}^{-5.266} \frac{x+5.3}{1.65}dx + \int_{-5.266}^{-2.980} 0.020407dx +}$$
$$\frac{+\int_{-2.980}^{-2.354} x\cdot(x+3)dx + \int_{-2.354}^{-1.646} x\cdot 0.646364dx +}{+\int_{-2.980}^{-2.354} (x+3)dx + \int_{-2.354}^{-1.646} 0.646364dx +}$$
$$\frac{+\int_{-1.646}^{-1.314} x\cdot(-x-1)dx + \int_{-1.314}^{-0.614} x\cdot 0.31407dx +}{+\int_{-1.646}^{-1.314} (-x-1)dx + \int_{-1.314}^{-0.614} 0.31407dx +} \quad (19)$$
$$\frac{+\int_{-0.614}^{-0.316} x\cdot(-x-0.3)dx + \int_{-0.316}^{-0.010} x\cdot 0.019157 +}{+\int_{-0.614}^{-0.316} (-x-0.3)dx + \int_{-0.316}^{-0.010} 0.019157\,dx +}$$
$$\frac{+\int_{-0.010}^{0} x\cdot(-2x)dx}{+\int_{-0.010}^{0} -2x\,dx} = -1.8481$$

$$HEP = 10^{-1.8481} = 0.0142 \quad (20)$$

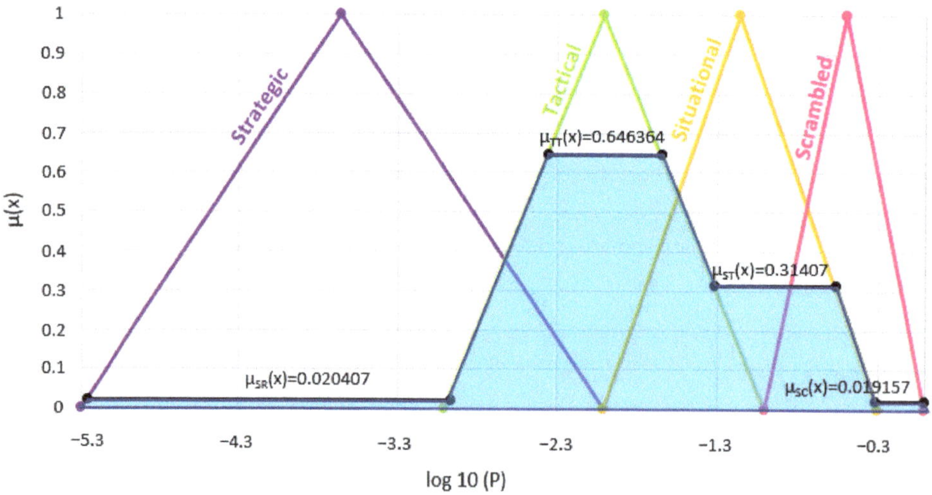

Figure 6. Graph of the probability distribution for the control mode of operator O1 for the water filtration process.

The results for the other operators and other tested scenarios are presented in Tables S3–S8 in the Supplementary Materials.

Table 5. Survey score for operator O1 for the water filtration process.

CPC Factor	Survey Score	Fuzzy Values
CPC1	50	0; 1; 0
CPC2	60	0.25; 0.75; 0
CPC3	50	0; 1; 0
CPC4	50	0; 1; 0
CPC5	60	0.25; 0,75
CPC6	60	0.25; 0.75; 0
CPC7	50	0; 1
CPC8	70	0.5; 0.5; 0
CPC9	60	0.25; 0.75; 0

Figure 7 shows the results of the HRA analysis for the tested group of operators in the form of Human Error Probability (HEP) values for six scenarios regarding operator work in the water intake subsystem (WI SubS), water treatment subsystem in the coagulation process (WT SubS: C), water treatment subsystem in the filtration process (WT SubS: F), water treatment subsystem in the disinfection process (WT SubS: D), water pumping subsystem (WP SubS), and water storage subsystem (WS SubS).

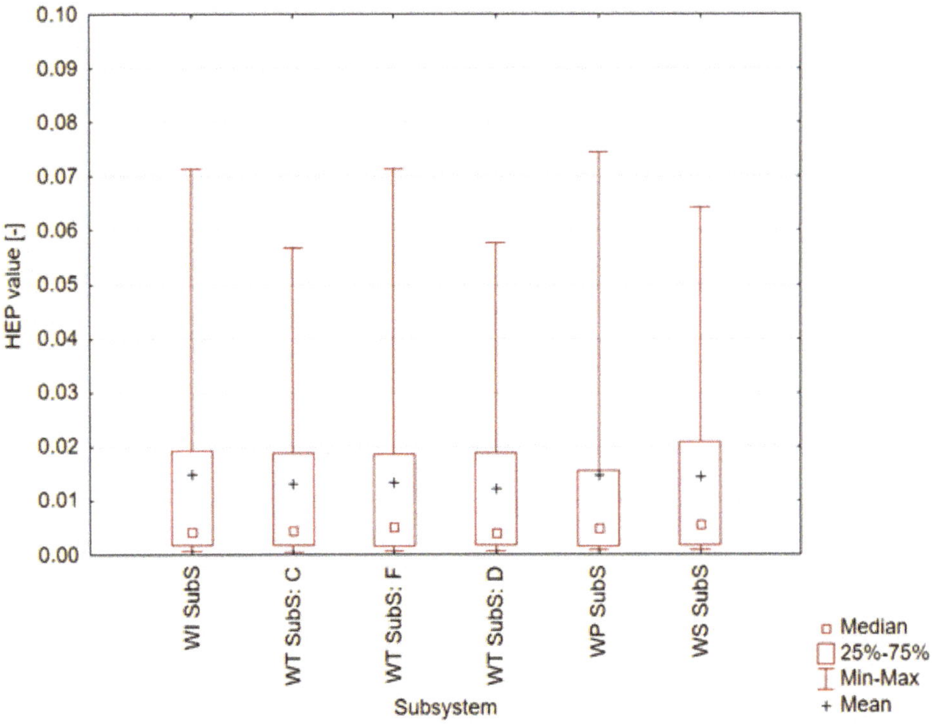

Figure 7. HEP values for the tested scenarios.

The highest average HEP value was determined for operators in the water intake subsystem (0.0148), while the lowest average HEP value was determined in the water treatment subsystem in the disinfection process (0.0122). The highest unit HEP value was determined for the operator in the water pumping subsystem (0.0746). The lowest unit HEP value was observed for the operator in the water treatment subsystem in the coagulation process (0.0005). The obtained values, in relation to the criteria of the classic CREAM method version, classified the operators' work in situational, tactical, and strategic control modes. The use of the Fuzzy-Bayes CREAM method allowed obtaining precise HEP values for each operator, which is necessary for using these data in further reliability analyses, e.g., with the fault tree analysis or the parametric method [4,25,38]. A detailed analysis of the obtained results is presented in the next chapter.

4. Discussion

For all processes in the water treatment subsystem, lower average HEP values are observed than in the other tested subsystems (average value of $HEP_{WT\ SubS:\ C} = 0.0132$; $HEP_{WT\ SubS:\ F} = 0.0147$; $HEP_{WT\ SubS:\ D} = 0.0122$). The coagulation process involves removing colloids and hard-to-fall particles from water, which are responsible for the turbidity and color of the water, as a result of adding chemicals to the water. In the coagulation process, the operator's main tasks include adjusting the doses of reagents to changes in water quality parameters as well as inspections and maintenance of devices and equipment of technological facilities [4]. In the filtration process, water is passed through a filter bed, where contaminants are removed in the form of suspension. Filter operation control consists of maintaining a constant filtration speed while increasing hydraulic pressure losses on the filter bed, determining the end of the filtration cycle, and then carrying out the procedure of rinsing the bed and putting it back into operation. In addition, the operator is also responsible for replacing the filling of the filter bed, checking the technical condition of the filter drainage, washing troughs, pipelines, and filter equipment [4]. The purpose of the disinfection process is to neutralize living and spore forms of pathogenic organisms and to protect water against secondary contamination in the water supply network. The most popular disinfectants include chlorine gas Cl_2, chlorine dioxide ClO_2, sodium hypochlorite, and UV radiation. Due to the use of hazardous chemicals, the operation of devices intended for water disinfection and the preparation of disinfectants must be carried out with particular care and in compliance with applicable occupational safety and health requirements. The main tasks of the operator in the disinfection process include adjusting the dose of disinfectant to the demand for chlorine, setting the power of UV radiators, and checking the technical condition of devices and technological facilities [4]. Improper operation of the water treatment subsystem poses a threat to human life and health. The operator's work in the water treatment subsystem, due to its specificity, has the greatest impact on the reliability and safety of the water production process; therefore, operators must have high competence. In the water treatment plants where the research was carried out, operators in the water treatment subsystem had a number of facilities that allowed them to minimize errors, including an extensive control system, an operating parameter monitoring system, an alarm system, workplace instructions, and periodic operator training. The obtained results confirm the effectiveness of such activities in reducing threats resulting from erroneous human activity. In other subsystems, the HEP values are higher, which may be due to the fact that the impact of the failure of these subsystems on the reliability and safety of the water production process is smaller than the failure of the water treatment subsystem [39]. The second most important subsystem influencing the reliability and safety of water production is the water pumping subsystem, followed by the water storage subsystem, while the water intake subsystem has the smallest impact [39]. The HEP values obtained in the remaining subsystems are proof of the statement above, where the highest average HEP value (0.0148) was obtained for the water intake subsystem, while for the pumping and storage subsystems, the obtained HEP values were lower at: 0.0146 and 0.0145. The main tasks of operators in the water storage subsystem are to monitor the operation of tanks and control valves on the water supply and discharge pipes and the drainpipe in the tanks. Other operational tasks include monitoring the water level in the tank, checking water quality, checking the condition of the tank and the valve chamber (leakage control), as well as maintaining the tank (washing, disinfection) [4]. In the water pumping subsystem, the operator's tasks include supervising the operation of pumps and their proper operation, including proper start-up and shutdown of the units, periodic tests of the efficiency of the units, and qualification of damaged or worn-out pumps for renovation or replacement [4]. The operator in the water intake subsystem is responsible for regulating the amount of water intake depending on the demand for water at a given moment, the hydrological or hydrogeological situation, or the observable threat to water quality [4]. Other operational tasks in the water intake subsystem include: for surface water intakes-observation of the lowest water levels during the year, observation of flood surges, movement of the bottom and debris, removal of coastal vegetation, protection of water against blooms, and control of the quality of the intake water; for groundwater intakes-water quality control, observation of well efficiency, observation of water table levels in the well and nearby of the intake, removal of sediments formed on the well filter, and, if necessary, renovation and desanding [4].

The differences between HEP values in individual subsystems are small, but they show the work priorities of WTP operators. The greatest threats, the effects of which are also of the greatest importance (threat to the health and life of the consumer), concern the operator's work in the water treatment subsystem. Maintaining the appropriate quality of treated water that meets quality standards is the most important task of WTP operators and requires the greatest level of skills and concentration when performing their work. The operation of the remaining subsystems, including the water intake subsystem, water pumping subsystem, and water storage subsystem, is more related to ensuring the appropriate amount of water supplied to consumers. Their operation seems to be more resistant to human errors from the technical side (e.g., the existence of reserve water intakes; raw water ponds; backup pumps; independent treated water tanks). Therefore, the work of operators in these subsystems is associated with less responsibility than in the water treatment subsystem, which is confirmed by the results obtained.

5. Conclusions

The water production process is an extensive process; it includes the operator's work in the water intake subsystem, water treatment subsystem, water pumping subsystem, and water storage subsystem. It is a key process to ensure the security and reliability of water supplies to consumers, both in terms of quantity and quality. For this reason, reliability and safety analyses for WSS are essential tools for maintaining the appropriate level of quality of water supply services. The paper presents the results of HRA research for a group of 38 operators working in 12 water treatment plants located in south-eastern Poland. The analysis was performed using the Fuzzy-Bayes CREAM method. The obtained results present the probability of making an error by the operator (HEP value) while working in the water production process, with the division of the operator's work into individual subsystems. The analysis allowed the following conclusions to be formulated:

- The probability of making an error by the WTP operator varies in the range of 0.0005–0.0746 depending on the analyzed subsystem.
- The lowest average probability of making an error by the WTP operator occurs for the water treatment subsystem in the disinfection process ($HEP_{WTSubS:\,D} = 0.0122$).
- Due to the nature of the operator's work in the water treatment subsystem, where chemicals are used in the water treatment process and precise technological processes are performed, the operator is required to have high competence and care for the correct execution of tasks. The failure of this subsystem is mainly related to maintaining appropriate water quality and may pose a threat to the health or life of water consumers.
- The highest average probability of making an error by the WTP operator occurs for the water intake subsystem ($HEP_{WISubS} = 0.0148$).
- Failures of the water intake, water pumping, or water storage subsystems are mainly related to maintaining the supply of an appropriate amount of water and may pose a threat to the continuity of its supply or maintaining appropriate hydraulic parameters of the network operation.

The obtained results allow taking into account the impact of the human factor on the operation of WTP and WSS, which has been ignored so far due to a lack of probability data. On their basis, reliability analyses, including the influence of human factors on the WSS, can be carried out in the future. Taking into account the influence of the human factor in the form of the system operator will allow for a better representation of the actual conditions of operation of the anthropotechnical system, such as WSS, which is also a critical infrastructure system.

Supplementary Materials: The following supporting information can be downloaded at: https://www.mdpi.com/article/10.3390/w16172399/s1, Table S1: O1–O19 operators survey scores; Table S2: O20–O38 operators survey scores; Table S3: Results of HEP calculations in the water intake subsystem (WI SubS); Table S4: Results of HEP calculations in the water treatment subsystem in the coagulation process (WT SubS: C); Table S5: Results of HEP calculations in the water treatment subsystem in the filtration process (WT SubS: F); Table S6: Results of HEP calculations in the water treatment subsystem in the disinfection process (WT SubS: D); Table S7: Results of HEP calculations in the water pumping subsystem (WP SubS); Table S8: Results of HEP calculations in the water storage subsystem (WS SubS).

Author Contributions: Conceptualization J.Ż. and B.T.-C.; data curation J.Ż.; formal analysis J.Ż.; methodology J.Ż. and K.S.; resources J.Ż.; supervision, B.T.-C.; writing—original draft preparation, J.Ż., B.T.-C. and K.S.; visualization, J.Ż. and K.S.; writing—review and editing, J.Ż. and B.T.-C. All authors have read and agreed to the published version of the manuscript.

Funding: This research received no external funding.

Data Availability Statement: The original contributions presented in the study are included in the Supplementary Materials, further inquiries can be directed to the corresponding author.

Conflicts of Interest: The authors declare no conflicts of interest.

References

1. Liu, W.; Song, Z. Review of studies on the resilience of urban critical infrastructure networks. *Reliab. Eng. Syst. Saf.* **2020**, *193*, 106617. [CrossRef]
2. French, S.; Bedford, T.; Pollard, S.J.T.; Soane, E. Human reliability analysis: A critique and review for managers. *Saf. Sci.* **2011**, *49*, 753–763. [CrossRef]
3. Zheng, X.; Bolton, M.L.; Daly, C.; Biltekoff, E. The development of a next-generation human reliability analysis: Systems analysis for formal pharmaceutical human reliability (SAFPH). *Reliab. Eng. Syst. Saf.* **2020**, *202*, 106927. [CrossRef]
4. Żywiec, J.; Tchórzewska-Cieślak, B.; Rak, J. Safety of the Water Supply System from the System Operator Perspective. In *Lecture Notes in Civil Engineering, Proceedings of the 18th International Conference Current Issues of Civil and Environmental Engineering Lviv—Košice—Rzeszów, CEE 2023, Rzeszów, Poland, 6–8 September 2023*; Springer Nature: Cham, Switzerland, 2024; Volume 438, pp. 551–561. [CrossRef]
5. Wu, S.; Hrudey, S.; French, S.; Bedford, T.; Soane, E.; Pollard, S. A role for human reliability analysis (HRA) in preventing drinking water incidents and securing safe drinking water. *Water Res.* **2009**, *43*, 3227–3238. [CrossRef]
6. Tang, Y.; Wu, S.; Miao, X.; Pollard, S.J.T.; Hrudey, S.E. Resilience to evolving drinking water contamination risks: A human error prevention perspective. *J. Clean. Prod.* **2013**, *57*, 228–237. [CrossRef]
7. Dhillon, B.S. *Human Reliability: With Human Factors*; Elsevier: Oxford, UK, 1986.
8. De Felice, F.; Petrillo, A. Human Factors and Reliability Engineering for Safety and Security. In *Critical Infrastructures—Decision Making, Theory, and Practice*; De Felice, F., Ed.; Springer International Publishing: Cham, Switzerland, 2018. [CrossRef]
9. Dsouza, N.; Lu, L. A Literature Review on Human Reliability Analysis Techniques Applied for Probabilistic Risk Assessment in the Nuclear Industry. *Adv. Intell. Syst. Comput.* **2017**, *495*, 41–54. [CrossRef]
10. Tao, J.; Qiu, D.; Yang, F.; Duan, Z. A bibliometric analysis of human reliability research. *J. Clean. Prod.* **2020**, *260*, 121041. [CrossRef]
11. Williams, J.C. A data-based method for assessing and reducing human error to improve operational performance. In Proceedings of the IEEE Fourth Conference on Human Factors and Power Plants, Monterey, CA, USA, 5–9 June 1988; IEEE: New York, NY, USA, 1988. [CrossRef]
12. Bello, G.C.; Colombari, V. The human factors in risk analyses of process plants: The control room operator model 'TESEO'. *Reliab. Eng. Syst. Saf.* **1980**, *1*, 3–14. [CrossRef]
13. US Nuclear Regulatory Commission. *NUREG-1624: Technical Basis and Implementation Guidelines for a Technique for Human Event Analysis (ATHEANA)*; US Nuclear Regulatory Commission: Washington, DC, USA, 2000.
14. Hollnagel, E. *Cognitive Reliability and Error Analysis Method (CREAM)*; Elsevier Science: Amsterdam, The Netherlands, 1998. [CrossRef]

15. Kirwan, B.; Gibson, H. CARA: A Human Reliability Assessment Tool for Air Traffic Safety Management—Technical Basis and Preliminary Architecture. In *Safety of Systems, Proceedings of the 15th Safety-Critical Systems Symposium, Bristol, UK, 13–15 February 2007*; Springer: London, UK, 2007. [CrossRef]
16. Tu, J.; Lin, W.; Lin, Y. A Bayes-SLIM based methodology for human reliability analysis of lifting operations. *Int. J. Ind. Ergon.* **2015**, *45*, 48–54. [CrossRef]
17. Li, D.; Li, L. Preliminary Study of Human Factor Reliability in Hydropower Station. *Adv. Mat. Res.* **2011**, *422*, 803–806. [CrossRef]
18. Tavakoli, M.; Nafar, M. Modification of the FFTA method for calculating and analyzing the human reliability of maintenance groups in power transmission grids. *Int. J. Syst. Assur. Eng. Manag.* **2021**, *12*, 1221–1234. [CrossRef]
19. Borgheipour, H.; Mohammadfam, I.; Narenji, M.A. Assessing and Comparing Human Errors in Technical operations in Petroleum Wells using Extended CREAM Technique. *J. Occup. Hyg.* **2017**, *9*, 132–141.
20. Wang, X.; Zhang, W.; Pan, X.; Liu, T. Method for Crew Human Reliability Analysis Based on CREAM. In *Advances in Human Error, Reliability, Resilience and Performance*; Springer: Cham, Switzerland, 2019; pp. 225–234. [CrossRef]
21. Guo, Y.; Sun, Y.; Yang, X.; Wang, Z. Flight Safety Assessment Based on a Modified Human Reliability Quantification Method. *Int. J. Aerosp. Eng.* **2019**, *2019*, 2812173. [CrossRef]
22. Jin, J.; Li, K.; Yuan, L. A Fuzzy and Bayesian Network CREAM Model for Human Error Probability Quantification of the ATO System. In Proceedings of the 4th International Conference on Electrical and Information Technologies for Rail Transportation (EITRT), Qingdao, China, 25–27 October 2019; Springer: Singapore, 2020; pp. 567–576. [CrossRef]
23. Akyuz, E.; Celik, E. A modified human reliability analysis for cargo operation in single point mooring (SPM) off-shore units. *Appl. Ocean Res.* **2016**, *58*, 11–20. [CrossRef]
24. Sezer, S.I.; Akyuz, E.; Arslan, O. An extended HEART Dempster–Shafer evidence theory approach to assess human reliability for the gas freeing process on chemical tankers. *Reliab. Eng. Syst. Saf.* **2022**, *220*, 108275. [CrossRef]
25. Zhou, Q.; Wong, Y.D.; Loh, H.S.; Yuen, K.F. A fuzzy and Bayesian network CREAM model for human reliability analysis—The case of tanker shipping. *Saf. Sci.* **2018**, *105*, 149–157. [CrossRef]
26. He, C.; Söffker, D. Human reliability analysis in situated driving context considering human experience using a fuzzy-based clustering approach. In Proceedings of the 2021 IEEE 2nd International Conference on Human-Machine Systems (ICHMS), Magdeburg, Germany, 8–10 September 2021; IEEE: New York, NY, USA, 2021; pp. 1–6. [CrossRef]
27. Lin, Q.L.; Wang, D.J.; Lin, W.G.; Liu, H.C. Human reliability assessment for medical devices based on failure mode and effects analysis and fuzzy linguistic theory. *Saf. Sci.* **2014**, *62*, 248–256. [CrossRef]
28. Zadeh, L.A. Fuzzy sets. *Inf. Control* **1965**, *8*, 338–353. [CrossRef]
29. Kluska, J. *Analytical Methods in Fuzzy Modeling and Control*; Springer: Berlin/Heidelberg, Germany, 2009. [CrossRef]
30. Zimmermann, H.J. *Fuzzy Set Theory—And Its Applications*; Springer: Dordrecht, The Netherlands, 2001. [CrossRef]
31. Dubois, D.; Prade, H. *Fuzzy Sets and Systems: Theory and Applications*; Academic Press: Cambridge, MA, USA, 1980.
32. Jensen, F.V. *An Introduction to Bayesian Networks*; UCL Press Ltd.: London, UK, 1996.
33. Habier, D.; Fernando, R.L.; Kizilkaya, K.; Garrick, D.J. Extension of the bayesian alphabet for genomic selection. *BMC Bioinform.* **2011**, *12*, 186. [CrossRef]
34. Nistal-Nuño, B. Tutorial of the probabilistic methods Bayesian networks and influence diagrams applied to medicine. *J. Evid.-Based Med.* **2018**, *11*, 112–124. [CrossRef]
35. Tchórzewska-Cieślak, B.; Rak, J.; Szpak, D. Bayesian Inference in the Analysis of the Failure Risk of the Water Supply Network. *J. KONBiN* **2019**, *49*, 433–450. [CrossRef]
36. Bishop, C. *Pattern Recognition and Machine Learning*; Springer: Singapore, 2006.
37. BayesFusion. Genie 3.0 Software. 2020. Available online: https://bayesfusion.com/2020/08/26/genie-3-0-released/ (accessed on 10 December 2023).
38. Boryczko, K.; Szpak, D.; Żywiec, J.; Tchórzewska-Cieślak, B. The Use of a Fault Tree Analysis (FTA) in the Operator Reliability Assessment of the Critical Infrastructure on the Example of Water Supply System. *Energies* **2022**, *15*, 4416. [CrossRef]
39. Żywiec, J.; Tchórzewska-Cieślak, B. Analysis of the impact of failures related to the operator's actions on the functioning of the collective water supply system. In *Problems of Reliability of Technical Systems: Theory and Applications*; Oficyna Wydawnicza Politechniki Warszawskiej: Warsaw, Poland, 2021. (In Polish)

water

MDPI

Article

Causes and Effects of Scale Deposition in Water Supply Pipelines in Surakarta City, Indonesia

Saiful Amin [1], Shinobu Kazama [2], Benyapa Sawangjang [1] and Satoshi Takizawa [1,*]

1 Department of Urban Engineering, Graduate School of Engineering, The University of Tokyo, Bunkyo-ku, Tokyo 113-8654, Japan; saifulamin@g.ecc.u-tokyo.ac.jp (S.A.); benyapa@env.t.u-tokyo.ac.jp (B.S.)

2 Department of Socio-Cultural Environmental Studies, Graduate School of Frontier Sciences, The University of Tokyo, 5-1-5 Kashiwanoha, Kashiwa 277-8563, Japan; kazama@edu.k.u-tokyo.ac.jp

* Correspondence: takizawa@env.t.u-tokyo.ac.jp

Abstract: Globally, scale deposition in water supply pipelines is one of the major problems faced by water utilities. This research aimed to determine the causes and effects of scale deposition in the water supply pipelines in Surakarta City, Indonesia. The total dissolved solids (TDS), hardness, manganese, and alkalinity in groundwater were higher than those in the surface water and spring water; thus, the supply areas from groundwater were identified using TDS at the taps. The three scaling indicators, i.e., the Langelier saturation index (LSI), the Ryznar stability index (RSI), and the Puckorius scaling index (PSI), indicated moderate calcium carbonate scaling. However, elemental analysis of eight scale samples using X-ray fluorescence (XRF) revealed that the major components of scale were either manganese (50.1–80.8%) or iron (45.6–63.8%), whereas calcium (3.0–7.8%) was a minor component. Because only five of twenty groundwater sources were chlorinated before distribution, it is estimated that dissolved manganese is oxidized by manganese-oxidizing bacteria. The manganese deposition rate in the networks was estimated to be 1660 kg/year using the manganese concentration at groundwater sources and in customers' taps. These results suggest the importance of the elemental analysis of scale and avoidance of overreliance on scale indicators.

Keywords: calcium carbonate; distribution pipes; elemental analysis; groundwater; iron; manganese

1. Introduction

Although access to clean water is a basic human need, maintaining a clean water supply has always been a major challenge in many countries [1]. Groundwater often contains high concentrations of iron and manganese; however, it is not easy to reduce their concentration to a sufficiently low level to prevent scale deposition in distribution pipelines. Therefore, many water utilities face scale deposition problems in water mains, leading to lower operating efficiency and degrading the quality of supplied water [2–4]. Scale deposition causes many other operational problems, including increased hydraulic resistance and decreased water pressure in networks, increased operational failures, and reduced energy efficiency [5]. Although old pipes are replaced every 30–50 years depending on their condition [6,7], scaling reduces the operational life of the water mains, leading to additional expenditure for water utilities when replacing pipes [8]. Some water utilities implement physical and/or chemical descaling using hydrochloric and polyacrylic acids [9]. However, this process incurs the high costs of using heavy machinery and water flushing for pipe cleaning [5]. In addition, the use of acidic chemicals for pipe cleaning poses a water safety risk [10]. Therefore, the precise forecasting and mitigation of pipe scaling is pivotal in reducing pipe maintenance costs and maintaining the quality of supplied water [5].

The accumulation of scale deposits also provides habitats for microorganisms. Actinomycetes, fungi, and pathogenic bacteria, such as coliforms, were identified in scale deposits [11,12], posing risks relating to the microbial contamination of drinking water. Although an enhanced chlorine (Cl_2) dosage can suppress microbial growth in pipelines, it is necessary to maintain a high chlorine concentration of over 1.8 mg/L for 24 h to achieve this [13]. Furthermore, this method is sometimes ineffective because loose scale deposits protect microbes from contact with chlorine. However, high dosages of chlorine lead to customer complaints due to the distinct smell and formation of disinfection byproducts.

Therefore, scale deposition in water mains is a common problem for many water utilities, including those in Mexico [8], China [14], Morocco [15,16], Tunisia [17], and Iran [18]. Previous studies have suggested that scale deposits in water mains are caused by the precipitation of poorly soluble salts, slowly attaching to the pipes over time and reacting with ions, especially iron and manganese, in water [19,20]. Predicting scale deposition in water mains is difficult because it is a slow process involving complex reactions of trace elements. The water quality influences the elemental composition of scale deposits in water mains; common elements found in scale deposits are calcium (Ca), manganese (Mn), iron (Fe), aluminum (Al), silica (Si), and magnesium (Mg) [4,5,21–23]. Hafid et al. [15] and Belattar et al. [16] studied scale deposition in downtown Taliouine and Agadir City, Morocco. They found that the main cause of scale deposition in those cities was the precipitation of calcium carbonate ($CaCO_3$). The scale samples consisted of 53% calcium oxide in Taliouine and 85.5% calcium carbonate in Agadir. This problem occurs mainly in summer when the water temperature rises, lowering the solubility of calcium carbonate [16]. Studies conducted in Mexico, Tunisia, and Iran obtained similar findings concerning the causes of scaling, including water flow reductions in distribution pipes, the scaling of cooling towers, and the scaling of sand filters in water treatment plants [8,17,18].

In water distribution systems with high concentrations of manganese in the distributed water, manganese scale deposition is prevalent [2,4,24]. Most of the manganese present in the distributed water comes from the source water [25]. Manganese is an abundant element in the Earth's crust, with levels of between 950 and 1000 mg/kg [26]. Manganese oxides are reduced to Mn(II) under anaerobic conditions and dissolve into groundwater, as documented in numerous aquifers worldwide [26]. High manganese concentrations were found to be 6.56 mg/L in Finland [27], 3.60 mg/L in Indramayu, Indonesia [28], 3.41 mg/L in southern China [29], 1.60 mg/L in Costa Rica [30], 3.44 mg/L in Bangladesh [31], >0.3 mg/L in the alluvial region of the upper Yamuna basin, India [32], and 4.00 mg/L in North Carolina, the United States of America [33]. Soluble manganese (Mn(II)) can be oxidized to insoluble Mn(III) and Mn(IV) via chlorination or manganese-oxidizing bacteria (MOB) [34,35]. After a certain period, a deposit layer forms in the water mains, which grows further through the adsorption and oxidation of manganese and other metal cations [11,36]. Some of the metals and metalloids in scale deposits, such as arsenic (As), lead (Pb), and cadmium (Cd), pose health risks to customers [3,37]. Manganese deposition can form at a low manganese concentration of 0.01 mg/L [34] and cause changes in color, odor, and/or taste at 0.05 mg/L [2].

The water resources in Indonesia are principally managed by municipally owned corporations known as Perusahaan Daerah Air Minums (PDAMs). There are currently 393 PDAMs in the country [38]. References on scale deposition in the water mains in Indonesia are very limited because, in most cases, scale depositions are not prioritized by PDAMs, as their main objective is to provide sufficient water to their customers 24/7 [39,40]. This is evident in the annual evaluation of operational management performed by the Directorate of Drinking Water, which only uses the parameters of production efficiency, non-revenue water, operating hours, water pressure, and water meter replacement [41]. Furthermore, scale deposition is a slow process; therefore, it takes time to affect water quality and the operation of water distribution networks, which is one of the reasons why many drinking water companies neglect this issue [42]. Scale accumulation, if not addressed properly, is a ticking time bomb, slowly clogging and reducing the effective diameter of distribution pipes [24]. Furthermore, most PDAMs in Indonesia fail to provide potable water to their customers [43]. The release of loose scale deposits makes it difficult for PDAMs to win customers' trust in their piped water as a reliable source of drinking water because of problems concerning color and taste [13].

Therefore, this study aimed to make water utility providers and consumers more aware of the problems of pipe scaling by elucidating the current state and issues relating to distribution pipes supplying water containing high concentrations of iron and manganese. The objective of this study is to find the main causes and effects of scale deposition in the water supply pipelines in PDAM Surakarta via water quality analyses of the pipelines and customers' taps, as well as via the elemental analysis of scale samples. We took water samples from different water sources and customers' taps to find the relationship between them. Three saturation indexes were applied to evaluate the tendency of calcium carbonate scaling, which were compared with the elemental analysis results of ten cut-out pipe samples taken from different locations of the water supply network.

2. Materials and Methods

2.1. Study Area

Surakarta is a city in Central Java and has an area of 46.72 km^2 (Figure 1). Surakarta has a tropical monsoon climate, with a wet season from October to May, and a relatively short dry season from June to September [44]. The average temperature in Surakarta is 26.85 °C [44], and the average annual precipitation is 2200 mm, with the most rainy months being December to February. Geologically, the main rock formation in Surakarta is alluvium consisting of clay, mud, sand, gravel, pebbles, and cobbles [45]. This rock formation is composed of the sediment of the Bengawan Solo River [46].

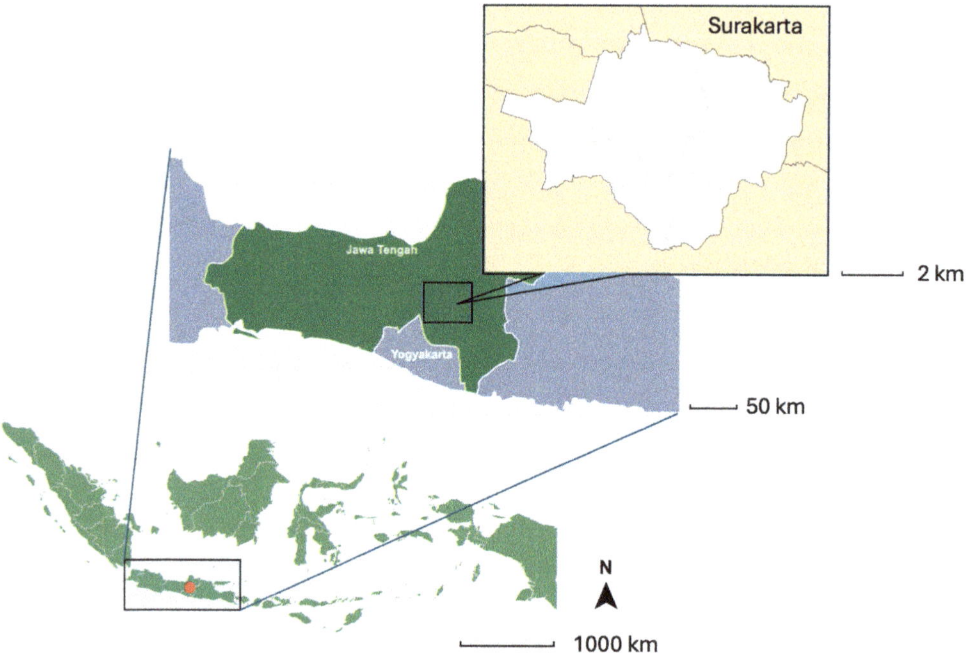

Figure 1. Maps of Surakarta.

Surakarta has a population of 523,000, with an average growth rate of 0.33% over the past 5 years [44]. The government of Surakarta established PDAM Surakarta to provide drinking water to its citizens. Currently, the PDAM covers 58.77% of the population, with an average water supply of 67,333 m^3/d and 57,772 house connections (HCs) [47]. On average, one HC serves 5.3 people.

PDAM Surakarta uses three water sources: groundwater, surface water from the Bengawan Solo River, and spring water from Cokro Tulung. Based on historical data obtained from PDAM Surakarta, scale deposition is prone to occur in pipelines carrying water from groundwater sources. Water treatment plants are installed only for the treatment of surface water and not for groundwater. The treatment process consists of pre-sedimentation, coagulation, flocculation, sedimentation, filtration, and chlorination. The groundwater and spring water are assumed to be clean and, thus, are directly supplied to customers after chlorination, although chlorination is installed only in five out of twenty groundwater sources. The largest water source is spring water, followed by groundwater and surface water. The average monthly water production is 2.02 million m^3/month, consisting of 0.78 million m^3/month of spring water, 0.67 million m^3/month of groundwater, and 0.57 million m^3/month of surface water [47]. Meanwhile, only 56.4% of produced water was billed every month; thus, non-revenue water arising from unbilled authorized consumption, apparent losses, and real losses are high at 43.6% in PDAM Surakarta.

Table 1 shows the materials, diameters, and lengths of the transmission and distribution mains. The transmission mains have a total length of 98.08 km, consisting of cast iron (CI), ductile cast iron (DCI), polyvinyl chloride (PVC), and asbestos cement (AC) pipes. The diameter of the transmission mains varies from 200 to 600 mm, with the majority of the pipes being CI with a total length of 45.67 km, while the diameter and the total length of the distribution mains are 40–150 mm and 724.16 km, respectively, for PVC and CI pipes. The distribution system consists of a reservoir and direct pumping to customers.

Table 1. The transmission and distribution pipes in PDAM Surakarta.

Usage	Pipe Type	Diameter (mm)	Length (km)
Transmission	PVC	200	14.50
	PVC	250	5.25
	AC	300	12.17
	CI	450	29.42
	CI	500	16.25
	DCI	600	20.49
		Sub-total	98.08
Distribution	PVC	40	0.07
	PVC	50	283.60
	PVC	75	107.82
	CI	80	33.29
	PVC	100	188.53
	CI	125	20.70
	PVC	150	90.15
		Sub-total	724.16
		Total	822.24

2.2. Data Acquisition and Analyses

2.2.1. Water Sampling and Water Quality Analyses

Field surveys were carried out three times in February 2023 (rainy season), August–September 2023 (dry season), and February 2024 to assess the water quality and to obtain cut-out samples of the water supply pipes. Figure 2 shows the sampling points. Water samples were collected from 27 sampling points consisting of 20 deep wells (groundwater sources), 3 water treatment plant (WTP) inlets and outlets (surface water sources), and 1 source of spring water in February 2023 and August–September 2023. The samples were also collected from customers' taps. During the sampling in February 2023, eight customers' tap samples were collected, and in August–September 2023, fourteen customers' tap samples were collected (eight samples from the same customer taps as the sampling in February 2023 and six new customers' taps). At every sampling point, two to four replicate samples were taken for analysis. Table 2 presents the water quality parameters and their analysis methods.

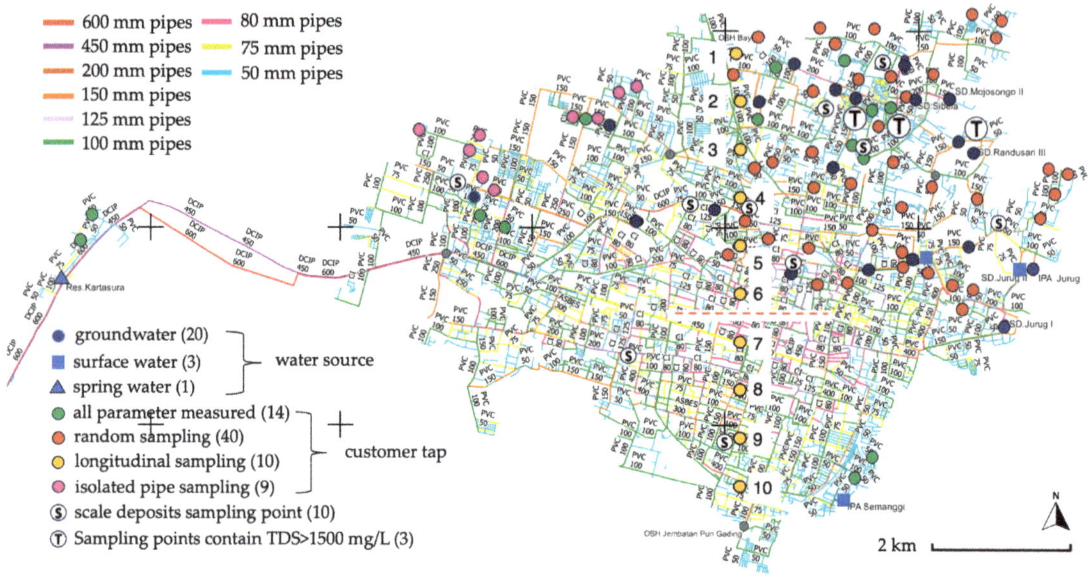

Figure 2. The sampling points across the PDAM Surakarta service area. The numbers inside the brackets indicate the number of samples. The red dashed line indicates the approximate border line between the areas supplied with groundwater (above) and river water (below) (the map was obtained from PDAM Surakarta and edited by the author).

Table 2. The water quality parameters and analysis methods.

Parameter	Unit	Method/Analytical Device	
pH	-	MP-6 Portable Meter; Hach Company, Loveland, CO, USA	
TDS	mg/L		
EC	μS/cm		
Temperature	°C		
Hardness	mg/L	Calmagite colorimetric method	DR900 Multiparameter Portable Colorimeter; Hach Company, Loveland, CO, USA
Total iron	mg/L	FerroVer® Method	
Ferrous iron	mg/L	1,10-Phenanthroline method	
Total chlorine	mg/L	USEPA DPT method	
Manganese	mg/L	PAN method (LR Mn)	
Silica	mg/L	Heteropoly blue method	
Alkalinity	mg/L as $CaCO_3$	Digital titration	Digital Titration Kit; Hach Company, Loveland, CO, USA
Dissolved oxygen (DO)	%	DO Meter DO-30N; Kasahara Chemical Instruments Corp., Saitama, Japan	
Ion concentration	mg/L	861 Advanced Compact IC; Metrohm, Herisau, Switzerland	

Further samples were taken from customers' taps in February 2024. The sampling points were divided into three categories: (1) random sampling of customers supplied with groundwater sources (40 sampling points), (2) longitudinal sampling from the north to the south at approximately every 700 m (10 sampling points), and (3) isolated pipes from a single groundwater source (nine sampling points from two different groundwater sources). (1) The areas supplied with groundwater were the main target of our study. Thus, we selected the sampling points randomly in the areas supplied with groundwater (random sampling points: red circles in Figure 2). (2) The longitudinal sampling points were selected from the north to the south to show how tap water quality changes from the areas supplied with groundwater and surface/spring water (longitudinal sampling: orange circles). (3) The isolated sampling points were also supplied with groundwater; however, because the water source and supply areas were isolated from the rest of the supply areas, we used these samples to analyze the influences of pipe diameter and the distance between groundwater and tap water on iron and manganese concentration reduction (isolated pipe sampling: pink circles). Furthermore, a survey was conducted regarding black water, taste, water pressure, and water flow at 57 customers' taps. At every sampling point, two to four replicate samples were taken for analysis. The parameters measured in the field survey were pH, total dissolved solids (TDS), electric conductivity (EC), temperature, total iron, and manganese.

2.2.2. Sampling and Analysis of Scale Deposits

To observe and analyze the elements in the scale deposits, 10 cut-out samples of pipe sections were obtained from the pipelines receiving water from different water sources (Figure 2). The pipe samples were collected randomly during the routine maintenance undertaken by PDAM Surakarta. The elemental compositions of the scale samples collected from the pipes were analyzed using the X-ray fluorescence (XRF) method (EA1400; Hitachi, Tokyo, Japan). Before the measurement, the scale samples were dried in an oven for 24 h at 105 °C. The samples were then ground using an alumina mortar and pestle to obtain a uniform sample size. The measurement used four different filters, lead (Pb), chlorine (Cl), cadmium (Cd), and chromium (Cr), to measure the different elements.

2.2.3. Saturation Indexes

Using the average water quality data of each sampling point, the Langelier saturation index (LSI), Ryznar saturation index (RSI), and Puckorius scaling index (PSI) were calculated to find the saturation and precipitation of calcium carbonate ($CaCO_3$). Table 3 presents the definition equations of the indexes and saturation categories. The LSI determines the potential of the water to precipitate calcium carbonate by calculating the difference between the pH of water samples and the saturation pH (pH_s). Water is considered in equilibrium when the value of the LSI is equal to 0. When the LSI value is positive, water is supersaturated and tends to precipitate calcium carbonate; at a negative value of the LSI, the water is undersaturated and tends to dissolve calcium carbonate [48]. The RSI resembles the LSI and uses the water saturation level to determine the tendencies to precipitate calcium carbonates. Meanwhile, the PSI considers the water's ability to act as a buffer and the maximum amount of precipitation required to restore the water to equilibrium [49].

Table 3. The saturation indexes, equations, and the implications of the saturation indexes.

Index	Equation	Index Value	Water Condition
LSI	$LSI = pH - pH_s$ If $pH \leq 9.3$, use $pH_s = A + B - \log[Ca^{2+}] - \log[alk]$ Else ($pH > 9.3$), use $pH_s = (9.3 + A + B) - (C + D)$ $A = \frac{\log[TDS]-1}{10}$ $B = -1.12[\log(°C + 273)] + 34.55$ $C = \log[Ca^{2+} + CaCO_3] - 0.4$ $D = \log[\text{alk as } CaCO_3]$	LSI > 0 LSI = 0 LSI < 0	$CaCO_3$ scale likely to be formed $CaCO_3$ in equilibrium $CaCO_3$ tends to be dissolved
RSI	$RSI = 2(pH_s) - pH$	RSI < 6 6 < RSI < 7 RSI > 7	$CaCO_3$ scale likely to be formed $CaCO_3$ in equilibrium $CaCO_3$ tends to be dissolved, corrosive
PSI	$PSI = 2(pH_s) - pH_{eq}$ $pH_{eq} = 1.465\log[alk] + 4.54$	PSI < 6 PSI > 6 PSI > 8	$CaCO_3$ scale likely to be formed $CaCO_3$ tends to be dissolved, corrosive Significantly corrosive

3. Results

3.1. Water Quality in Source Waters and at Customer Taps

Figure 3 shows the TDS, hardness, total iron, and manganese concentrations in different water sources and at customers' taps in the supply area of PDAM Surakarta. The red lines are the water quality standards stipulated in the Ministry of Health decree No. 2 Year 2023 [50]. The groundwater contains higher levels of TDS than the surface and spring water sources; additionally, all samples from the groundwater sources and customers' taps supplied with groundwater contained higher amounts TDS than the water quality standards of 300 mg/L because no treatment was completed for the groundwater (Figure 3a). There were water samples containing extremely high levels of TDS above 1500 mg/L: four samples from one groundwater source and six samples from two groundwater taps (indicated in Figure 2 with a T symbol). The TDS in the customers' taps were lower than the TDS in groundwater, indicating that the level of TDS was reduced in the distribution networks owing to the precipitation of insoluble salts, adsorption to pipe walls, or the mixing of water distributed from other sources.

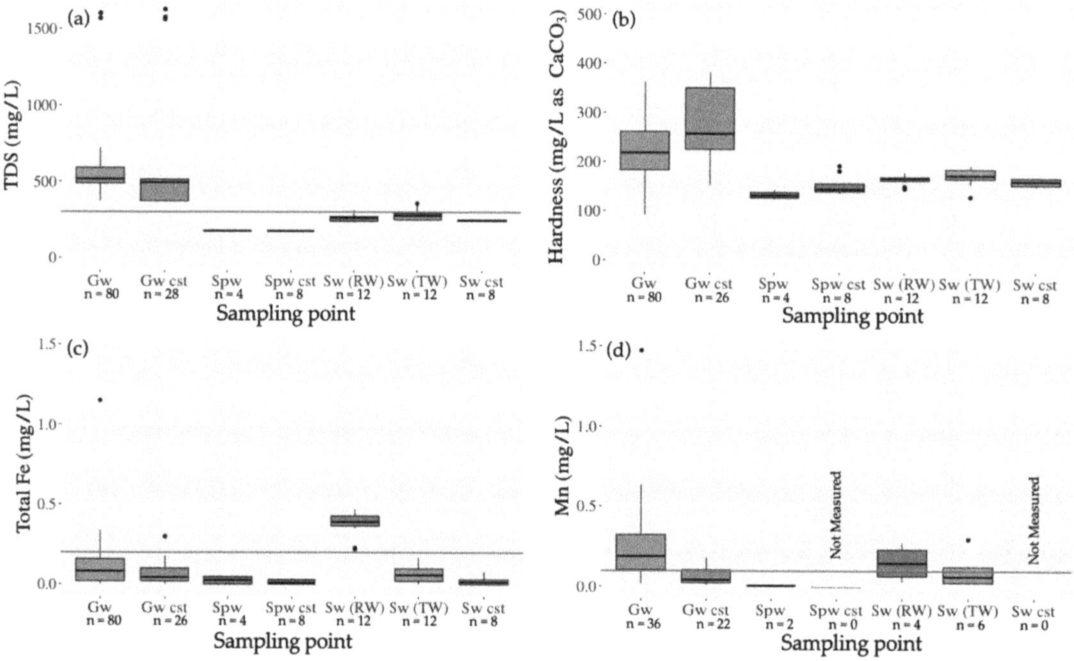

Figure 3. The water quality of different water sources: (**a**) TDS; (**b**) hardness; (**c**) total iron; (**d**) manganese. The red lines are the water quality standards in the Ministry of Health decree No. 2 Year 2023. Note: Gw: groundwater; Spw: spring water; Sw: surface water; RW: raw water; TW: treated water; cst: customers' taps.

Figure 3b shows the degree of hardness in water samples taken from different water sources. Based on the WHO guidelines for hardness in drinking water [51], the spring water and surface water sources are moderately hard, while groundwater is very hard, which indicates a high potential for calcium carbonate precipitation in the water supply pipes. Other parameters, such as pH, temperature, and alkalinity, also influence precipitation [51]. The hardness in customers' taps should be noted as being higher than in groundwater, indicating that calcium and magnesium were dissolved into water in the distribution networks. Thus, calcium scaling in the water mains is not likely to occur.

Figure 3c shows the total iron concentration in the samples from different water sources. A total of 13 samples from groundwater sources and one tap water sample out of 106 groundwater samples contained higher iron levels than the threshold stipulated in the water quality standards of 0.2 mg/L. All spring water samples and customer tap samples using spring water contained a low concentration of iron below 0.2 mg/L. However, the surface water contained a high concentration of total iron above 0.2 mg/L, which was reduced below 0.2 mg/L in the water treatment processes. In all of the samples taken from different sources, the iron concentration decreased from the water sources to the customers' taps, indicating iron removal via deposition in the distribution pipes and scale formation.

Figure 3d shows that the manganese concentration was higher in the order of groundwater (average concentration 0.26 mg/L), surface water (0.025 mg/L), and spring water (0.0045 mg/L). Twenty-eight out of thirty-six samples exceeded the water quality standards of 0.1 mg/L. The manganese concentrations were lower in the customers' taps than in the groundwater, indicating that manganese oxidation and deposition occurred in the distribution pipes before reaching customers' taps. Because only five out of twenty groundwater sources had chlorination facilities, the process of manganese oxidation was facilitated not only by chlorine but also possibly by MOB. Meanwhile, in the raw water of the surface water sources, high manganese potentially originates from textile wastewater discharge, decreasing surface water quality [52]. Furthermore, high manganese concentrations were recorded in the treated surface water due to mixing with untreated groundwater sources in the clear water tanks in two water treatment plants.

We measured the general water quality parameters, such as the pH, temperature, alkalinity, and sulfate concentration, because these could influence manganese release from loose scale deposits in the pipelines; low pH, low alkalinity, high temperatures, and high sulfate concentrations promote manganese release from deposits [53]. Table 4 shows the results of the water quality analyses. The pH level was between 7.13 and 7.45, while the temperatures ranged from 24.70 to 29.47 °C. Meanwhile, the average alkalinity in the groundwater samples was high at 375.2 mg/L, reducing to 297.0 mg/L at customers' taps. Scale deposits detach from the pipes as the temperature rises. Zhang et al. [53] observed that manganese was released into a solution over time at different water temperatures. The time needed for the solution to exceed 100 μg/L of manganese was twice as fast when the temperature rose from 15 °C to 25 °C (4 h at 15 °C compared to 2 h at 25 °C). The high temperatures in the study area suggest that the scale deposits in Surakarta are prone to detaching, causing discoloration and other problems.

The total chlorine concentrations were below the national standards for drinking water of 0.2 mg/L in all customer tap samples despite the fact that surface water and spring water sources, as well as five out of twenty groundwater sources, pass through a chlorination facility. The decreasing chlorine concentrations found in customers' taps fed by groundwater were significant compared to spring water and surface water customers' taps. We selected water sources and nearby customers' taps for the analyses of residual chlorine. The total chlorine was 0.46 mg/L in a groundwater source and 0.12 mg/L in a customer's tap supplied by groundwater from a deep well in Plesungan. The total chlorine was 0.17 mg/L in surface water sources (WTP Semanggi) and 0.14 mg/L and 0.16 mg/L in customers' taps supplied by surface water. Meanwhile, in spring water, it was recorded as being 0.14 mg/L at the source and 0.12 mg/L and 0.09 mg/L at customers' taps. Chlorine was rapidly consumed in the pipeline from groundwater sources before reaching customers' taps. Scale deposition is one of the main causes of lower chlorine concentrations in distribution pipes due to reactions with chlorine [24].

Table 4. Water quality of different water sources.

Parameter	Unit	Gw Ave. Min–Max	Gw cst Ave. Min–Max	Spw Ave. Min–Max	Spw cst Ave. Min–Max	Sw Ave. Min–Max	Sw cst Ave. Min–Max
pH	-	7.45 7.0–8.07	7.38 7.31–7.50	7.19 7.15–7.26	7.14 7.10–7.18	7.13 6.93–7.23	7.39 7.36–7.42
Temp.	°C	28.25 24.70–29.30	28.51 27.30–29.47	28.51 27.30–29.47	26.93 26.80–27.10	26.90 26.40–27.53	29.01 29.00–29.03
EC	μS/cm	806.3 567.7–2241.0	658.9 571.9–2240.0	276.2 275.5–276.7	276.3 276–277.1	416.7 340.8–542.1	379.8 374.6–384.7
Total Cl_2	mg/L	0.08 NA–0.50	0.02 NA–0.12	0.14 0.11–0.16	0.12 0.08–0.12	0.57 0.08–1.41	0.15 0.14–0.16
Alkalinity	mg/L	375.2 226.7–533.3	297.0 250.7–382.0	103.0 102.0–104.0	115.0 113.7–116.3	87.9 73.3–102.7	82.3 80.0–85.3
Silica	mg/L	66.0 44.0–84.0	68.0 52.0–90.0	74.0 74.0–74.0	73.0 72.0–74.0	27.0 18.0–44.0	18.5 18.0–19.0
Ferrous iron	mg/L	0.03 BDL–0.32	0.01 BDL–0.03	0.04 BDL–0.09	0.02 BDL–0.02	0.03 BDL–0.11	0.01 BDL–0.03
DO	mg/L	1.76 0.50–2.70	2.34 0.80–3.70	3.12 2.80–3.50	3.10 2.60–3.60	2.50 0.40–4.00	3.50 2.90–4.60

Note: Gw: groundwater; Spw: spring water; Sw: surface water; cst: customers' taps; BDL: below detection limit, NA: not applicable.

The silica in groundwater and spring water was higher than in surface water due to silica dissolution from rocks and soils in the aquifer [54]. Meanwhile, the DO in the groundwater sources was lower due to the limitation of contact with the air in the atmosphere.

Furthermore, major cations found in the water samples were Na^+, Ca^{2+}, K^+, Mg^{2+}, and NH_4^+, and major anions were Cl^-, SO_4^{2-}, HCO_3^-, NO_3^-, and PO_4^{3-}. The results of the analyses are shown in Table A1 of Appendix A, including evidence that the very high TDS concentration in samples from customers' taps was caused by high Na^+ and Cl^- concentrations in groundwater sources. As shown in Figure A1, the concentration of TDS in some wells becomes extremely high in the dry season because the wells dry up.

The water quality in customers' taps in the service area supplied from groundwater sources is shown in Figures 4–7. Figure 4 shows the results of isolated sampling points in two groundwater sources across different pipe diameters. In water supply systems, we use large-diameter pipes in areas closer to water sources, such as groundwater, and the diameters of pipes gradually decrease as we move away from water sources because water flow gradually decreases. Thus, small-diameter pipes are located in areas further away from water sources, whereas large-diameter pipes are installed in areas closer to water sources. Therefore, water in small-diameter pipes is likely to have a longer hydraulic retention time than water in large-diameter pipes as they are further away from the water sources. Thus, we compared iron and manganese concentrations in small- and large-diameter pipes to see how hydraulic retention times influence iron concentrations in water supply pipes. Figure 4a,b shows the sampling locations, and Figure 4c–e shows the TDS, total iron, and manganese, respectively. The TDS in Banyuanyar were stable from the source across different pipe diameters, indicating that the water in the pipes came from the same source. Meanwhile, in Karangasem, TDS were lower in pipes with 100 and 75 mm diameters, possibly because the pipes were located further away and were reduced in the distribution networks owing to the precipitation of insoluble salts and/or adsorption to pipe walls.

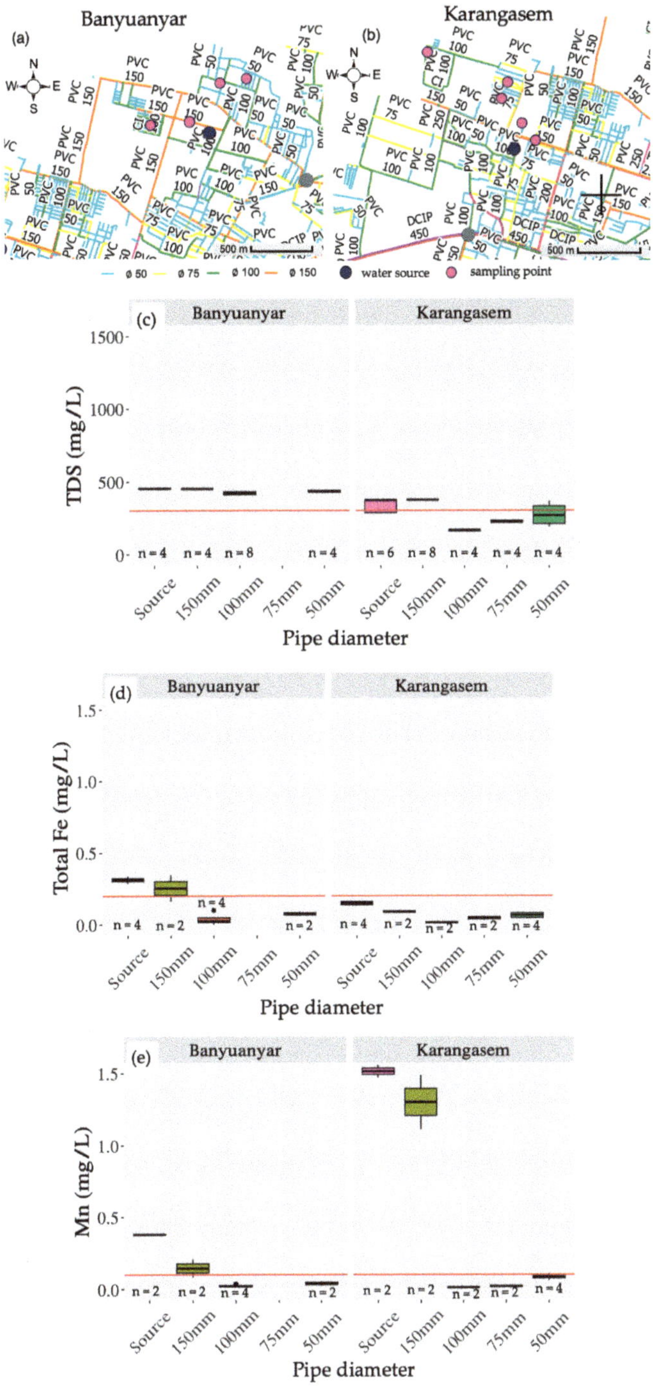

Figure 4. The water quality changes in distribution pipes with different diameters and distances: (**a**) sampling points in the Banyuanyar deep well; (**b**) sampling points in the Karangasem deep well; (**c**) TDS; (**d**) total iron; and (**e**) manganese. The red lines indicate the water quality standard.

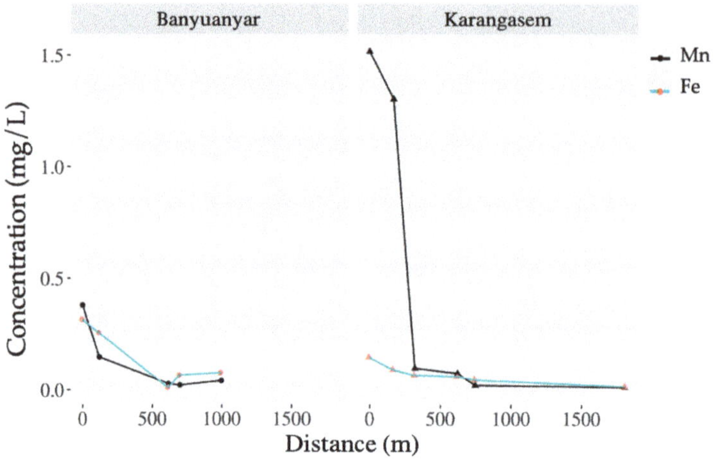

Figure 5. Manganese and iron concentrations along the pipeline from water sources.

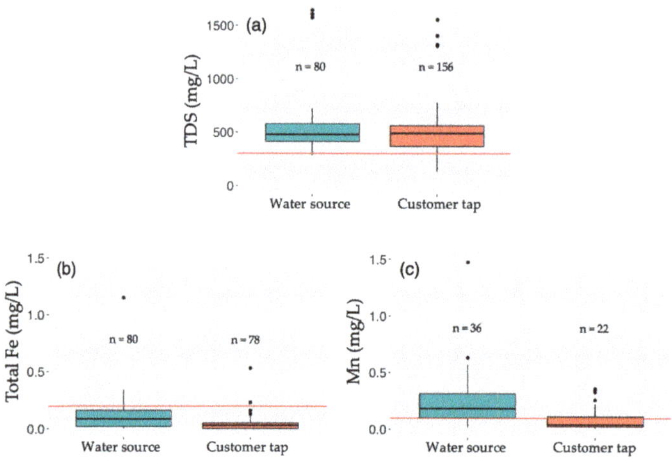

Figure 6. The water quality in groundwater sources and customers' taps: (**a**); TDS; (**b**) total iron; (**c**) manganese.

The iron concentration decreased from water sources along with decreasing pipe diameters in both Banyuanyar and Karangasem (Figure 4d). These results indicate that iron was gradually removed along the pipelines owing to scale deposition. Ferrous iron can be readily oxidized by dissolved oxygen in supplied water and removed by adsorption onto the deposits on the pipe walls. In addition, iron-oxidizing bacteria attached to the pipe walls might have contributed to iron oxidation to some extent, according to references [11,36].

Figure 4e shows that the manganese concentration also decreased from the source along with decreasing pipe diameter, which is more obvious in Karangasem than in Banyuanyar. As an exception, manganese concentrations in 50 mm pipes were higher than in pipes with larger diameters because the sampling points were located closer to the water source compared to the 75 and 100 mm pipes.

Meanwhile, the pH and temperature in both Karangasem and Banyuanyar were stable. The pH ranged from 6.96 to 7.61 in Karangasem, with an average value of 7.32, and from 7.53 to 7.96 in Banyuanyar, with an average value of 7.57. The temperature ranged from

27.0 to 29.7 °C in Karangasem, with an average value of 28.64 °C, and from 27.1 to 31.1 °C in Banyuanyar, with an average value of 29.17 °C.

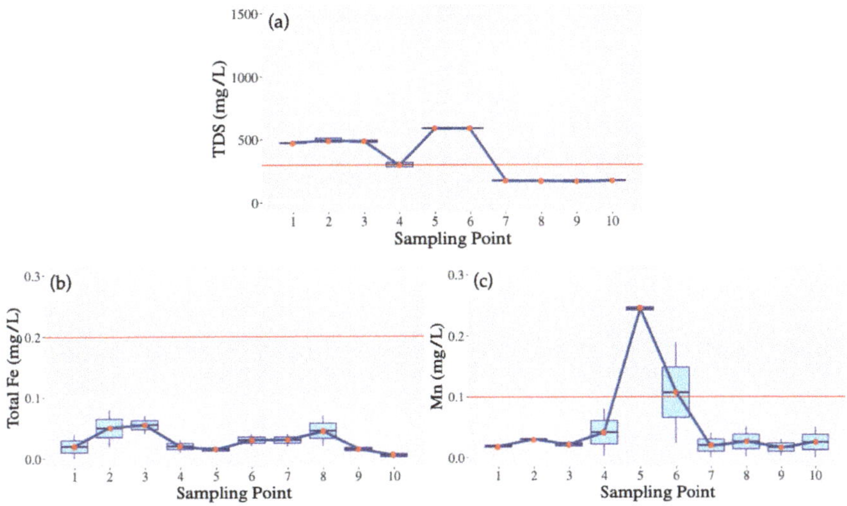

Figure 7. The longitudinal sampling results of customers' taps: (**a**) TDS; (**b**) total iron; (**c**) manganese.

Figure 5 shows the iron and manganese concentrations along with the distance from the water source. It must be noted that the distance does not necessarily indicate the hydraulic retention time, and it could be possible that the water is mixed with water coming from a different pipeline route. Nevertheless, these results clearly indicate that iron and manganese decreased along with the distance from the water sources, which provides evidence of iron and manganese scale deposition in the pipelines.

Figure 6 shows the water quality of groundwater sources and 40 customers' taps selected randomly from the areas supplied by groundwater sources (see Figure 2 for sampling locations). Figure 6a shows that the TDS values in customers' taps were not significantly different from the water sources (Wilcoxon rank-sum test, $p > 0.05$), excluding the outliers and the sampling points influenced by surface water sources. Meanwhile, a seasonal variation in TDS is shown in Figure A1. The higher TDS levels at every groundwater source in the dry season compared to the rainy season indicate the reduced infiltration of rainwater in the dry season in the study area. The three extremely high levels of TDS of approximately 1600 mg/L were only found in the dry season.

Figure 6b,c show the total iron and manganese in groundwater sources and customers' taps. The iron concentration in customer taps was significantly lower than the iron concentration in the source water (Wilcoxon rank-sum test, $p < 0.05$), providing evidence of iron deposition in the pipelines (Figure 6b). The manganese concentration in taps was also lower than the manganese concentration in the source water (Figure 6c), providing evidence of manganese deposition within the pipelines. As a result, the pipelines de facto remove iron and manganese from the source water. All the tap water samples contained iron lower than water quality standards, except for two samples, while sixteen out of twenty-two tap water samples contained manganese lower than the water quality standards.

Figure 7a shows the variation in TDS along with the longitudinal sampling points (see Figure 2). The north area (Sampling Points #1–3) is supplied by groundwater sources; thus, the level of TDS was high. At Sampling Point #4, the lower level of TDS is due to proximity to a surface water treatment plant and a pipeline from spring water sources. Meanwhile, the high level of TDS in Sampling Points #5 and #6 is due to a groundwater source located in proximity to the sampling points. TDS were lower at the southern Sampling Points #7–10 than in the north, indicating that the water in this area came from either surface water or spring water. The dashed line shown in Figure 2 between Sampling Points #6 and #7 indicates the boundary between an area supplied with groundwater in the north and an area supplied with surface or spring water in the south.

Figure 7b shows the total iron levels from north to south, taken approximately every 700 m. All of the samples were below the water quality standards. The groundwater customers have a concentration of total iron slightly higher than the surface water and spring water customers because the iron concentration in the source water is high. The low concentration in all customers' taps is further proof that iron was oxidized in the distribution pipelines. Figure 7c shows the manganese concentration along with the longitudinal sampling points. In the north part, the manganese concentrations were below 0.05 mg/L, indicating that these sampling points are further away from groundwater sources and manganese was removed in the pipelines. Meanwhile, in the middle section, Sampling Points #4–6 are mostly sourced from the nearby Banjarsari deep well, which has a high manganese concentration of 0.60 mg/L. Because Sampling Point #5 is the closest to the water source, the manganese concentration was the highest at 0.25 mg/L. The manganese concentration decreased rapidly in the pipelines from the sources and then gradually with distance from the water source (Figure 5). In the southern part, the water is sourced from either surface water or spring water; hence, the manganese concentrations were low.

3.2. Saturation Indexes

The LSI, RSI, and PSI were calculated for different water sources and customers' taps to estimate the saturation levels of calcium carbonate (Figure 8). The LSI of groundwater and its customer samples produced an LSI > 0, which indicates that the water samples were supersaturated. Meanwhile, other water sources were below saturation with an LSI < 0. Undersaturated water tends to dissolve calcium carbonate, while supersaturated water tends to form calcium carbonate scale deposits in the pipeline. Even though the values were positive, the LSI values were below 1, indicating that a change in temperature or another ambient factor could shift the LSI value.

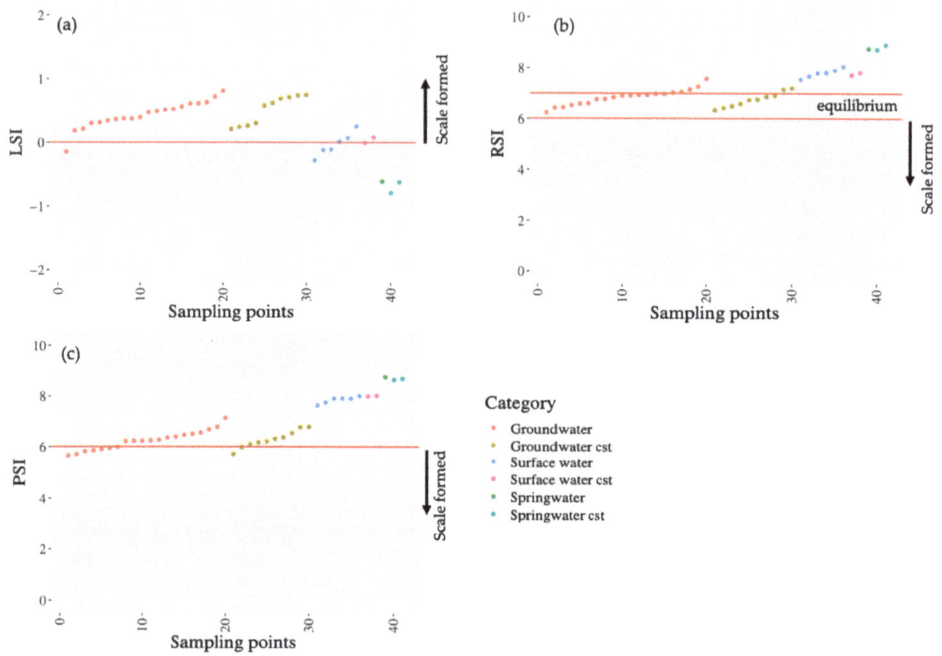

Figure 8. The saturation indexes: (**a**) LSI; (**b**) RSI; (**c**) PSI. The red lines indicate equilibrium.

Meanwhile, most of the RSI values, especially from groundwater sources, were between 6 and 7, indicating that the water samples were in equilibrium. All the RSI values from the surface water and spring water samples were above 7, indicating that water tends to dissolve calcium carbonate and has a corrosive property. On the contrary, six groundwater samples had PSI values <6, indicating the possibility of calcium carbonate precipitation. Other samples had PSI values >6, indicating that most samples were likely to dissolve calcium carbonate. Furthermore, surface water and spring water samples had corrosive tendencies based on the PSI values. These three saturation indices are not in clear agreement because most of the data were close to equilibrium; namely, the LSI indicated that groundwater samples had a weak tendency to form a calcium carbonate scale, while the RSI and PSI indicate equilibrium and undersaturation, respectively. However, we should note that the degree of hardness at customers' taps was higher than in water sources (Figure 3b), indicating the dissolution of calcium and magnesium in the pipelines.

3.3. Elemental Composition of Scale Samples

Figure 9 shows the scale in the cut-out pipe samples. Samples #1–7 were collected from pipelines supplying groundwater from different groundwater sources. Meanwhile, Sample #8 was collected from a pipeline transporting a mixture of surface water and groundwater. Scale deposits were observed in Samples #1–#8.

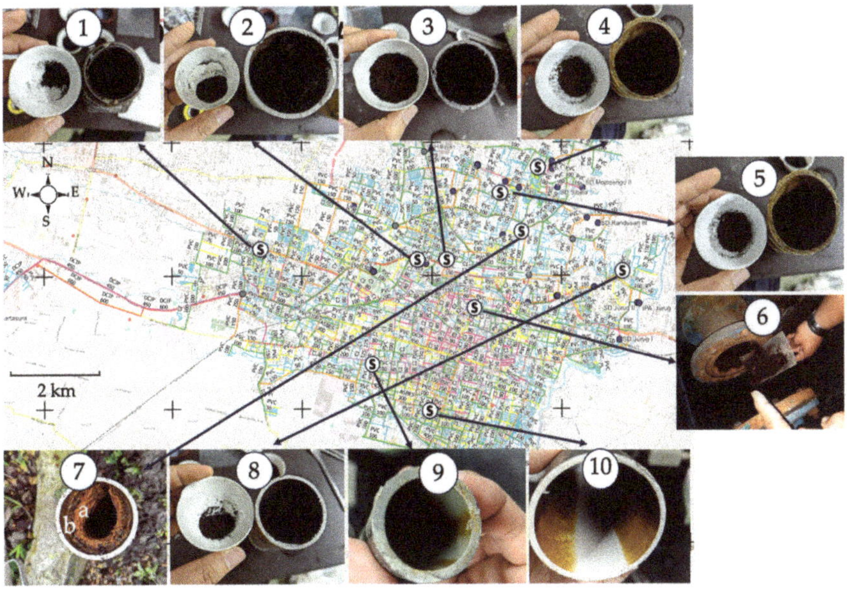

Figure 9. The scale samples collected from the distribution pipelines (map obtained from PDAM Surakarta and edited by the author). The numbers in the photos are sample numbers.

Sample #9 was obtained from a pipeline supplying spring water, and Sample #10 from a pipeline supplying spring water and surface water. The average manganese and iron concentrations were 0.0045 mg/L and 0.025 mg/L in spring water and 0.06 mg/L and 0.08 mg/L in treated surface water located in the southern part (no mixing with groundwater). Therefore, we did not find any scale deposits in Samples #9 and #10.

Table 5 shows the XRF measurement results. In Sample #7, XRF analysis of the surface layer (Sample #7a) and the inner layer (Sample #7b) was performed because of the thick layer of scale. The predominant elements in the scale samples were Mn (Samples #1–#6, 50.15–80.82%) or Fe (Samples #7a, #7b, and #8, 45.60–63.82%), and Ca was a minor component (2.98–7.83%) despite the results relating to calcium scaling indicators.

Table 5. The major elements found in scale samples.

Sample	Composition (%wt.)							Water Source
	Mn	Fe	Al	Si	Ca	Zn	Others	
1	**55.24**	15.51	13.62	8.17	2.98	2.44	2.04	Gw
2	**52.46**	7.04	16.69	14.72	6.41	1.74	0.95	Gw
3	**50.15**	9.54	12.49	17.52	6.57	1.45	2.27	Gw
4	**59.15**	11.00	6.66	13.06	6.35	0.98	2.80	Gw
5	**64.57**	4.31	11.18	10.20	7.66	1.33	0.75	Gw
6	**80.82**	4.92	ND	0.77	7.83	0.06	5.59	Gw
7a	3.29	**63.82**	3.10	19.25	5.36	ND	5.18	Gw
7b	11.21	**58.66**	1.79	17.25	4.68	0.05	6.37	Gw
8	31.22	**45.60**	4.88	12.09	3.53	0.19	2.50	Gw + Sw

Note: Gw: groundwater; Sw: surface water; ND: not detected; bold letters: the highest composition.

In Samples #7a, #7b, and #8, the iron composition was higher than that found in other samples. The water sources of these pipes contained high iron concentrations of 0.18 and 0.21 mg/L for #7 and #8, respectively, and the Fe/Mn ratios were much higher at 10.3 and 1.7 for #7 and #8, respectively, than those found in other water sources. A total of 15 water sources out of 20 had Fe/Mn ratios of less than 1.0. Thus, a high iron concentration and high Fe/Mn ratios might be the cause of high iron contents in the scale of Samples #7 and #8. The manganese content was lower in the surface section (#7a) than in the inner section (#7b). This suggests that the manganese scale formed initially, and then iron was adsorbed onto the manganese deposits, forming a composite scale. Furthermore, two customers' tap samples taken near scale Sample #7 showed an iron concentration of 0.02 mg/L and BDL. This finding suggests that iron was fully oxidized in the pipelines.

Furthermore, the scale samples contain small amounts of heavy metals. Samples #1, #2 #3, #5, and #8 contain 0.26%, 0.14%, 0.15%, 0.11%, and 0.07% of lead (Pb), respectively. The release of scale deposits can cause lead (Pb) to enter the distribution system. Lead (Pb) poisoning can result in neurodevelopmental, cardiovascular, renal, and reproductive health effects [55,56]. These health effects are thought to occur at blood lead levels as low as 1–2 μg/dL [55].

3.4. *Effects of Iron and Manganese Scale in PDAM Surakarta*

Other than the effect of reducing the effective diameter of the pipelines, the release of loose deposits causes problems such as high turbidity, color changes, taste changes, odor, and microbial contamination. Out of 57 customers' taps, black water was found in 41 taps (71.9%), with seven taps more frequently affected. The black water is caused by changes in water velocity during maintenance processes; the loose deposits have porous structures and can be easily detached at a velocity of 0.5–0.8 m/s [13]. While the majority of the respondents were satisfied with their water, in seven out of fifty-seven of the sampling taps, the water pressure was not high enough because the water flow was possibly being impeded through pipes that were clogged by scaling. Therefore, it is important to remove iron and manganese from groundwater, using similar methods as used in surface water treatments, not only to avoid scale deposition and the clogging of pipes but also to win the confidence of customers.

4. Discussion

4.1. *Scale Deposition and Saturation Indexes in PDAM Surakarta*

Manganese and iron were found to be the major components of scale deposits via elemental analysis using XRF. Because manganese is released into groundwater in a soluble form, namely Mn(II), in the aquifer, it needs to be oxidized to Mn(III) or Mn(IV) to become insoluble in water and form scale. Although the oxidation and deposition reactions are slow and invisible, it is very important to monitor pipelines to avoid the multiple problems that arise from scale deposition. The oxidation of Mn(II) by dissolved oxygen is thermodynamically favorable; however, the reaction rate is considerably slow at a near-neutral pH of (<8.5) [57]. Because PDAM Surakarta only chlorinates five out of twenty groundwater sources, manganese oxidation is likely to be facilitated by MOB. Once a manganese film forms on the inner surface of a pipeline, the oxidation rate is dramatically increased due to the propagation of the MOB population and the high absorption capacity of biogenic MnOx [11]. These biogenic MnOx also exhibit high adsorption capacities for other metal cations such as Al, Fe, and Ca, which, over time, form scale in pipelines [24,58]. The adsorption of metal cations, especially Al, is high at pH levels of 7.7–9 [58]. Non-chlorinated conditions are more favorable for the adsorption of particulate Al deposition [58]. Meanwhile, chlorinated conditions are more favorable for the adsorption of Cr(IV) deposition [59]. The pipe material used also affects scale formation. In PDAM Surakarta, 92.54% of the distribution pipes are PVC pipes. This material has a high affinity to manganese deposits [24,35]; therefore, the distribution pipelines in Surakarta are rendered prone to manganese oxide scale formation. However, there was a limitation in this study in that it was not possible to evaluate the contribution of multiple factors that influenced scale deposition in water supply pipes in Surakarta via our survey of the distribution networks because these factors have varied hourly depending on water consumption for more than ten years.

The three saturation indexes indicated different results concerning calcium carbonate saturation in groundwater, namely, supersaturation (LSI), equilibrium (RSI), and undersaturation (PSI) for $CaCO_3$. The LSI was proposed first as a simple comparison between water pH and the saturation pH levels of calcium carbonate. Then, the RSI and PSI were proposed to take saturation in various water matrices into consideration. Scaling indexes, such as the LSI, PSI, and RSI, have limitations; most importantly, water flowing in pipes is in a dynamic state and could be different from the equilibrium state of static water. Therefore, they should be combined with the elemental analysis of scale deposits. $CaCO_3$ scaling is a complex process, and many aspects need to be considered. These limitations arise from the complex nature of the system, which involves multiple equilibrium interactions at the interfaces between solid/liquid and liquid/gas phases [60]. Dependable thermodynamic equilibrium data are scarce for these interactions. The major weakness of saturation indexes is that the crystalline form of $CaCO_3$ only considers calcite [61] and the presence of other forms of $CaCO_3(s)$ are neglected. Calcium carbonate has six allotropic forms, consisting of three anhydrous forms (calcite, aragonite, and vaterite) and three hydrated forms (amorphous calcium carbonate, monohydrate calcium carbonate ($CaCO_3.H_2O$), and hexahydrate calcium carbonate ($CaCO_3.6H_2O$)) [62]. Amorphous calcium carbonate has been demonstrated to be a precursor for rapid $CaCO_3$ nucleation under conditions of high supersaturation, whereas monohydrate calcium carbonate has been identified as a precursor for $CaCO_3$ nucleation under low supersaturation [62]. Another limitation is that Ca^{2+} and HCO_3^- complexation is not considered. When polyphosphates are present, the equations relating to pH levels tend to overstate $CaCO_3$ saturation unless adjustment factors are incorporated to accommodate for the complexation [61]. Despite the limitations of saturation indices mentioned above, the results in this study show that the elemental composition of calcium in the scale samples was much lower than the manganese and iron contents (Table 5) and that the degree of hardness in customers' taps was increased when coming from groundwater sources (Figure 3b). Thus, it can be concluded that the PSI represents the saturation condition of water sources and supplied water in Surakarta the best among the three saturation indices.

4.2. Analyses of Water Sources and Scale Deposition in the Distribution Network Using Water Quality Data

TDS measurements at customers' taps are useful for examining possible scaling in pipelines. TDS data at customers' taps provide useful information to delineate the supply areas of different water sources, such as groundwater and surface water (Figures 2–7). The groundwater sources contain higher TDS than surface water and spring water; thus, we can use TDS data to identify the water supply areas requiring more detailed monitoring of scale deposition.

The decreasing trends in iron and manganese concentrations along the pipeline distance from water sources to customers' taps provide evidence of iron and manganese deposition in the pipelines. Furthermore, we can estimate the amounts of iron and manganese scale deposition in the pipelines using these water quality monitoring data. Using the average manganese and iron concentrations at the water source and customers' taps and the average water production, the rate of scale deposition was calculated. Among the 20 groundwater sources, iron and manganese deposition from one source was calculated separately from others owing to the high iron and manganese concentrations. As a result, an average of 1660.1 kg/year of manganese and 559.5 kg/year of iron are estimated to be deposited in the pipelines supplying groundwater. These deposits continually accumulate for years and reduce the effective diameter. In other cases, these deposits can become detached from the pipelines, causing problems such as black water, turbidity, color changes, taste changes, odor, and microbial contamination.

4.3. Prevention of Manganese Scale Deposition in PDAM Surakarta

Controlling manganese concentrations before the start of the distribution pipeline is the most important factor in preventing manganese scale deposition. If the treated water contains manganese, manganese precipitation is unavoidable, especially in Surakarta, where manganese deposits have accumulated in the pipes for a long time. The removal of other metal cations, such as iron and aluminum, is also crucial because these cations can be absorbed by the current scale layer deposited in the pipes.

PDAM Surakarta needs to consider installing a manganese removal facility and prioritize this installation, even under budget limitations. Among 20 groundwater sources, 14 had manganese concentrations above 0.1 mg/L, and one was exceptionally high, with an average manganese concentration of 1.47 mg/L. One feasible manganese removal method can be achieved via adsorption. This method is widely used because it is highly efficient, simple to operate, and cost-effective. Mn(II) can be removed by sorption to a solid surface, commonly using manganese oxide in a pH range of 6–9. Some of the common materials used for this method are glauconite (greensand), pyrolusite (a natural mineral containing $MnO_2(s)$), ceramic materials coated with a MnO_x surface, and anthracite coal or silica sand coated with MnO_x [63]. When the adsorption capacity of the MnO_x-coated media is exhausted, a breakthrough of dissolved Mn(II) occurs. This problem can be solved by adding oxidants, commonly permanganate (MnO_4^-), to the backwash water.

5. Conclusions

The elemental analysis of scale deposits combined with the analysis of iron and manganese at water sources and customers' taps provided evidence that manganese and iron are the main causes of scale deposition in the PDAM Surakarta network. Iron and manganese concentrations gradually decreased in the pipelines with distance from the water sources, providing evidence of the causes of scale deposition in pipelines. Manganese and iron are mainly sourced from groundwater; in addition, other metal cations are incorporated into the scale due to the high adsorption capacity of MnOx. The fact that PDAM Surakarta directly distributed groundwater to the customers without any treatment exacerbated the problem of scale deposition, and further provides evidence that iron and manganese can be oxidized and deposited in pipelines in the absence of chlorine. Concerning current pipe conditions, where scale deposits have already developed, the continuous precipitation of manganese and iron is unavoidable. Manganese removal to as low as 0.05 mg/L is pivotal for avoiding manganese precipitation in the pipes.

Measuring iron and manganese at customers' taps and comparing them with water sources is an easy and rapid method of verifying scale deposition in pipelines. This method can also help identify the water distribution pipelines that are accumulating more iron and manganese scale deposition and provide data for estimating the rate of scale deposition in the whole distribution network. TDS data at taps can be used to delineate the water supply areas receiving water from different sources.

Although the saturation indexes showed the potential for calcium carbonate precipitation to occur in pipelines conveying groundwater sources, less than 8% of calcium carbonate was detected in the scale sample. In addition, the degree of hardness at customers' taps was higher than in groundwater. These results show the limitation of saturation indexes, such as the LSI, RSI, and PSI, in predicting and finding the causes of scale deposition. The elemental analysis of scale samples provides direct evidence that can be used to identify the causes of scale deposition.

Author Contributions: Conceptualization, S.T. and S.K.; methodology, S.A. and S.T.; formal analysis, S.A.; investigation, S.A.; data curation, S.A.; writing—original draft preparation, S.A.; writing—review and editing, S.K., B.S. and S.T.; supervision, S.K., B.S. and S.T.; project administration, S.T.; funding acquisition, S.T. All authors have read and agreed to the published version of the manuscript.

Funding: This research was supported by the Japan International Cooperation Agency (JICA) through the collaborative program with the University of Tokyo and the scholarship provided to Saiful Amin for his graduate studies. This study was also supported by a Grant-in-Aid for Scientific Research (No. 22H01621) provided by the Japan Society for the Promotion of Sciences (JSPS).

Data Availability Statement: The data are available upon request.

Acknowledgments: The authors extend their gratitude to the Ministry of Public Works and Housing of the Republic of Indonesia, Perusahaan Daerah Air Minum (PDAM) Surakarta, Fujimura Kazuyoshi from the Department of Urban Engineering, the University of Tokyo, and Yamagishi Takayuki from the Department of Applied Chemistry, the University of Tokyo.

Conflicts of Interest: The authors declare no conflicts of interest.

Appendix A

Table A1. The cations and anions in different water sources (ND; not detected).

Parameter	Unit	Gw Mean Min–Max	Gw cst Mean Min–Max	Spw Mean Min–Max	Spw cst Mean Min–Max	Sw Mean Min–Max	Sw cst Mean Min–Max
Cation							
Na^+	mg/L	110.86 25.78–350.58	128.29 26.85–360.28	15.39	15.43 15.42–15.44	29.43 20.58–50.01	21.55 21.52–21.58
Ca^{2+}	mg/L	48.46 13.85–75.09	52.92 32.53–70.60	20.78	20.75 20.73–20.78	37.06 35.99–38.38	37.29 37.25–37.32
Mg^{2+}	mg/L	16.45 7.02–29.99	17.31 8.68–24.01	10.08	10.13 10.13–10.14	8.99 8.39–10.04	8.56 8.56–8.57
K^+	mg/L	8.38 4.19–12.52	7.61 3.12–9.89	4.69	4.6 4.54–4.66	4.24 3.48–5.33	3.42 3.40–3.45
NH_4^+	mg/L	0.23 ND–1.54	0.67 ND–1.44	ND	0.01 ND–0.02	0.36 ND–0.99	ND
Anion							
HCO_3^-	mg/L	451.89 305.0–561.81	403.15 297.98–492.88	130.54	133.89 132.37–135.42	170.66 151.89–193.57	149.24 146.4–152.93
Cl^-	mg/L	55.19 9.49–511.94	116.86 8.55–507.61	8.95	9.01 8.99–9.03	32.93 21.70–51.86	32.91 32.78–33.06
SO_4^{2-}	mg/L	7.74 1.47–16.42	7.94 1.51–15.98	5.67	5.79 5.79–5.80	16.10 14.53–17.56	14.64 14.56–14.72
PO_4^{3-}	mg/L	0.68 ND–2.16	0.31 ND–1.56	ND	ND	0.5 0.00– 1.51	ND
NO_3^-	mg/L	2.05 0.69–9.68	2.00 0.69–8.87	10.27	10.45 10.45–10.45	5.24 3.24–7.38	3.84 3.81–3.86

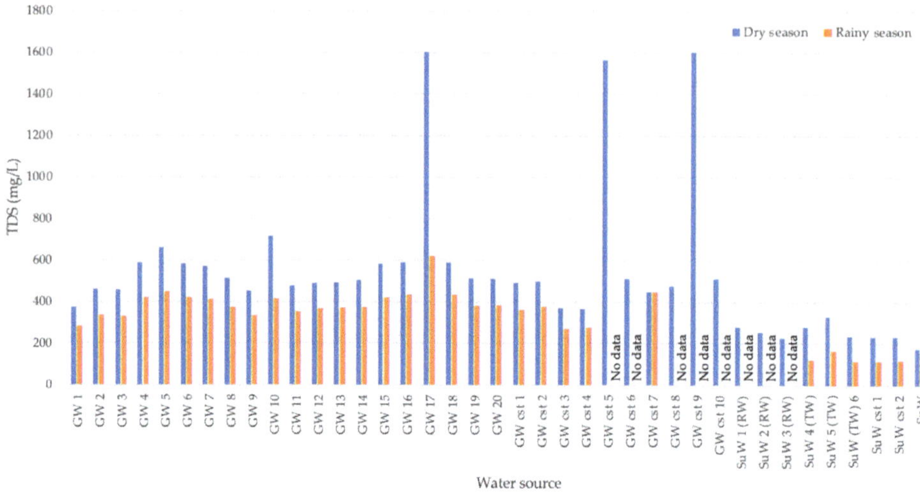

Figure A1. The level of TDS in the dry and rainy seasons. Some data are unavailable because of limitations during the field survey.

References

1. Zhang, H.; Liu, D.; Zhao, L.; Wang, J.; Xie, S.; Liu, S.; Lin, P.; Zhang, X.; Chen, C. Review on Corrosion and Corrosion Scale Formation upon Unlined Cast Iron Pipes in Drinking Water Distribution Systems. *J. Environ. Sci.* **2022**, *117*, 173–189. [CrossRef]
2. Gerke, T.L.; Little, B.J.; Barry Maynard, J. Manganese Deposition in Drinking Water Distribution Systems. *Sci. Total Environ.* **2016**, *541*, 184–193. [CrossRef] [PubMed]
3. Pan, L.; Li, G.; Li, J.; Gao, J.; Liu, Q.; Shi, B. Heavy Metal Enrichment in Drinking Water Pipe Scales and Speciation Change with Water Parameters. *Sci. Total Environ.* **2022**, *806*, 150549. [CrossRef] [PubMed]
4. Li, G.; Ma, X.; Chen, R.; Yu, Y.; Tao, H.; Shi, B. Field Studies of Manganese Deposition and Release in Drinking Water Distribution Systems: Insight into Deposit Control. *Water Res.* **2019**, *163*, 114897. [CrossRef] [PubMed]
5. Li, C.; Liu, C.; Xu, W.H.; Shan, M.G.; Wu, H.X. Formation Mechanisms and Supervisory Prediction of Scaling in Water Supply Pipelines: A Review. *Water Res.* **2022**, *222*, 118922. [CrossRef] [PubMed]
6. Ghobadi, F.; Jeong, G.; Kang, D. Water Pipe Replacement Scheduling Based on Life Cycle Cost Assessment and Optimization Algorithm. *Water* **2021**, *13*, 605. [CrossRef]
7. Malm, A.; Ljunggren, O.; Bergstedt, O.; Pettersson, T.J.R.; Morrison, G.M. Replacement Predictions for Drinking Water Networks through Historical Data. *Water Res.* **2012**, *46*, 2149–2158. [CrossRef] [PubMed]
8. González, P.P.; Bautista-Capetillo, C.; Ruiz-Canales, A.; González-Trinidad, J.; Júnez-Ferreira, H.E.; Rodríguez, A.R.C.; Rovelo, C.O.R. Characterization of Scale Deposits in a Drinking Water Network in a Semi-Arid Region. *Int. J. Environ. Res. Public Health* **2022**, *19*, 3257. [CrossRef]
9. Tut Haklidir, F.; Haklidir, M. Fuzzy Control of Calcium Carbonate and Silica Scales in Geothermal Systems. *Geothermics* **2017**, *70*, 230–238. [CrossRef]
10. Alahmad, M. Factors Affecting Scale Formation in Sea Water Environments—An Experimental Approach. *Chem. Eng. Technol.* **2008**, *31*, 149–156. [CrossRef]
11. Li, G.; Su, Y.; Wu, B.; Han, G.; Yu, J.; Yang, M.; Shi, B. Initial Formation and Accumulation of Manganese Deposits in Drinking Water Pipes: Investigating the Role of Microbial-Mediated Processes. *Environ. Sci. Technol.* **2022**, *56*, 5497–5507. [CrossRef] [PubMed]
12. Torvinen, E.; Suomalainen, S.; Lehtola, M.J.; Miettinen, I.T.; Zacheus, O.; Paulin, L.; Katila, M.-L.; Martikainen, P.J. Mycobacteria in Water and Loose Deposits of Drinking Water Distribution Systems in Finland. *Appl. Environ. Microbiol.* **2004**, *70*, 1973–1981. [CrossRef] [PubMed]
13. Zhang, Y.; Hou, H.; Qiu, M.; Zhou, L. Control of Loose Deposits in a Simulated Drinking Water Distribution System Using Ultrafiltration. *Water* **2023**, *15*, 2210. [CrossRef]
14. Ma, X.; Li, G.; Yu, Y.; Chen, R.; Zhang, Y.; Tao, H.; Zhang, G.; Shi, B. Spatial Variation of Loose Deposit Characteristics in a 40 Km Long Operational Drinking Water Distribution System. *Environ. Sci. Water Res. Technol.* **2019**, *5*, 1689–1698. [CrossRef]
15. Hafid, N.; Ben-Aazza, S.; Hadfi, A.; Ezahri, M.; Driouiche, A. Characterization of Scale Formed in Drinking Water and Hot Water Pipes in the Taliouine Downtown—Morocco. *Am. J. Anal. Chem.* **2015**, *6*, 677–686. [CrossRef]
16. Belattar, M.; Hadfi, A.; Said, B.; Mohareb, S.; Hafid, N. Characterization of Scale Deposits Formed in Sanitary Hot Water Pipelines in the Northern Tourist Zone of Agadir City. *Mediterr. J. Chem.* **2018**, *7*, 86–92. [CrossRef]

17. Touati, K.; Alia, E.; Zendah, H.; Elfil, H.; Hannachi, A. Sand Filters Scaling by Calcium Carbonate Precipitation during Groundwater Reverse Osmosis Desalination. *Desalination* **2018**, *430*, 24–32. [CrossRef]
18. Tavanpour, N.; Noshadi, M.; Tavanpour, N. Scale Formation and Corrosion of Drinking Water Pipes: A Case Study of Drinking Water Distribution System of Shiraz City. *Mod. Appl. Sci.* **2016**, *10*, 166–177. [CrossRef]
19. Mahmoud, B.; Yosra, M.; Nadia, A. Effects of Magnetic Treatment on Scaling Power of Hard Waters. *Sep. Purif. Technol.* **2016**, *171*, 88–92. [CrossRef]
20. Dobersek, D.; Goricanec, D. An Experimentally Evaluated Magnetic Device's Efficiency for Water-Scale Reduction on Electric Heaters. *Energy* **2014**, *77*, 271–278. [CrossRef]
21. Belattar, M.; Hadfi, A.; Ben-Aazza, S.; Mohareb, S.; Karmal, I.; Hafid, N.; Driouiche, A. Kinetic Study of the Scaling Power of Sanitary Water in the Tourist Area of Agadir. *Mater. Today Proc.* **2020**, *22*, 61–63. [CrossRef]
22. Yanagisawa, N. Case Study of Calcium Carbonate Scale at EGS and Hot Spring Binary System. In Proceedings of the World Geothermal Congress, Melbourne, Australia, 19–25 April 2015; pp. 1–5.
23. Peng, C.-Y.; Korshin, G. V Speciation of Trace Inorganic Contaminants in Corrosion Scales and Deposits Formed in Drinking Water Distribution Systems. *Water Res.* **2011**, *45*, 5553–5563. [CrossRef] [PubMed]
24. Zhou, X.; Kosaka, K.; Nakanishi, T.; Welfringer, T.; Itoh, S. Manganese Accumulation on Pipe Surface in Chlorinated Drinking Water Distribution System: Contributions of Physical and Chemical Pathways. *Water Res.* **2020**, *184*, 116201. [CrossRef] [PubMed]
25. Dickinson, W.H.; Wiatr, C. Manganese-Related Corrosion and Fouling in Water Systems. *Analyst* **2013**, *20*, 20–35.
26. Gerke, T.L.; Little, B.J. Manganese. In *Encyclopedia of Geochemistry: A Comprehensive Reference Source on the Chemistry of the Earth*; White, W.M., Ed.; Springer International Publishing: Cham, Switzerland, 2018; pp. 864–867, ISBN 978-3-319-39312-4.
27. Kousa, A.; Komulainen, H.; Hatakka, T.; Backman, B.; Hartikainen, S. Variation in Groundwater Manganese in Finland. *Environ. Geochem. Health* **2021**, *43*, 1193–1211. [CrossRef] [PubMed]
28. Rusydi, A.; Onodera, S.-I.; Saito, M.; Ioka, S.; Maria, R.; Ridwansyah, I.; Delinom, R. Vulnerability of Groundwater to Iron and Manganese Contamination in the Coastal Alluvial Plain of a Developing Indonesian City. *SN Appl. Sci.* **2021**, *3*, 1–12. [CrossRef]
29. Hou, Q.; Zhang, Q.; Huang, G.; Liu, C.; Zhang, Y. Elevated Manganese Concentrations in Shallow Groundwater of Various Aquifers in a Rapidly Urbanized Delta, South China. *Sci. Total Environ.* **2020**, *701*, 134777. [CrossRef] [PubMed]
30. De Joode, B.V.W.; Barbeau, B.; Bouchard, M.F.; Mora, A.M.; Skytt, Å.; Córdoba, L.; Quesada, R.; Lundh, T.; Lindh, C.H.; Mergler, D. Manganese Concentrations in Drinking Water from Villages near Banana Plantations with Aerial Mancozeb Spraying in Costa Rica: Results from the Infants' Environmental Health Study (ISA). *Environ. Pollut.* **2016**, *215*, 247–257. [CrossRef] [PubMed]
31. Khan, K.; Factor-Litvak, P.; Wasserman, G.A.; Liu, X.; Ahmed, E.; Parvez, F.; Slavkovich, V.; Levy, D.; Mey, J.; van Geen, A. Manganese Exposure from Drinking Water and Children's Classroom Behavior in Bangladesh. *Environ. Health Perspect.* **2011**, *119*, 1501–1506. [CrossRef]
32. Thakur, D.; Sharma, A.; Goel, P.; Thakur, A.; Raturi, M. Groundwater Quality Assessment in the Alluvial Region of Upper Yamuna Basin, India. *Groundw. Sustain. Dev.* **2023**, *22*, 100969. [CrossRef]
33. Gillispie, E.C.; Austin, R.E.; Rivera, N.A.; Bolich, R.; Duckworth, O.W.; Bradley, P.; Amoozegar, A.; Hesterberg, D.; Polizzotto, M.L. Soil Weathering as an Engine for Manganese Contamination of Well Water. *Environ. Sci. Technol.* **2016**, *50*, 9963–9971. [CrossRef] [PubMed]
34. Sly, L.I.; Hodgkinson, M.C.; Arunpairojana, V. Deposition of Manganese in a Drinking Water Distribution System. *Appl. Environ. Microbiol.* **1990**, *56*, 628–639. [CrossRef] [PubMed]
35. Cerrato, J.M.; Reyes, L.P.; Alvarado, C.N.; Dietrich, A.M. Effect of PVC and Iron Materials on Mn(II) Deposition in Drinking Water Distribution Systems. *Water Res.* **2006**, *40*, 2720–2726. [CrossRef] [PubMed]
36. Spiro, T.G.; Bargar, J.R.; Sposito, G.; Tebo, B.M. Bacteriogenic Manganese Oxides. *Acc. Chem. Res.* **2010**, *43*, 2–9. [CrossRef] [PubMed]
37. Watanabe, J.; Tani, Y.; Miyata, N.; Seyama, H.; Mitsunobu, S.; Naitou, H. Concurrent Sorption of As(V) and Mn(II) during Biogenic Manganese Oxide Formation. *Chem. Geol.* **2012**, *306–307*, 123–128. [CrossRef]
38. Center of Data and Information Technology Kinerja BUMD Air Minum (Regional Drinking Water Enterprise Performance) (In Indonesian Language). Available online: https://data.pu.go.id/visualisasi/kinerja-bumd-air-minum (accessed on 30 June 2024).
39. Intani, D.; Suripin; Suharyanto; Soetrisno, Y.A.A. Multi Criteria Decision Making for Integrated Assessment of Indonesian Water Company. *J. Adv. Res. Dyn. Control Syst.* **2020**, *12*, 1050–1060. [CrossRef] [PubMed]
40. Sutrisno, E. Air Sebagai Prioritas Utama Pembangunan Di Indonesia (Water Is the Main Development Priority in Indonesia) (In Bahasa Indonesia). Available online: https://indonesia.go.id/kategori/editorial/7000/air-sebagai-prioritas-utama-pembangunan-di-indonesia?lang=1 (accessed on 28 April 2024).
41. Directorate of Drinking Water Kinerja BUMD Air Minum 2022 (Regional Drinking Water Enterprise Performance 2022) (In Bahasa Indonesia); Jakarta. 2023. Available online: https://www.perpamsi.or.id/publikasi/buku (accessed on 30 June 2024).
42. Liu, G.; Zhang, Y.; Knibbe, W.-J.; Feng, C.; Liu, W.; Medema, G.; van der Meer, W. Potential Impacts of Changing Supply-Water Quality on Drinking Water Distribution: A Review. *Water Res.* **2017**, *116*, 135–148. [CrossRef]
43. Prasetiawan, T.; Nastiti, A.; Muntalif, B.S. 'Bad' Piped Water and Other Perceptual Drivers of Bottled Water Consumption in Indonesia. *WIREs Water* **2017**, *4*, e1219. [CrossRef]

44. Statistics of Surakarta Municipality Kota Surakarta Dalam Angka (Surakarta Municipality in Figure) 2023 (In Bahasa Indonesia); Surakarta. 2023. Available online: https://surakartakota.bps.go.id/publication/2023/02/28/8da64ce70793fbb9886e147f/kota-surakarta-dalam-angka-2023.html (accessed on 30 June 2024).
45. Puslitbang Geologi Peta Geologi Lembar Surakarta—Giritontro, Jawa. Available online: https://geologi.esdm.go.id/geomap/pages/preview/peta-geologi-lembar-yogyakarta-jawa (accessed on 11 May 2023).
46. Patriadi, A.; Soemitro, R.A.A.; Warnana, D.D. Preliminary Assessment to Sediment Evaluation On The Estuary of Bengawan Solo River. In Proceedings of the International Conference on Problematic Soils and Ground Improvement (Soft Soils 2016), Bandung, Indonesia, 27–28 September 2016; pp. X1-1–X1-4.
47. PDAM Surakarta Rencana Bisnis 2020–2024 Perumda Air Minum Kora Surakarta (Business Plan 2020–2024 of Surakarta Municipal Water Supply) (In Bahasa Indonesia); Surakarta, 2020.
48. Gholizadeh, A.; Mokhtari, M.; Naimi, N.; Shiravand, B.; Ehrampoush, M.; Miri, M.; Ebrahimi, A. Assessment of Corrosion and Scaling Potential in Groundwater Resources; a Case Study of Yazd-Ardakan Plain, Iran. *Groundw. Sustain. Dev.* **2017**, *5*, 59–65. [CrossRef]
49. El Baroudi, H.; Ouazzani, C.; Moustaghfir, A.; Er-ramly, A.; El Baroudi, Y.; Dami, A.; Balouch, L. Assessing the Corrosion and Scaling Potential of Drinking Water in Morocco Using Water Stability Indices. *Ecol. Eng. Environ. Technol.* **2024**, *25*, 130–139. [CrossRef]
50. Ministry of Health Ministry of Health Decree Number 2 Regarding Regulations on Implementing Government Regulation Number 66 of 2014 Concerning Environmental Health; Indonesia. 2023. Available online: https://peraturan.bpk.go.id/Details/245563/permenkes-no-2-tahun-2023 (accessed on 30 June 2024).
51. World Health Organization. Hardness in Drinking-Water: Background Document for Development of WHO Guidelines for Drinking-Water Quality. Available online: https://iris.who.int/bitstream/handle/10665/70168/WHO_HSE_WSH_10.01_10_Rev1_eng.pdf (accessed on 23 May 2024).
52. Mulato. After Being Shut Down Twice Due to Waste Exposure, the Semanggi Water Treatment Plant (IPA Semanggi) Has Been Reoperated. (In Bahasa Indonesia). Radio Republik Indonesia. 2024. Available online: https://www.rri.co.id/daerah/707415/sempat-2-kali-off-terpapar-limbah-ipa-semanggi-dioperasikan-kembali (accessed on 30 June 2024).
53. Zhang, S.; Tian, Y.; Guo, Y.; Shan, J.; Liu, R. Manganese Release from Corrosion Products of Cast Iron Pipes in Drinking Water Distribution Systems: Effect of Water Temperature, PH, Alkalinity, SO_4^{2-} Concentration and Disinfectants. *Chemosphere* **2021**, *262*, 127904. [CrossRef]
54. Khan, M.; Fadzil, F.; Eva, H.; Nazaruddin, D.; Shah, Z. Significance of Silica Analysis in Groundwater Studies of Domestic Shallow Wells in Parts of Jeli District, Kelantan, Malaysia. In *Water Resources in Arid Areas: The Way Forward*; Springer: Berlin/Heidelberg, Germany, 2017; pp. 103–114, ISBN 978-3-319-51856-5.
55. World Health Organization. Lead in Drinking Water: Background Document for Development of WHO Guidelines for Drinking-Water Quality. Available online: https://cdn.who.int/media/docs/default-source/wash-documents/wash-chemicals/lead-background-feb17.pdf?sfvrsn=fc50727b_4 (accessed on 23 May 2024).
56. Levallois, P.; Barn, P.; Valcke, M.; Gauvin, D.; Kosatsky, T. Public Health Consequences of Lead in Drinking Water. *Curr. Environ. Health Rep.* **2018**, *5*, 255–262. [CrossRef] [PubMed]
57. Zhou, H.; Fu, C. Manganese-Oxidizing Microbes and Biogenic Manganese Oxides: Characterization, Mn(II) Oxidation Mechanism and Environmental Relevance. *Rev. Environ. Sci. Bio/Technol.* **2020**, *19*, 489–507. [CrossRef]
58. Li, G.; Chen, Q.; Zhou, Y.; Su, Y.; Wu, B.; Yu, J.; Yang, M.; Shi, B. Manganese and Iron Oxides on Pipe Surface Promote Dissolved Aluminum Accumulation in Drinking Water Distribution Systems. *Sci. Total Environ.* **2024**, *924*, 171606. [CrossRef] [PubMed]
59. Tian, Y.; Wei, L.; Yu, T.; Shen, H.; Zhao, W.; Chu, X. Adsorption of Cr(VI) and Cr(III) on Layered Pipe Scales and the Effects of Disinfectants in Drinking Water Distribution Systems. *J. Hazard. Mater.* **2024**, *474*, 134745. [CrossRef]
60. Ferguson, R.J. Chapter 30—Scaling Indices: Types and Applications. In *Mineral Scales and Deposits*; Amjad, Z., Demadis, K.D., Eds.; Elsevier: Amsterdam, The Netherlands, 2015; pp. 721–735, ISBN 978-0-444-63228-9.
61. Ozair, G. An Overview of Calcium Carbonate Saturation Indices as a Criterion to Protect Desalinated Water Transmission Lines from Deterioration. *Nat. Environ. Pollut. Technol.* **2012**, *11*, 203–212.
62. Hannachi, A.; Naimi, I.; Zinoubi, R.; Elfil, H. A New Index for Scaling Assessment. In Proceedings of the IDA World Congress, Gran Canaria, Spain, 21–26 October 2007.
63. Tobiason, J.E.; Bazilio, A.; Goodwill, J.; Mai, X.; Nguyen, C. Manganese Removal from Drinking Water Sources. *Curr. Pollut. Rep.* **2016**, *2*, 168–177. [CrossRef]

water

MDPI

Article

To Feel the Spatial: Graph Neural Network-Based Method for Leakage Risk Assessment in Water Distribution Networks

Wenhong Wu [1,2], Xinyu Pan [1,2,*], Yunkai Kang [1,2,*], Yuexia Xu [3] and Liwei Han [1]

[1] School of Information Engineering, North China University of Water Resources and Electric Power, Zhengzhou 450046, China; wuwenhong@ncwu.edu.cn (W.W.); cvvvd7140@gmail.com (L.H.)
[2] Henan Province Water Distribution Network Intelligent Management Engineering Research Center, Zhengzhou 450046, China
[3] Zhengzhou Water Supply Investment Holdings Co., Ltd., Zhengzhou 450046, China; yuexia.xu@163.com
* Correspondence: pan19760928@gmail.com (X.P.); kykinfdarksoul@gmail.com (Y.K.); Tel.: +86-152-2603-7361 (X.P.)

Abstract: As water distribution networks expand, evaluating pipeline network leakage risk has become increasingly crucial. Contrary to traditional evaluation methods, which are often hampered by subjective weight assignment, data scarcity, and high expenses, data-driven models provide advantages like autonomous weight learning, comprehensive coverage, and cost-efficiency. This study introduces a data-driven framework leveraging graph neural networks to assess leakage risk in water distribution networks. Employing geographic information system (GIS) data from a central Chinese city, encompassing pipeline network details and historical repair records, the model achieved superior performance compared to other data-driven approaches, evidenced by metrics such as precision, accuracy, recall, and the Matthews correlation coefficient. Further analysis of risk factors underscores the importance of factors like pipe age, material, prior failures, and length. This approach demonstrates robust predictive accuracy and offers significant reference value for leakage risk evaluation.

Keywords: water distribution network; risk assessment; graph neural network; supervised learning

1. Introduction

China's urban water distribution networks (WDNs) play a vital role in connecting water sources to end-users and are essential for the survival and development of cities. However, pipeline networks often experience wear and tear in practical operations, leading to leaks that can disrupt daily life and impact the economy. To prevent significant losses and ensure the smooth operation of WDNs, it is vital to establish a robust leakage risk assessment system and implement appropriate early warning measures for high-risk pipelines (Pietrucha-Urbanik et al., 2023) [1].

Leakage risk prediction for WDNs has been the focus of research in the industry. Aiming at the complex influence factors of WDN pipelines and the surrounding environment, some scholars have proposed utilizing statistics and fuzzy mathematics to comprehensively consider the risk of pipeline leakage. For example, Peng et al. (2020) [2] used the fuzzy mathematical comprehensive evaluation method to predict the safety of WDNs, while Song et al. (2023) [3] used the multilevel fuzzy comprehensive evaluation method to clarify the different leakage risk influencing factors and their weights. Although these methods achieve a quantitative analysis of the risk of WDN pipelines, the subjectivity exposed by these methods due to their reliance on expert opinion reduces their credibility. With the construction of a database for pipeline attribute features based on a Geographic Information System (GIS) (Ho et al., 2010) [4] and the development of machine learning technology, some scholars have proposed that artificial intelligence models can be applied to the field of leakage risk prediction to improve the accuracy of risk prediction through data-driven learning of the nonlinear relationship between pipeline complexity factors and leakage probability (Bubtiena et al., 2011) [5]. Data-driven machine learning models can create relatively objective and effective risk assessment methods (Zhou et al., 2019) [6]. For example, Taiwo et al. (2024) [7] achieved a reliable assessment of pipeline leakage risk using characteristics of the pipeline itself and a logistic regression (LR) model optimized by a genetic algorithm, and Robles-Velasco et al. (2020) [8] developed predictive models using LR and support vector machine (SVM) algorithms for the prediction of water pipe failure probability. Chen et al. (2022) [9] developed predictive models using random forest (RF), boosted trees (BT), and extreme gradient boosting (XGBoost) algorithms with data collected from six WDNs.

The above machine learning-based method achieved a high identification rate of leaky pipelines in different cities or areas, but the description of pipeline topology in the GIS system is neglected and the black-box modeling makes it difficult to distinguish which risk factors on which pipelines have a more significant impact on leakage. We summarize the shortcomings of existing methods as follows:

1. Existing methods lack spatial considerations or cannot intuitively model the topological space of WDNs.
2. Existing methods utilize a limited number of pipeline features and cannot accurately capture the complex interactions between features at different domain levels, thus reducing the model's accuracy.
3. Existing methods can conduct quantitative analysis of pipelines but not qualitative analysis of the various risk factors of pipelines, and therefore cannot provide practical solutions.

We propose a novel solution to address the challenges of predicting pipeline risk in WDN. The proposed solution involves using a GIS pipeline feature database to build a data-driven model based on graph neural network (GNN) that has spatial sensing ability. This model considers the pipeline's characteristics and spatial factors to make quantitative predictions about pipeline risk. Furthermore, the model provides risk factor ranking for different pipelines, which can guide real-world decision-making. The study's main contributions are as follows:

1. This study proposes the utilization of node embedding learning in the domain of leakage risk estimation for WDNs via the dual-stream network optimization embedding-multilayer perceptron framework.
2. This study introduces two novel techniques, the spatial perception block and the dynamic graph optimization feature enhancement block, to capture the complex nonlinear relationship between pipeline properties, spatial topology, and the likelihood of leakage.
3. This study uses the SHAP method to investigate further the risk factors that may lead to pipeline leakage based on the quantitative risk prediction results.

2. Related Work

2.1. Leakage Risk Research for WDNs

Leakage risk research in WDNs has been a major focus of the industry. Leakage detection methods based on data-driven models are becoming mainstream. Konstantinos et al. (2017) [10] leveraged evolutionary polynomial regression and K-means clustering to identify abnormal pipelines through unsupervised learning. Wang et al. (2017) [11] implemented leakage prediction of pipelines using support vector machines by using pipeline operating water pressure as the basis of judging. Jensen et al. (2018) [12] predicted pipeline leakage incidents using neural networks and pipeline water volume as a metric. Sousa et al. (2023) [13] utilized a learning vector quantitative classifier to identify abnormal water pressures, which serve as indicators of leakage incidents, aiding in determining their occurrence within WDNs. Leakage prediction schemes for WDNs based on data-driven models have become a research priority. However, there is still a need for enhancement in quantitative risk analysis and extraction of spatial characteristics of WDN pipelines.

2.2. Graphical Neural Network

GNNs have become an important tool for analyzing and processing graph data. They have been widely used in various fields, including e-commerce, biology, and communication (Tang et al., 2023, Jin et al., 2023) [14,15], for graph node classification and anomaly detection tasks. For instance, Huang et al. (2022) [16] harnessed a real-world financial dataset, employing graph convolutional neural networks to detect financial fraudsters within a network. Choi et al. (2020) [17] merged the transformer encoder with GNNs to enhance the accuracy of heart failure prediction. Ahmed et al. (2019) [18] extensively reviewed the applications of graph neural networks in social media networks, showcasing their significant advantage and heightened accuracy in predicting fake news. Li et al. (2020) [19] integrated graph neural networks with spatio-temporal convolutional networks, achieving superior performance in predicting neighborhood crime rates. Yu et al. (2018) [20] employed graph neural networks for urban traffic computation, leveraging spatial information on connected roads to improve traffic flow predictions.

This study suggests incorporating GNNs into WDN leakage risk research. This approach would consider pipeline features and spatial factors to understand better and mitigate the risk of water leaks in WDNs.

3. Materials and Methods

3.1. Preliminary

3.1.1. Description of Tasks

The study was conducted in the B2 district metering area of a city in central China, and the data were derived from the pipeline GIS data and inspection and maintenance records covering the period from 2015 to 2021. Our task is to predict the probability of a leakage incident in this section of the network using the available pipeline properties. There is a complex nonlinear relationship between the pipeline properties of WDNs and the leakage risk, so it is necessary to process the pipeline properties to the embedding space, and then obtain the risk probability of this pipeline after nonlinear calculation within the embedding space, which is described in Equation (1):

$$\{X_1, X_2 \ldots, X_M\} \overset{f_1(\cdot)}{\Rightarrow} \{E_1, E_2 \ldots, E_D\} \overset{f_2(\cdot)}{\Rightarrow} \hat{Y} \quad (1)$$

where X denotes the natural features that the pipeline has, M denotes the number of features, E denotes the embedded representation of the pipeline in embedding space that has D dimensions, and $\hat{Y}$ denotes the result of predicting pipeline leakage.

3.1.2. Modeling of WDNs with Graph

A graph is a data structure with node and connection relationships in the form of edges. Graph structures are useful for recording spatial topological relationships between data elements. This study aims to model WDNs using graph structures to better understand spatial distribution. Recent research by Cen et al. (2023) [21] emphasizes the link between WDN leakage incidents and the network's intrinsic characteristics. Therefore, the modeling process is divided into two main parts: the semantic features of the pipeline itself and the spatial topological distribution. The semantic features of the pipeline itself are constructed as the feature matrix $X \in \mathbb{R}^{N \times M}$, where N is the number of pipelines to be predicted, and each pipeline has a feature vector of $\{X_1, X_2 \ldots, X_M\}$. The spatial distribution of pipelines is modelled as $G = (V, E, X)$; pipelines are defined as nodes in the form of $V \in \mathbb{R}^{N \times 1}$, which have corresponding feature vectors in X; and their connectivity is defined by edges in the form of $E \in \mathbb{R}^{N \times 2}$.

Due to the spatial proximity of pipelines with similar water quality, soil properties, and other external factors, this model uses neighbor node sampling to avoid disturbing the factors' interference with the model results. The neighbor node sampling process is denoted as $N(v)$ (Hamilton et al., 2018) [22]. It represents the search for pipelines with spatial connectivity relationships in the node set V using the edge set E. The above modeling approach provides the possibility of utilizing the spatial distribution of pipelines, as shown in subsequent models.

3.2. Model Framework

The framework is shown in Figure 1; the model adopts the dual-stream network design through the spatial perception block (SPB) and dynamic graph optimization feature enhancement block (DGB). The pipeline feature matrix and graph structure are inputs, and the stacked spatial domain graph convolution model learns the potential embedding of each pipeline. The results of the dual-stream network are aggregated and fed into a multilayer perceptron (MLP) optimized with BatchNorm layers (Ioffe et al., 2015) [23] to generate the final pipeline node representation vectors, then the final leakage probability of the pipeline nodes is generated. The components of the model are described in detail in this section.

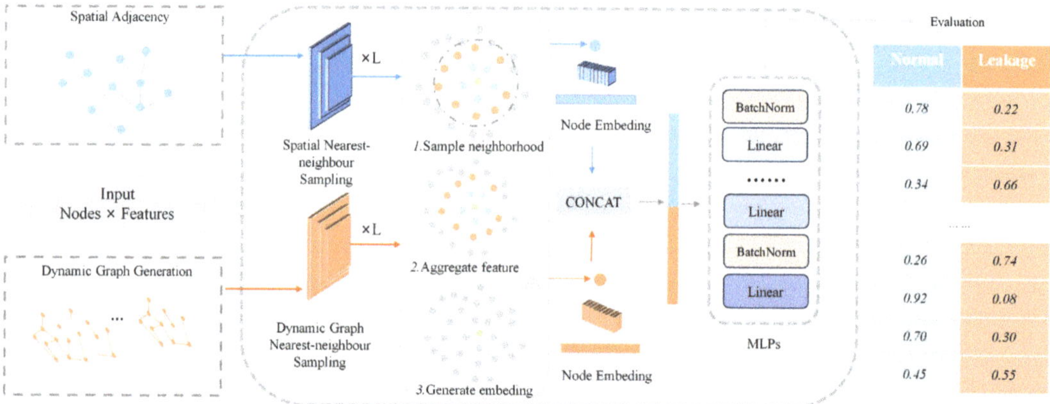

Figure 1. Framework of the model.

3.2.1. Spatial Perception Block

This block explicitly utilizes the spatial topology of WDNs after graph modeling, specifically applying graph convolution operations in the spatial domain to perform feature averaging and aggregation operations on the target pipeline and its surrounding pipeline nodes, which are then subjected to nonlinear transformations to ensure dimensional invariance (Hamilton et al., 2018). After the multi-layer graph convolution operation, this block realizes the synchronous embedding of pipeline features with spatial topology (Yang et al., 2023) [24]. The spatial embedding of the pipeline node v is generated as shown in Equation (2):

$$E_{1v}^{k} = \sigma\left(W_k \cdot mean\left(\{E_{1v}^{k-1}\} \cup \{E_{1u}^{k-1}\}\right)\right) \in \mathbb{R}^{1 \times D} \tag{2}$$

where $\forall u \in N(v)$, $E_{1v}^{(0)} = X(v)$, mean is the average computation, and σ represents a nonlinear activation function such as the ReLU.

This part achieves differentiated spatial embedding by aggregating the features of neighboring pipelines and their pipelines. Also, it enables the final generated pipeline embedding node to perceive the spatial structure of WDNs.

3.2.2. Dynamic Graph Optimization Feature Enhancement Block

This block focuses on the features of the pipeline itself in the embedding generation of the pipeline nodes. Since there are pipes of a close nature in the WDN system, the feature enhancement of individual pipes is achieved by aggregating the feature vectors of the more relative pipes. This module introduces a dynamic graph mechanism that allows the model to spontaneously learn potentially similar nodes to enhance the richness of the embedding representation (Ma et al., 2021) [25]. The dynamic graph A_{apt} construction process is shown in Equation (3):

$$\begin{gathered} M^k = tanh\left(\beta E^k \Gamma_{gc}^k\right) \in \mathbb{R}^{N\times D} \\ A_{apt}(i,j) = ReLU\left(tanh\left(\beta\left(M_{ij}^1 {M_{ij}^2}^T - M_{ij}^2 {M_{ij}^1}^T\right)\right)\right) \in \mathbb{R}^{N\times N} \end{gathered} \quad (3)$$

where $M_1, M_2 \in R^{N\times D}$ is described by two neural networks with randomly initialized embedding $E_1, E_2 \in \mathbb{R}^{N\times D}$ and trainable parameters Γ_{gc}^1, $\Gamma_{gc}^2 \in \mathbb{R}^{N\times D}$ computed via the Tanh activation function. β is hyper-parameter that adjusts the saturation rate of activation.

After constructing the dynamic graph, this module uses the maximum pooling method to achieve the feature enhancement of the pipeline node v, as shown in Equation (4):

$$E_{2v}^k = max\left(\sigma\left(W_p E_{2u}^{k-1} + b\right)\right) + W_p E_{2v}^{k-1} \in \mathbb{R}^{1\times D} \quad (4)$$

where $\forall u \in N_{A_{apt}}(v)$, $E_{2u}^{(0)}$=X(u), $E_{2v}^{(0)}$=X(v), max is the maximizing function, W_p is the pool of learnable weights, and b is the learnable bias constant.

After feature-enhanced embedding through this block, it is spliced with the spatial embedding obtained from the spatial perception block and fed into the MLP optimized by BatchNorm to finally obtain the representation vectors of the pipeline node. This process can be described by Equation (5):

$$z_v{}^{(l)} = Dropout\left(w^{(l)} z_v{}^{(l-1)} + b^{(l)}\right) \quad (5)$$

where $z_v{}^{(0)} = \frac{CONCAT(E_{1v}, E_{2v})}{1+e^{(-CONCAT(E_{1v}, E_{2v}))}} \in \mathbb{R}^{2\times D}$, $CONCAT$ denotes the dimension splicing operation, $w^{(l)}$ and $b^{(l)}$ are all linear layer learning parameters, and $Dropout$ introduces randomness in the training process by randomly masking some neurons (Hinton et al., 2012) [26].

3.2.3. Loss Functions and Leakage Probability Generation

The node representation z_v output after stacking the MLP needs to be mapped to the probability space and the final leakage probability obtained. To ensure the non-negativity and discriminability of the leakage probability and treat it as a probability distribution, the model maps the final generated node representations into probability space using the sigmoid function defined in Equation (6). This study minimizes the loss using the cross-entropy function in Equation (7) during model training.

$$\hat{Y} = \frac{1}{1+e^{-z_v}} \quad (6)$$

$$L = -\frac{1}{N}\sum_{i=1}^{N}\left(Y_i log(\hat{Y}_i) + (1-Y_i) log(1-\hat{Y}_i)\right) \quad (7)$$

where $\hat{Y}_i$ is the predicted label of pipeline i, and Y_i is the actual label of pipeline i.

3.3. *Evaluation Indicators and Model Interpretation*

3.3.1. Evaluation Indicators

In this model, the commonly used model evaluation metrics in binary classification tasks were chosen, such as *precision*, *accuracy*, *recall*, and the *Matthews correlation coefficient* (MCC):

$$Precision = \frac{TP}{TP+FP} \quad (8)$$

$$Accuracy = \frac{TP+TN}{TP+FP+TN+FN} \quad (9)$$

$$Recall = \frac{TP}{TP + TN} \quad (10)$$

$$Mcc = \frac{TP \times TN - FP \times FN}{\sqrt{(TP + FP)(TP + FN)(TN + FP)(TN + FN)}} \quad (11)$$

where true positive (TP), false positive (FP), true negative (TN), and false negative (FN) denote normal samples predicted to be normal, leakage samples predicted to be normal, leakage samples predicted to be leakage, and normal samples predicted to be leakage, respectively.

The first three metrics all range from 0% to 100%, and the MCC metrics range from −100% to 100%, with higher metrics being better. Different metrics reflect the ability of the model with different focuses. For example, higher recall and precision metrics indicate that the model is more capable of recognizing positive samples, and a higher accuracy indicates that the model is more capable of recognizing the current data. Due to the imbalance of leakage samples relative to normal samples, we additionally introduce the MCC metric, which is more robust in the category imbalance scenario, and a value of0% means that the model carries out a random prediction, a value of −100% means the model predicts completely wrongly, and a value of 100% means the model predicts all samples perfectly. To better evaluate the quality of a model, this study introduces the receiver operating characteristic curve (ROC) to visualize and analyze the prediction results. The model's predictions for test samples generate the ROC using different probability thresholds for the two classification outcomes (Carrington et al., 2021) [27]. This curve plots the actual positive rate against the false positive rate. The area under the curve (AUC) value summarizes the model's performance. In general, a larger AUC indicates better model performance.

3.3.2. SHAP Risk Factor Explanatory Mechanism

To determine the effect of different risk factors of a particular pipeline on the pipeline's leakage situation, this study introduces SHAP (Lundberg et al., 2017) [28] ranking to analyze the effect of different risk factors. The SHAP method employs the Shapley value in game theory, quantifying each feature's contribution to the model's final predicted outcome. The SHAP values represent the expected change in the model prediction when the attribute value of each feature is assigned to adjust for that feature, as shown in Equation (12):

$$\phi_i(x) = \sum_{S \in \{1,\ldots,p\}/\{i\}} \frac{|S|!(p - |S| - 1)!}{p!} [f(x_S \cup i) - f(x_S)] \quad (12)$$

where x_S denotes the input after removing the features in feature set S in x, p denotes the number of features in the input, $|S|$ denotes the size of S, $f(x_S \cup i)$ denotes the output of the model with feature i included, and $f(x_S)$ denotes the output of the model without feature i included.

The results of SHAP analysis are generally expressed through overall feature violin plots. Within the overall feature plot, the order of the vertical coordinates corresponds to the level of influence ability of different features. In contrast, the width of the horizontal coordinates of the violin plot corresponding to different feature values corresponds to the degree of influence of the feature on the predicted results of the model. The violin plot's width corresponds to the samples' distribution, and its color also represents the level of feature values. The SHAP methodology allows for the analysis of pipelines on a case-by-case basis and guides the repair of risky pipelines in practice. Specifically, if we find that a pipeline has a high probability of incurring a risk of leakage, and the SHAP analysis yields that the highest contributing factor to the high probability of the pipeline is the pipe material, we can replace the material of the pipeline during the repair work. Similarly, if the risk factor contributing more to the risk of the line is the age of the pipe and the water pressure in the pipe, then we can renew the pipe or reduce its operating pressure.

4. Results and Discussion

To verify the feasibility of the model, extensive comparison and visualization experiments were conducted on real-world pipeline datasets based on the GIS pipeline system of a managed area in a city in central China.

4.1. Dataset Description

This study selected pipeline materials, pipeline diameters, pipeline length, pipeline age, pipeline burial depth, operating pressure, and previous failures as seven critical pipeline network characteristics influencing pipeline network leakage. Pipelines where leakage occurs were labeled with a separate label item: "leakage". Specifically for the missing values, we supplemented the missing pipeline features in the GIS pipeline database and the maintenance work order with each other according to the section number. If the relevant records were missing in both data sources, we deleted the pipeline from the dataset considering the reliability of the model; after cleaning the data, we needed to max-min normalize the numerical type features in the pipeline, especially the pipeline length, pipeline age, pipeline burial depth, and operating pressure. Data preprocessing is a crucial step in preparing data for machine learning models. It involves various tasks, such as handling missing values and converting data types.

To clean up the GIS database, irrelevant indicators such as pipeline numbers, update times, and duplicate records were removed. Categorical features such as pipeline material and previous failures were expanded to nine dimensions using one-hot encoding. This resulted in seven pipeline types with two markers for previous failure, making the total dimensions 14. Numerical features such as pipeline length and age were scaled to a minimum–maximum range for model training. The dataset contained 30,612 edges. Out of the 11,489 data entries in the dataset, random sampling was used to divide them into training, validation, and test sets in a 7:2:1 ratio. To avoid the risk of potential data leakage (Chen et al., 2019) [29], a BFS optimized stratified sampling method was used when dividing the dataset. The pipeline nodes were divided into three regions with as few intersections as possible to avoid the model aggregating the test node information in advance. In the experiment, the model was trained on the training set, and the trained parameters were validated on the validation set. The optimal model on the validation set was finally tested on the test set.

4.2. Visual Comparison Experiment

This section compares the model proposed with five commonly used models for WDN leakage risk prediction to assess its performance. These models include BP networks, random forest, decision tree, AdaBoost (Suárez et al., 2023) [30], and logistic regression. In this study, we conducted 20 experiments with the baseline model and our model and subsequently calculated the resultant averages for the various evaluation metrics. The results of the combined experiments are shown in Figure 2 and detailed in Table 1.

Figure 2 and Table 1 show that the novel model introduced within our model exhibited superior performance metrics in comparison to extant models, achieving an accuracy of 93.23%, a precision of 92.18%, and a recall of 92.05%. These figures represent enhancements of 2.85%, 1.514%, and 4.000%, respectively, over a previously identified sub-optimal model. The graphical representations in Figure 2 underscore the model's supremacy across all evaluated metrics and the remarkable stability of the results compared to other machine learning models. Despite the enhanced classification stability attributed to tree models predicated on ensemble learning, their failure to adequately capture the nonlinear dynamics between feature sets and leakage and the spatial interrelations amongst pipeline structures rendered their accuracy subpar. Conversely, while BP-based models consider nonlinear interactions, the variability of their evaluative outcomes across different metrics signaled a deficiency in their generalization capacity. Through its incorporation of spatial considerations, GCN manifested superior performance indicators on the dataset, highlighting the criticality of spatial elements in predictive accuracy. A common trend among most models is a higher accuracy relative to recall, potentially indicative of a data imbalance within the test set. The ROC curves furnish a more illustrative depiction of the model's predictive capabilities, particularly within an imbalanced dataset. Given an equilibrated dataset and the ambition to generate a more precise anomaly score, both the ROC and MCC offer a more accurate reflection of the model performance. The proposed model achieved an optimal ROC and AUC of 0.952 and an MCC of 87.14%, underscoring its exceptional efficacy. To substantiate the effectiveness of the methodology delineated in this study, an ablation study was conducted, obfuscating the SPB and the dynamic DGB within the model and assessing their respective performances. The results from these ablation experiments affirm that both modules substantially enhanced the model's accuracy in predicting leakage probability. Specifically, the incorporation of the DGB, predicated on the dynamic graph's feature augmentation, significantly bolstered the model's accuracy, with improvements of 2.6319%, 1.7727%, and 4.5708% observed in accuracy, recall, and MCC metrics, respectively, compared to the model devoid of the DGB.

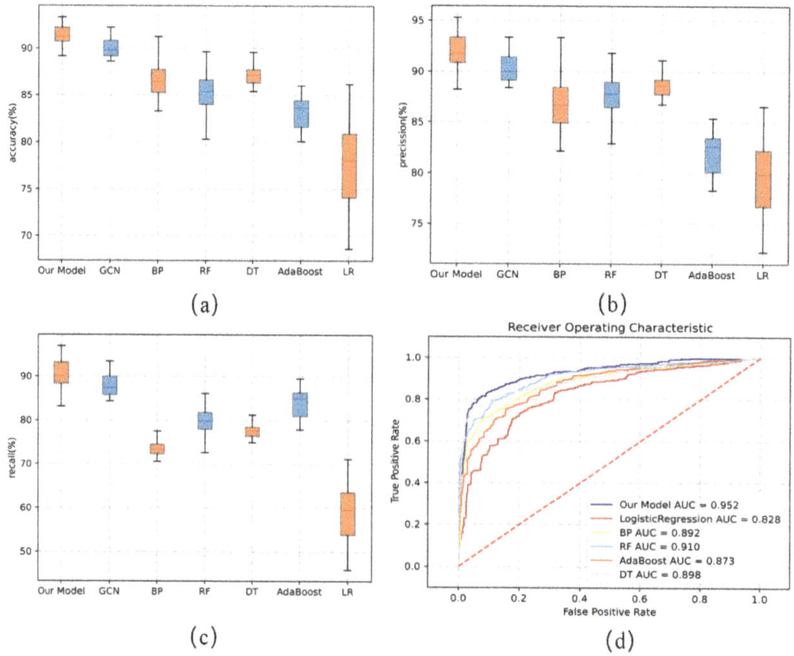

Figure 2. Results of the comparison experiments on the test set: (**a**) precision result, (**b**) accuracy result (**c**) recall result, and (**d**) ROC and AUC results.

Table 1. Comparison of experimental results.

Model Type	Training Set Precision (%)	Test Set Precision (%)	Test Set Accuracy (%)	Test Set Recall (%)	Test Set Mcc (%)
Proposed method	96.8996	93.2346	92.1823	92.0526	87.1423
GCN	94.4862	90.3806	90.6678	88.0526	83.6821
BP	99.7746	86.1159	86.1271	73.0392	70.2132
RF	93.4798	87.2053	84.8039	79.0697	72.0127
DT	96.8715	88.3217	86.9109	77.2093	71.9531
AdaBoost	96.2484	82.3374	83.4951	84.7041	79.4239
LR	81.2648	79.1739	77.8501	58.5784	53.4952
No-SPB	92.4824	90.5982	92.6342	91.4586	86.2581
No-DGB	95.4237	91.3249	90.0023	89.6859	81.6873

Reference to the GCN model's outcomes elucidates that an in-depth understanding of the interplay between pipeline features and a generic representation of analogous pipelines can markedly augment the model's predictive capabilities. Consequently, the TSNE-KDE joint graph approach was adopted to elucidate the model's node representation learning process. This methodology posits that their similarity is enhanced when high-dimensional representations of pipeline nodes are nonlinearly projected and proximally aligned in the high-dimensional continuum.

In Figure 3a–c, the results of the TSNE-KDE analyses performed on the training, validation, and test datasets are presented, demonstrating the model's adeptness at internalizing pipeline representations with clear class demarcation. Within the TSNE-KDE visualizations of the training dataset, it is evident that the model adeptly conglomerated disparate pipeline representations, engendering clusters with distinct nuclei. This phenomenon underscores the model's exemplary learning aptitude in the training phase. Conversely, the TSNE-KDE visualizations for the validation and test sets, while still delineating between leakage and normative pipeline representations, did not manifest pronounced clustering within the leakage representations. This observation intimates a potential diminution in the model's generalization capabilities across the validation and test datasets. However, it is imperative to note that this limitation predominantly impinges upon the model's interpretative capacity rather than its predictive accuracy. Furthermore, as depicted in Figure 3d, there was a nominal decrease in leakage prediction accuracy when transitioning from the training set to the test set, with the decrease being less than 5%. Notwithstanding this slight attrition, the model's ultimate prediction efficacy substantially exceeded competing models. This resilience in performance is ascribed to the disparity between the quantities of leakage and normal samples within the dataset. Consequently, enhancing the model's generalization capabilities is a critical area for further investigation to overcome the identified shortcomings while maintaining or improving its predictive accuracy.

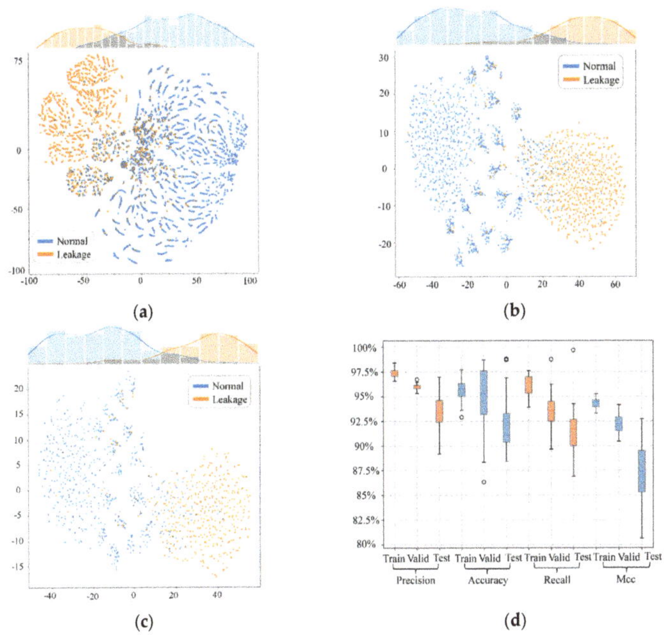

Figure 3. Visualization experimental results: (**a**) TSNE−KDE on the training set, (**b**) TSNE−KDE on the validation set, (**c**) TSNE−KDE on the test set, and (**d**) comparison of model performance on the training set, test set, and validation set.

4.3. Leakage Risk Factor SHAP Analysis

As indicated earlier, the SHAP algorithm complemented our model's interpretability. In this section, we used the SHAP method to visualize pipeline leakage risk factors. Specifically, we used a violin plot to analyze the risk factors for pipeline leakage at the full pipeline level. Then, we further used a waterfall plot to analyze two pipelines with different model identification results. The results of the evaluation are shown in Figure 4.

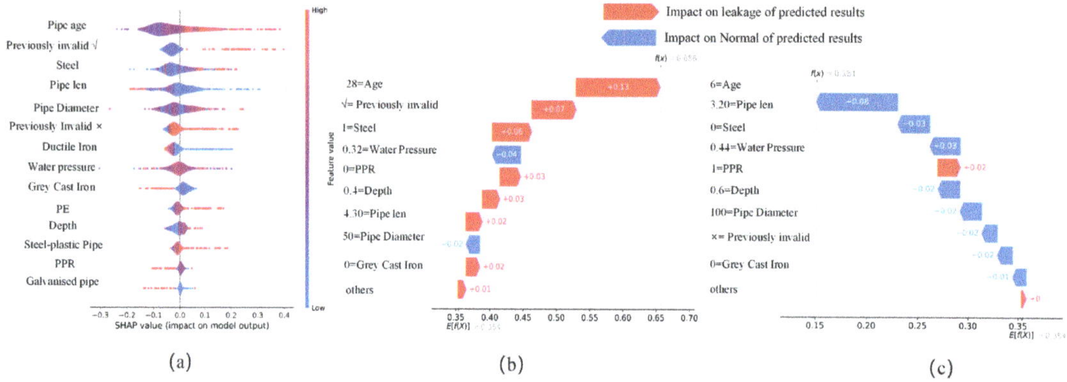

Figure 4. Result of SHAP analysis: (**a**) analysis results of pipeline leakage at the full pipeline level and (**b**,**c**) waterfall analysis for two separate samples.

Figure 4a describes the ranking of leakage risk factors obtained from analyzing the results of all pipeline forecasts. The pipeline's age was a key factor in assessing the risk of leakage, as shown by the overall characteristic violin plot ranking results. This may be due to the corrosive effects of soil and water quality on the pipeline, which increase over time and distance. Additionally, steel pipes are less resistant to corrosion, and previous damage or degradation of pipeline material can contribute to risk. It is worth noting that the previously invalid factor also significantly increases the risk of pipeline leakage because repair work on the pipeline destroys the strength of the original pipeline. Figure 4b shows the contribution of a pipeline identified by the model as having a risk of leakage, in which the age of the pipeline being 28 years was the largest contributor to the risk of leakage, and the previous invalid and the quality of its steel greatly influenced the model's judgment. Figure 4c shows the contribution of a pipeline identified by the model as not having a risk, in which the newer, 6-year-old pipeline reduced the risk of leakage. The PPR material and the absence of previous invalid also reduced the risk of leakage, and it is worth noting that the relatively high pressure increased the probability of leakage. The results of these two specific pipelines analyses are consistent with the analysis of all pipelines. In addition, the influence of pipeline operation water pressure on the risk of leakage is unclear for now. Although the contribution of water pressure to the prediction results was more common when analyzing the overall pattern, we found that the leakage loss of the pipeline was more sensitive to the change in water pressure by analyzing the prediction results of the specific pipeline. The interaction between the characteristics of the water pressure and the other characteristics still needs to be further clarified. Conversely, it was observed that steel–plastic, PPR, and galvanized features held minimal significance in the predictive models, which can be partially attributed to the inherent safety of these materials in terms of leakage risks. Consequently, this insight paves the way for strategic interventions such as prioritizing the replacement of obsolete pipelines, advocating for a diminished reliance on steel pipe in pipeline repair protocols, and advocating for reducing operating water pressure within pipelines deemed to be at high risk. This approach enhances the safety and efficiency of the pipeline infrastructure and aligns with best practices for sustainable and risk-averse management.

4.4. Practical Applications

In this section, we utilized pipeline coordinates from the GIS system and treated them as nodes in the B2 area. The assessment uses model-predicted leakage probability as an indicator to classify nodes. Nodes with a leakage probability of 0.50 or below are represented by blue, indicating normal conditions. Orange nodes signify medium-risk situations, with a leakage probability of between 0.50 and 0.75, and red nodes indicate high-risk situations, with a leakage probability of 0.75 or above. This criterion was used to evaluate the status of the local pipeline network, as illustrated in Figure 5b below.

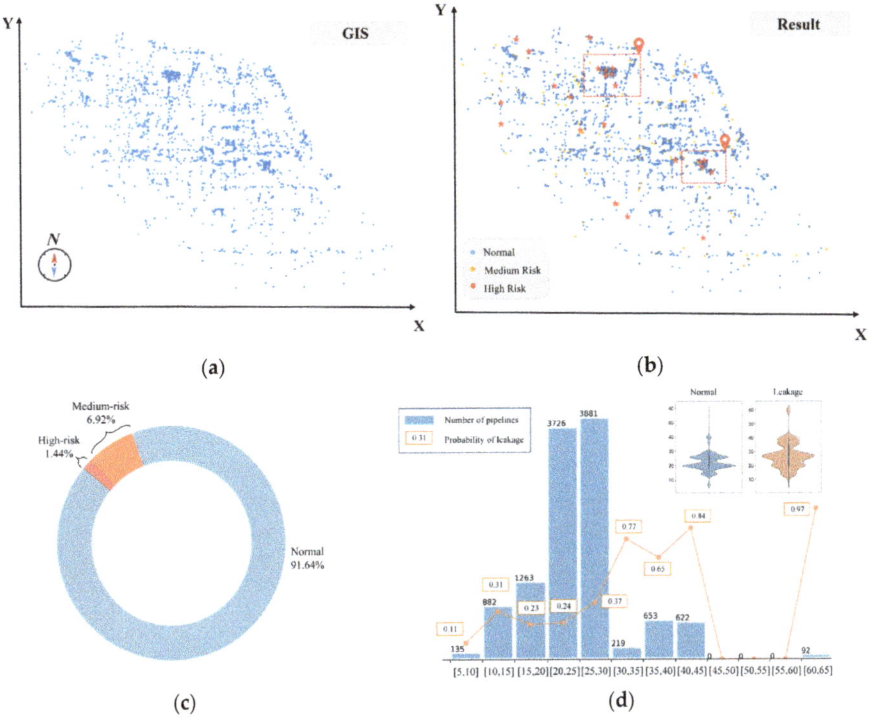

Figure 5. Coordinate mapping of model prediction results: (**a**) GIS system pipeline coordinate mapping results, (**b**) assessment results, (**c**) circular assessment results, and (**d**) results of heuristic data analysis using pipeline age as a reference.

In the results delineated in Figure 5c, a comprehensive analysis of 11,489 pipelines contained within the database was conducted. The findings revealed that 91.64% of these pipelines were categorized as usual, 6.92% as medium-risk, and 1.44% as high-risk. To augment the robustness of the SHAP analysis outcomes, the pipeline's age was employed as a principal reference feature to conduct detailed data analysis. This examination is visually represented in the violin plot depicted in Figure 5d, where a discernible distinction in the age distribution between leaky and regular pipelines was observed. Notably, as the age of the pipeline advanced, there was a palpable escalation in the likelihood of leakage, a trend that aligns with the SHAP analysis results. This correlation underscores the critical impact of pipeline age on its integrity and the imperative of incorporating age considerations into pipeline risk assessment frameworks.

5. Conclusions

The failure in WDNs is always unprecedented, facilitating numerous unpalatable consequences, for which previous studies have developed physical, statistical, and machine learning models to predict the probability of failure of individual pipes in a network. In this study, a novel embedding–MLP model is conducted to address the challenges of predicting the risk of water supply leakage in WDN pipelines. The proposed model incorporates a spatial perception block and a dynamic graph optimization feature enhancement block, introducing SHAP into the model prediction. This approach enhances the accuracy of leakage risk prediction by overcoming the shortcomings of existing methods, including insufficient spatial perception ability, lack of feature learning ability, and inadequate quantitative analysis of risk factors.

In the present inquiry, pipeline GIS data alongside maintenance archives from an urban area in China's heartland were meticulously aggregated. Following an extensive process of data cleansing and the fine-tuning of parameters for optimal performance, the predictive model demonstrated commendable efficacy, as evidenced by its achievement of *accuracy*, *precision*, *recall*, *MCC*, and AUC values of 93.23%, 92.18%, 92.05%, 87.14%, and 0.95, respectively. The robustness of the model was further corroborated through a series of detailed comparative analyses and visualization techniques. This verification underscores the model's potential to significantly aid water utility administrators in making well-informed decisions regarding the prioritization and identification of pipeline repair, which needs to be predicated on the calculated probability of failure. Moreover, the elucidation of model determinants via SHAP value experiments revealed that the factors of "pipeline age," "previous invalid," and "material steel" were predominant influencers in the model's predictive accuracy. This finding aligns seamlessly with the insights from heuristic data analyses, reinforcing the model's credibility. The congruence of these results with heuristic analyses validates the model's predictive prowess and establishes it as an indispensable instrument for management teams. By leveraging such analytical tools, decision-makers are empowered to enact strategic interventions aimed at enhancing the operational efficiency and reliability of WDNs.

Despite the significant insights provided by this study, there are acknowledged limitations that pave the way for future research. Currently, our model still has high requirements for the quality of the data, we need to have topological–spatial maps that record the connection relationship between pipelines in GIS data, and more pipeline features that can be used for model training and inference are required, which leads to the geographic–scenario migration capability of the model still being a big challenge. In our subsequent research, we will propose a novel way of selecting pipeline features to cope with model training under different data qualities. Additionally, the scarcity of pipeline leakage events within the dataset highlights the necessity for data augmentation strategies to achieve a more balanced sample. Moving forward, plans include expanding the scope to encompass various operating conditions and background characteristics pertinent to WDN pipelines. This will involve building a more comprehensive database of pipeline attributes. Furthermore, by leveraging the foundations laid by this study, the ultimate goal is to develop an early warning platform specifically designed to assess and mitigate the risk of WDN pipeline leakages.

Author Contributions: Conceptualization, W.W.; methodology, W.W. and Y.K.; validation, X.P.; formal analysis, X.P.; resources, W.W.; data curation, Y.X.; writing—original draft preparation, X.P. and Y.K.; writing—review and editing, L.H. and Y.K.; visualization, X.P. and Y.X.; supervision, Y.K. and L.H.; project administration, W.W. All authors have read and agreed to the published version of the manuscript.

Funding: This research received no external funding.

Data Availability Statement: Data available on request due to restrictions, e.g., privacy or ethical.

Acknowledgments: Thanks to Zhengzhou Water Supply Company for the data support.

Conflicts of Interest: Author Yuexia Xu was employed by the company Zhengzhou Water Supply Investment Holdings Co., Ltd. The remaining authors declare that the research was conducted in the absence of any commercial or financial relationships that could be construed as a potential conflict of interest.

References

1. Pietrucha-Urbanik, K.; Rak, J. Water, Resources, and Resilience: Insights from Diverse Environmental Studies. *Water* **2023**, *15*, 3965. [CrossRef]
2. Peng, L.; Liu, Y.F.; Zheng, J.C. Safety assessment and analysis for urban water supply distribution network. *Water Purif. Technol.* **2020**, *39*, 140–148.
3. Song, Y.; Liu, S.G.; Qi, Y.Y. Safety assessment and analysis for urban municipal pipeline networks. *Water Purif. Technol.* **2023**, *42*, 151–160+175.
4. Ho, C.; Lin, M.; Lo, S. Use of a GIS-based hybrid artificial neural network to prioritize the order of pipe replacement in a water distribution network. *Environ. Monit. Assess.* **2010**, *166*, 177–189. [CrossRef]
5. Bubtiena, A.M.; Elshafie, A.H.; Jafaar, O. Application of Artificial Neural networks in modeling water networks. In Proceedings of the 2011 IEEE 7th International Colloquium on Signal Processing and Its Applications, Penang, Malaysia, 4–6 March 2011; pp. 50–57.
6. Zhou, X.; Tang, Z.; Xu, W.; Meng, F.; Chu, X.; Xin, K.; Fu, G. Deep learning identifies accurate burst locations in water distribution networks. *Water Res.* **2019**, *166*, 115058. [CrossRef] [PubMed]
7. Taiwo, R.; Zayed, T.; Ben Seghier, M.E.A. Integrated Intelligent Models for Predicting Water Pipe Failure Probability. *Alex. Eng. J.* **2024**, *86*, 243–257. [CrossRef]
8. Robles-Velasco, A.; Cortés, P.; Muñuzuri, J.; Onieva, L. Prediction of pipe failures in water supply networks using logistic regression and support vector classification. *Reliab. Eng. Syst. Saf.* **2020**, *196*, 106754. [CrossRef]
9. Chen, T.Y.; Vladeanu, G.; Yazdekhasti, S.; Daly, C.M. Performance evaluation of pipe break machine learning models using datasets from multiple utilities. *J. Infrastruct. Syst.* **2022**, *28*, 05022002. [CrossRef]
10. Konstantinos, K.; Kourosh, B.; Raz1yeh, F. Pipeline failure prediction in water distribution networks using evolutionary polynomial regression combined with K means clustering. *Urban Water J.* **2017**, *14*, 737–742.
11. Wang, X.; Chen, Z.; Zhong, X.; Lu, N. PSO-SVM based leakage diagnosis method of water supply pipeline. *Mod. Electron. Tech.* **2018**, *41*, 156–159.
12. Jensen, T.N.; Puig, V.; Romera, J.; Kallesøe, C.S.; Wisniewski, R.; Bendtsen, J.D. Leakage localization in water distribution using data-driven models and sensitivity analysis. *IFAC-Pap.* **2018**, *51*, 736–741. [CrossRef]
13. Sousa, D.P.; Du, R.; Mairton Barros da Silva, J., Jr.; Cavalcante, C.C.; Fischione, C. Leakage detection in water distribution networks using machine-learning strategies. *Water Supply* **2023**, *23*, 1115–1126. [CrossRef]
14. Tang, J.; Hua, F.; Gao, Z.; Zhao, P.; Li, J. GADBench: Revisiting and Benchmarking Supervised Graph Anomaly Detection. *arXiv* **2023**. [CrossRef]
15. Jin, G.; Liang, Y.; Fang, Y.; Shao, Z.; Huang, J.; Zhang, J.; Zheng, Y. Spatio-Temporal Graph Neural Networks for Predictive Learning in Urban Computing: A Survey. *arXiv* **2023**. [CrossRef]
16. Huang, X.; Yang, Y.; Wang, Y. DGraph: A large-scale financial dataset for graph anomaly detection. *Adv. Neural Inf. Process. Syst.* **2022**, *35*, 22765–22777.
17. Edward, C.; Xu, Z.; Li, Y.; Dusenberry, M.; Flores, G.; Xue, E.; Dai, A. Learning the Graphical Structure of Electronic Health Records with Graph Convolutional Transformer. In Proceedings of the AAAI Conference on Artificial Intelligence, New York, NY, USA, 7–12 February 2020; Volume 34, pp. 606–613.
18. Ahmed, S.; Hinkelmann, K.; Corradini, F. Combining machine learning with knowledge engineering to detect fake news in social networks—A survey. In Proceedings of the AAAI Conference on Artificial Intelligence, Honolulu, HI, USA, 27 January–1 February 2019; Volume 12, p. 8.
19. Li, Z.; Huang, C.; Xia, L.; Xu, Y.; Pei, J. Spatial-Temporal Hypergraph Self-Supervised Learning for Crime Prediction. In Proceedings of the IEEE 38th International Conference on Data Engineering (ICDE), Kuala Lumpur, Malaysia, 9–12 May 2022; pp. 2984–2996.
20. Yu, B.; Yin, H.; Zhu, Z. Spatio-Temporal Graph Convolutional Networks: A Deep Learning Framework for Traffic Forecasting. In Proceedings of the 27th International Joint Conference on Artificial Intelligence, Stockholm, Sweden, 13–19 July 2018; pp. 3634–3640.
21. Cen, H.; Huang, D.; Liu, Q.; Zong, Z.; Tang, A. Application Research on Risk Assessment of Municipal Pipeline Network Based on Random Forest Machine Learning Algorithm. *Water* **2023**, *15*, 1964. [CrossRef]
22. Hamilton, W.L.; Ying, R.; Leskovec, J. Inductive Representation Learning on Large Graphs. *arXiv* **2017**. [CrossRef]
23. Ioffe, S.; Szegedy, C. Batch Normalization: Accelerating Deep Network Training by Reducing Internal Covariate Shift. In Proceedings of the 32nd International Conference on Machine Learning, Lille, France, 6–11 July 2015; pp. 448–456.
24. Yang, C.; Wu, Q.; Wang, J.; Yan, J. Graph Neural Networks are Inherently Good Generalizers: Insights by Bridging GNNs and MLPs. *arXiv* **2023**. [CrossRef]
25. Ma, X.; Wu, J.; Xue, S.; Yang, J.; Zhou, C.; Sheng, Q.Z.; Xiong, H.; Akoglu, L. A Comprehensive Survey on Graph Anomaly Detection with Deep Learning. *IEEE Trans. Knowl. Data Eng.* **2021**, *35*, 12012–12038. [CrossRef]
26. Hinton, G.E.; Srivastava, N.; Krizhevsky, A.; Sutskever, I.; Ruslan, R. Salakhutdinov Improving neural networks by preventing co-adaptation of feature detectors. *arXiv* **2012**. [CrossRef]
27. Carrington, A.M.; Manuel, D.G.; Fieguth, P.W. Deep ROC analysis and AUC as balanced average accuracy to improve model selection, understanding, and interpretation. *IEEE Trans. Pattern Anal. Mach. Intell.* **2021**, *45*, 329–341. [CrossRef] [PubMed]

28. Lundberg, S.M.; Lee, S.I. A Unified Approach to Interpreting Model Predictions. *Adv. Neural Inf. Process. Syst.* **2017**, *30*, 4768–4777.
29. Chen, D.; Lin, Y.; Li, W.; Li, P.; Zhou, J.; Sun, X. Measuring and Relieving the Over-smoothing Problem for Graph Neural Networks from the Topological View. In Proceedings of the AAAI Conference on Artificial Intelligence, Honolulu, HI, USA, 27 January–1 February 2019; Volume 34.
30. Suárez, M.; Martínez, R.; Torres, A.M.; Ramón, A.; Blasco, P.; Mateo, J. Personalized Risk Assessment of Hepatic Fibrosis after Cholecystectomy in Metabolic-Associated Steatotic Liver Disease: A Machine Learning Approach. *J. Clin. Med.* **2023**, *12*, 6489. [CrossRef] [PubMed]

water

MDPI

Article

Case Study for Predicting Failures in Water Supply Networks Using Neural Networks

Viviano de Sousa Medeiros [1], Moisés Dantas dos Santos [2,*] and Alisson Vasconcelos Brito [3]

1 TRIL Lab, Graduate Program in Mechanical Engineering, Center of Technology, Federal University of Paraíba, João Pessoa 58058-600, Brazil; vivianomedeiros@gmail.com
2 TRIL Lab, Scientific Computing Department, Center of Informatics, Federal University of Paraíba, João Pessoa 58058-600, Brazil
3 LASER Lab, Scientific Computing Department, Center of Informatics, Federal University of Paraíba, João Pessoa 58058-600, Brazil; alisson@ci.ufpb.br
* Correspondence: mdantas@ci.ufpb.br

Abstract: This study deals with the prediction of recurring failures in water supply networks, a complex and costly task, but essential for the effective maintenance of these vital infrastructures. Using historical failure data provided by Companhia de Água e Esgotos da Paraíba (CAGEPA), the research focuses on predicting the time until the next failure at specific points in the network. The authors divided the failures into two categories: Occurrences of New Faults (ONFs) and Recurrences of Faults (RFs). To perform the predictions, they used predictive models based on machine learning, more specifically on MLP (Multi-Layer Perceptron) neural networks. The investigation unveiled that through the analysis of historical failure data and the consideration of variables including altitude, number of failures on the same street, and days between failures, it is possible to achieve an accuracy greater than 80% in predicting failures within a 90-day interval. This demonstrates the feasibility of using fault history to predict future water supply outages with significant accuracy. These forecasts allow water utilities to plan and optimize their maintenance, minimizing inconvenience and losses. The article contributes significantly to the field of water infrastructure management by proposing the applicability of a data-driven approach in diverse urban settings and across various types of infrastructure networks, including those pertaining to energy or communication. These conclusions underscore the paramount importance of systematic data collection and analysis in both averting failures and optimizing the allocation of resources within water utilities.

Keywords: fault prediction; water supply networks; machine learning; predictive modeling; infrastructure management

1. Introduction

Water distribution networks are essential infrastructures for the development of a city and its productive activities [1]. Failures in these networks can bring inconvenience and losses to the population and economic activities, industry, and agriculture [2,3]. In this context, predicting future failures or breaks in a water supply network allows the management of this network to carry out planned interventions, anticipating failures, which can reduce inconvenience and losses caused by water supply interruption [4–7].

According to information from the National Sanitation Information System (in Portuguese: SNIS—Sistema Nacional de Informações sobre Saneamento), Brazil has a Water Distribution Loss Index (WDLI) of 40.3%. This number indicates that more than 40% of treated water is lost in the distribution process. In this context, the Water and Sewage Company of Paraíba (in Portuguese: CAGEPA—Companhia de Água e Esgotos da Paraíba) presents a WDLI of 35.4%, a value below the national average, but expressing a concerning figure: more than one-third of the water produced by the company is lost in the distribution process [8].

Predicting future failures in water supply networks constitutes a complex and computationally intensive endeavor, as well involves processing data with high volume and dimensionality [9]. In this context, three approaches stand out for predicting failures in this type of network: predictions based on physical models of the pipeline, predictions based on statistical models, and predictions based on machine learning models. Of these three approaches, the use of machine learning has gained garnered significance due to its capacity for automated pattern recognition and complex relationships among the variables linked to a water supply network [6,10].

Machine learning-based prediction models have been employed by water supply companies to forecast failures in their distribution system. However, this approach requires high-dimensional and highly representative data. Such data may be scarce in terms of quantity and variety (diversity of variables) because they are highly sensitive for companies, and access to them may be limited. Data on network structure, historical failure records, and spatial and meteorological data, for example, are private to water supply companies and are considered strategic for their business models, which complicates access [9,11].

According to [1,10], there are few studies involving failure prediction and optimization of corrections in water supply infrastructures. This type of operation is common in the energy sector, but few applications are known for water supply. Moreover, ref. [3] also considers statistical results from research evaluating the condition or reliability of water distribution systems to be scarce. Additionally, according to [12], research about water distribution system data using machine learning techniques is scarce or rare.

Amidst this scenario, the present research aims to predict the occurrence of failures in a water distribution network using a machine learning-based model and real network data, including the history of failures. This, combined with other variables, allows estimating how many days will elapse until a new failure occurs at a point in the system where previous failures have been recorded.

The present research work is organized as follows: first, we begin with a brief introduction to the topic, followed by a detailed presentation of the proposal and theoretical foundation. Next, this work present the methodology and the development of the research. Finally, the results achieved and their implications are discussed.

2. Related Works

Failures in water supply networks can occur due to the influence of various factors, as described in the work of [13], which provides a detailed description of the main factors influencing failure mechanisms in that systems. Along the same lines, the study presented by [14] also considers elements that play a determinant role in network breakdowns, taking into account information on physical, mechanical, environmental, and social components. Both works highlight the influence of failure history as an important factor for predicting new failures. This information forms the basis of our research, which considers failure history as our main predictive variable.

In the literature, some studies explore the prediction of failures within water supply networks employing various methodologies, yet the predominant approach involves leveraging failure history as a significant predictive variable. Some of these works employ predictive models based on machine learning to forecast the remaining useful life for a single pipeline in the network, as seen in [15,16]. Additionally, the authors of [17] uses statistical relationships between failure frequencies and weather conditions to assess the effect of climate change on future breakdowns in water supply networks.

On the other hand, this research also found other works that aim to predict the probability of a failure occurrence. In this line of research, is possible observe that the work of [18], which uses an adapted model from electrical networks to predict the risk of failure in water supply networks. Similarly, the work of [19] utilizes a predictive model to obtain a failure probability associated with each sample, i.e., each pipeline in the network. Based on the predicted information, the replacement of parts of the water supply network is planned.

The work of [9] aims to predict the risk of failures using AutoML based on the failure history. Similarly, the work [7] seeks to predict the probability of network failure. Finally, the study by [20] aims to predict the frequency of failure (failure per kilometer of the network), but it uses limited data in its model, restricting itself to data such as material, length, diameter, and installation year.

Other works deal with predicting the failure rate in water distribution mains (networks that typically distribute water between two different points with long length and diameter). This is the case of the study presented by [21]. A similar approach is taken by [10,22], although the latter apply their work to predicting failures in long-length water mains.

There are also studies that deal with the analysis of failure risk using hydraulic models of real supply networks. For example, the work by [23,24] employs models built on EPANET to analyze failure risk. Both studies use real data from Polish cities in their analysis. However, while the first study analyzes the consequences of failures occurring in individual pipes and classifies areas according to their vulnerability to failures caused by pressure variations, considering statistical factors such as seasonality, the second study examines failure risk indicators and proposes the adoption of new failure indicators for pipelines. Both studies deal with failure risk indicators based on network structure data, which were not available for the research presented in this manuscript. However, the present research has a different objective (predicting the number of days until the next failure) and uses different data (mostly historical failure data).

Unlike what is found in the literature, our work analyzes points in an urban network where there is already a history of past failures, and our prediction consists of a regression task in which the returned value indicates the number of days until the occurrence of the next failure at a particular point.

3. Problem Definition

The present work purposes to predict failures in a water systems based on its historical records of previous failures. A failure is understood as any occurrence recorded in the water supply network related to leaks, indicating a potential risk of temporary water shortage (water outage).

Accordingly, this research analyzes points in the water network where there is already a history of failures, meaning points in the network where failures are recurrent. These points were mapped based on the historical records of registered complaints, and the data from these points were subjected to a machine learning algorithm, a model based on a neural network, aimed at predicting a numerical value indicating the number of days between the last registered failure and the occurrence of a new failure in the future.

Armed with the estimated prediction of the forthcoming failure, the water utility can plan maintenance activities so that this future failure can be repaired preemptively, anticipating the leak, avoiding water shortages, and minimizing inconvenience for the potentially affected population. Therefore, the predictive maintenance work can contribute to creating a positive image for the water utility among its customers.

The data used in this research were provided by the Water Supply Company CAGEPA (Water and Sewage Company of Paraíba). These are real data provided by the company regarding the recorded occurrences. These occurrences include information such as date, geographical coordinates, type, address, and other data.

It is essential to emphasize that this research was conducted through an institutional partnership between CAGEPA and the Federal University of Paraíba. Consequently, this study is tailored to the specific context provided by the company. Hence, the company was actively involved in defining the research objectives and supplying the relevant data for analysis. Consequently, the company's primary objective is to forecast future failures, necessitating the provision of data for the predictive model.

For this research, data recorded in the city of Guarabira–PB were selected. The choice of this city was motivated by a recommendation from CAGEPA, as this city has the highest availability of data, with a greater number of variables available, as well as the longest data interval, extending from November 2017 to April 2023.

For clarification purposes, it is important to emphasize that failures may recur over time, meaning a failure occurring at a point in the network may be recorded again at that same point in the future. Thus, occurrences recorded within a radius of 50 m from each other are considered a recurrence of the same failure, i.e., a repetition of a failure over time, while occurrences outside this radius are considered new independent failures. A representation of this information can be visualized in Figure 1 below:

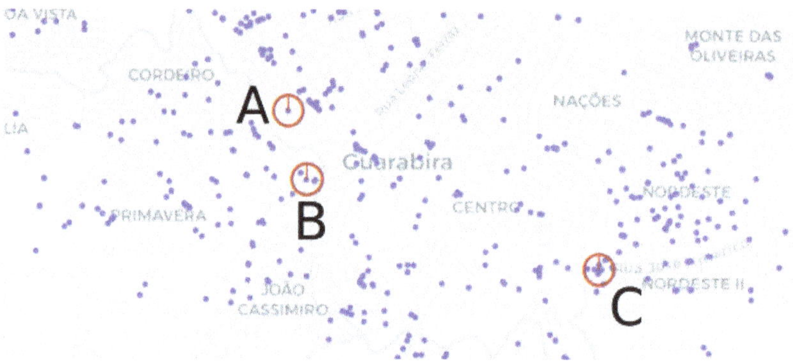

Figure 1. Recurring failures.

Observing Figure 1, it is possible to visualize the occurrences recorded in the city of Guarabira–PB. The figure shows the map of the city and several blue dots representing the location of the failures. On the other hand, is possible see three highlighted points, where it is possible to perceive a central occurrence circled by a radius. This radius indicates whether other occurrences around it are new failures or recurrences of the main failure.

In Figure 1 that Point A is a failure that does not have any recurrences, while Point B has only two recurrences. Finally, Point C, unlike the previous ones, has several recurrences around the main failure. It is worth noting that the main failure is the oldest one, meaning that it was recorded before the others.

Failure History

To enhance understanding of the research problem addressed in this study, Figure 2 below presents an analysis of a failure case and the arrangement of its failure history. Additionally, the figure also presents the forecast to be estimated by the neural network developed in this study.

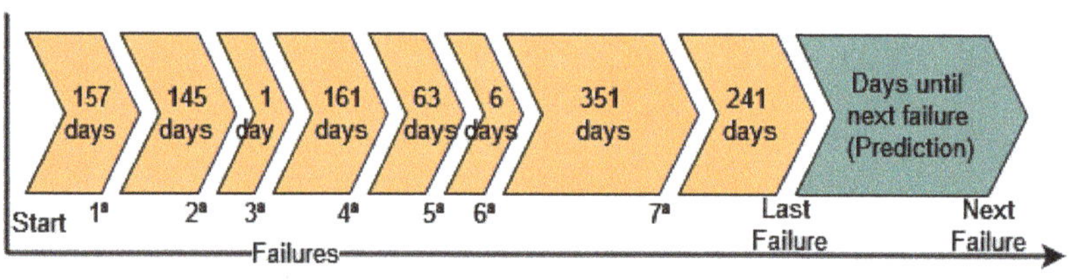

Figure 2. Failure predictions—region of Guarabira–Brazil.

Analyzing Figure 2, it is possible to visualize a point in the water supply network that experienced a recurrence of eight failures. In the figure, the beginning of the monitoring occurred in November 2017 and the failure history is presented in blocks separated by the failures that occurred at this point. For this specific case, the first failure occurred 157 days after the start of monitoring, the second failure occurred 145 days after the first one, the third failure occurred only 1 day after the second one, and so on, presenting all the failures that occurred at this point in the network.

Finally, on the rightmost part of the figure, a blue block indicates the number of days between the last recorded failure and a new potential failure that may occur in the future. Therefore, this is the prediction objective of this study: to forecast the number of days between the last recorded occurrence and an estimated future failure.

4. Materials and Methods

This research was developed to predict failures in a water supply network using real data from the same network applied to machine learning models. Consequently, the predictive data analysis process presented in [25] was followed. Therefore, this work is structured into six main stages with their respective characteristics and artifacts generated in each one. These stages are described below in Figure 3:

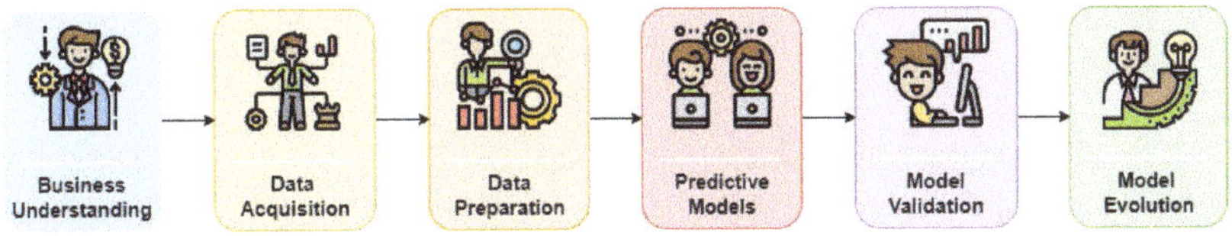

Figure 3. Stages of work development.

According to Figure 3, the work begins with understanding the business context in which the data are embedded. It is during this stage that the objectives of analysis and prediction are delineated. Next, the data collection and understanding work are carried out, with exploratory analyses aiding in understanding the information and phenomena present in the data. The third stage of the work consists of preparing the data to be used in predictive models. It is in this phase that the processes of data cleaning, variable selection, and data preparation with necessary transformations take place.

The fourth stage of the process involves the final selection of data attributes and the construction of predictive models. The fifth stage, in turn, consists of applying the constructed models to the data to obtain predictions. With the predictions made, it is possible to validate the accuracy of the model. Finally, the last stage consists of using the model with real data and its evolution.

Below, this manuscript describes how each stage described above was applied in the context of our work:

The initial two stages described in Figure 3 were carried out together, and both were developed in an immersion context within the company CAGEPA, the data provider. Understanding the business enabled a better grasp of the phenomena encountered in the exploratory analysis stage, where company experts were consulted to help uncover trends and patterns present in the datasets. After a preliminary exploration of the data, the data cleaning stage was conducted to remove attributes and values that had the potential to impair subsequent analyses.

The attribute selection stage was carried out in several phases. The CAGEPA databases used in this study have dozens of attributes, with the main one used in this work dealing with network failure occurrences and the services performed to carry out the operations of correction and recovery of these failures.

The primary attributes utilized in this study were related to the network failure history of the company, from which the date of occurrence registration, its geographical location, its type (only occurrences of the Leak Removal type were considered), and finally, its registered address were extracted.

In addition to attributes present in the datasets provided by the company, other attributes were incorporated during the development of the work, most of which were statistical values calculated from previously existing data, such as mean, standard deviation, and range, among others described below.

Below are described all the attributes or variables used in this research work. Table 1 and the following enumerated list present each of the variables used. The attributes were divided into two types: predictor attributes and the target attribute, according to the indication of [26].

1. Target attribute (variable).
   1.A. Number of days until the occurrence of the next failure for a specific point in the network.
2. Predictor attributes (variables).
   2.A. Distance to the nearest neighbor (meters).
   2.B. Terrain elevation at the location where the failure was registered relative to sea level (meters).
   2.C. Number of recurrences of the failure in question.
   2.D. Number of days since the last failure in the network.
   2.E. Number of days between failures in the network at the same point.
   2.F. Number of occurrences on the same street.
   2.G. Mean of the number of days between registered failures at the same point.
   2.H. Variance in registered failures for the same point.
   2.I. Standard deviation of the number of days between failures for the same point.

Table 1. Description of the attributes/variables used by the predictive models.

Attribute/ Variable	Name	Attribute Use	Data Type	Min Value	Max Value	Metric Unit
1.A	days_until_ next_failure	Target	integer	1	1081	days
2.A	dist_ nearest_neighbor	Predictor	float	~0.01	0.126	meters
2.B	elevation	Predictor	float	27.63	132.5	meters
2.C	qtd_recurrences	Predictor	integer	2	95	failures
2.D	days_since_ last_failure	Predictor	integer	4	1174	days
2.E	qtd_recurs_street	Predictor	integer	1	131	failures
2.F	avg_days_ btw_failures	Predictor	float	1.5	785.5	days
2.G	variance	Predictor	float	4.25	5.2×10^5	days
2.H	standard_deviation	Predictor	float	2.06	722	days
2.I	qtd_days_ btw_failures	Predictor	integer	5.0	1668	days

The target attribute consists of a numeric value that expresses the number of days until the occurrence of the next failure for a specific point in the network. Furthermore, the data interval used corresponds to all occurrence records in the supply network starting from January 2018 and extending until April 2023.

Conversely, data such as terrain elevation and distances were calculated using the Google Maps API through the Elevation API [27].

The attributes or variables used in this work are listed above. In regard to the attributes or variables used in this study, it is important to analyze the existence of correlations between the variables. The correlation values between the target attribute and each of the predictor attributes are expressed in Table 2 below, which shows the values obtained for Pearson's Correlation [28] between the variables.

Observing this figure, it is possible to perceive that the variables do not show strong correlation with the target attribute. Of all the analyzed attributes, only the number of recurrences and the number of days between failures show some correlation (0.4); all the others present values below this measure. These values indicate that variation in the values of these variables alone does not explain the behavior of the target attribute. In other words, the number of days until the occurrence of the next failure in the network is not directly explained by the predictor variables. It is important to highlight that the colors expressed in the table indicate the intensity of the correlation: positive (lighter colors) and negative (darker colors).

Table 2. Correlation between attributes or variables.

ATTRIBUTES	2.A	2.B	2.C	2.D	2.E	2.F	2.G	2.H	2.I	1.A
2.A	1.0	−0.3	−0.1	0.1	0.2	0.0	−0.1	−0.1	−0.2	−0.3
2.B	−0.3	1.0	−0.2	0.1	−0.6	0.1	0.1	0.1	0.0	−0.1
2.C	−0.1	−0.2	1.0	−0.3	0.4	−0.5	−0.2	−0.3	0.4	0.4
2.D	0.1	0.1	−0.3	1.0	−0.2	−0.1	−0.1	−0.1	−0.5	−0.2
2.E	0.2	−0.6	0.4	−0.2	1.0	−0.4	−0.2	−0.4	−0.1	0.0
2.F	0.0	0.1	−0.5	−0.1	−0.4	1.0	0.7	0.8	0.3	0.2
2.G	−0.1	0.1	−0.2	−0.1	−0.2	0.7	1.0	0.9	0.3	0.1
2.H	−0.1	0.1	−0.3	−0.1	−0.4	0.8	0.9	1.0	0.4	0.2
2.I	−0.2	0.0	0.4	−0.5	−0.1	0.3	0.3	0.4	1.0	0.4
1.A	−0.3	−0.1	0.4	−0.2	0.0	0.2	0.1	0.2	0.4	1.0

From the attributes listed above, predictive models based on Multi-Layer Perceptron (MLP) neural networks for the regression problem were used. In this context, this work used manually configured models, meaning that the arrangement of layers, and the number of neurons in each layer, were defined manually. Additionally, for comparison purposes, models with automatic configuration and linear regression were used, which can be used to compare the accuracy of the results.

In this stage, the process of training the predictive model occurred, meaning that the model was configured and trained using labeled data with the correct response that the model should seek to achieve by correcting its training error. In this context, training occurred with a dataset of 1727 failure samples. This dataset was divided into training and testing data, the former consisting of a total of 1175 samples (corresponding to the period from January 2018 to December 2021), while the testing set consisted of 552 samples spaced in time between January 2022 and April 2023.

The models used were obtained from the Scikit-Learn [29] and Keras [30] libraries, where the MLP Regressor and Keras DNN (Deep Neural Network) models were, respectively, utilized. Additionally, for comparison purposes, the same data were subjected to linear regression models provided by both libraries.

The predictive models used had different configurations: while the MLP model built using the Keras DNN library was automatically configured, the same applies to linear regression in both libraries. Conversely, a manually configured model was built using the MLP Regressor library. This model had the following hyperparameter configuration.

Four hidden layers contained, respectively, 128, 128, 648, and 550 neurons in each. The activation function used was 'relu' or rectified linear unit function. The training process occurred over 370 iterations and 100 epochs.Below, the convergence graph of the error in the training process is represented in Figure 4.

Some error metrics were used to evaluate the accuracy of the model. The first error metric used was the Mean Absolute Error (MAE), which indicates the difference between the expected value and the value predicted by the model in absolute terms, that is, non-negative values. MAE is calculated using Equation (1), where $\hat{y}$ is the value predicted by the model for the n-th sample and y is the expected value or true corresponding value. All other error metrics used in this work were calculated considering the same meanings for these same variables.

$$\mathrm{MAE}(y,\hat{y}) = \frac{1}{n}\sum_{i=0}^{n-1}|y_i - \hat{y}_i|. \tag{1}$$

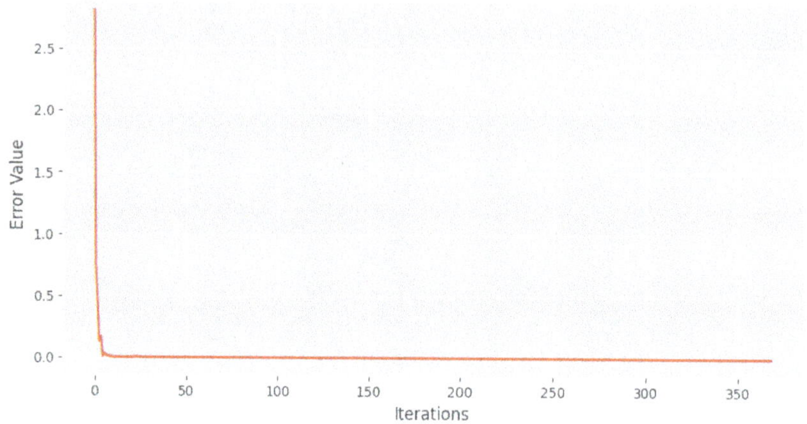

Figure 4. Training convergence error.

In conjunction with MAE, other error metrics such as MSE (Mean Squared Error) and RMSE (Root Mean Squared Error) were also used. They show the error value between the expected value and the predicted value, but MSE has the ability to highlight small dimension error values, while RMSE, in turn, seeks to keep the error value in the same dimension as the target variable. MSE is calculated using Equation (2), while RMSE is calculated using Equation (3).

$$\mathrm{MSE}(y, \hat{y}) = \frac{1}{n} \sum_{i=0}^{n-1} (y_i - \hat{y}_i)^2. \tag{2}$$

$$\mathrm{RMSE}(y, \hat{y}) = \sqrt{\frac{1}{n} \sum_{i=0}^{n-1} (y_i - \hat{y}_i)^2}. \tag{3}$$

A fourth error metric used was the Mean Absolute Percentage Error (MAPE). Unlike the previous ones, this metric is not altered by the global scale of the target variable. The best expected value for MAPE is 0.0. Equation (4) is used to calculate this metric. The fifth metric used was the Median Absolute Error (MedAE), which is considered robust for outlier values. The error value is calculated from the median of all absolute errors between the predicted and expected values. MedAE is calculated from Equation (5), and its result consists of a non-negative value, with the best possible value being 0.0.

$$\mathrm{MAPE}(y, \hat{y}) = \frac{1}{n} \sum_{i=0}^{n-1} \frac{|y_i - \hat{y}_i|}{\max(\epsilon, |y_i|)} \tag{4}$$

$$\mathrm{MedAE}(y, \hat{y}) = \mathrm{median}(\mid y_1 - \hat{y}_1 \mid, \ldots, \mid y_n - \hat{y}_n \mid). \tag{5}$$

Finally, the Max. Error, or Maximum Error, was also calculated, capturing the highest error value, i.e., the worst case of error between the predicted and expected values. Max Error is calculated using Equation (6).

$$\text{Max Error}(y, \hat{y}) = \max(|y_i - \hat{y}_i|) \tag{6}$$

From the models used, the last stage consisted of applying new unknown data to the model and verifying the results. After this stage, the use and evolution of the predictive models used could be subsequently performed .

5. Results and Discussion

The Table 3 shows the error values calculated for each of the algorithms. Within it, one can observe each of the models employed, along with the corresponding error values derived from their respective predictions. It is worth mentioning that the error values used in this work were obtained using the ScikitLearn Metrics library [31].

Table 3. Comparison of models in terms of accuracy.

Error (Days)	MLP Manual	MLP Auto.	Linear Regr.	KERAS DNN	KERAS LR
MAE	33.84935	37.06102	45.34312	236.10285	64.07651
MSE	1981.12961	1994.35068	3762.60283	67,591.47127	5740.66704
RMSE	44.50988	44.65815	61.34006	259.9836	75.76719
MAPE	3.63132	4.10978	7.99508	35.77144	1.93596
MedAE	28.37494	30.69204	34.8845	248.3089	74.3719
Max. Erro	158.47765	150.88351	251.06084	492.20184	217.66253

Observing the data in the table, is possible see that the Manual MLP model obtained the best accuracy values for Mean Absolute Error (MAE), Mean Squared Error (MSE), Root Mean Squared Error (RMSE), and Median Absolute Error (MedAE), while the Maximum Error (MAX. ERROR) value was obtained by the Automatic MLP. However, the values are very close, indicating similar accuracy in both cases.

The error values obtained by the predictive models used in this work are expressed in the table. The error metric column presents each of the metrics used in this work. The errors are presented in days. For example, an MAE of 33.84 for the Manual MLP means that the model is able to predict, on average, a pipeline failure with a margin of error of 33.84 days.

Given the above, the models based on Manual MLP and Automatic MLP obtained the best results. Furthermore, the error values are close in both cases. Despite being close, the performance of the Manual MLP model achieved superior performance, and for this reason, the results obtained by this model are presented below.

To evaluate the error obtained in the predictions of our model, let us observe Figure 5, which presents the histogram of the error for the proposed model.

The Figure 5 shows a histogram of the error obtained in the predictions of the Manual MLP model. The error values indicate the number of days of discrepancy between the predicted value and the expected value. In this graph, the majority of error values are below 40 days, and only five error values above 80 days were recorded.

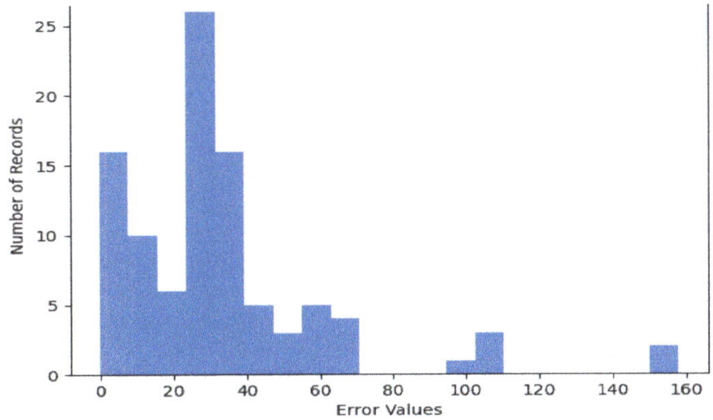

Figure 5. Error histogram (in days).

In order to provide further clarification regarding the accuracy of the model, the obtained error values were analyzed and we found the following:

- 12.37% of predictions have an error of less than 5 days;
- 21.64% of predictions have an error of less than 10 days;
- 26.80% of predictions have an error of less than 15 days;
- 57.73% of predictions have an error of less than 30 days;
- 80.41% of predictions have an error of less than 45 days;
- 87.62% of predictions have an error of less than 60 days;
- 93.81% of predictions have an error of less than 90 days.

Considering the presented data, over 20% of the predictions have an error of less than 10 days, and 80% of these predictions have an error of less than 45 days. Therefore, it is possible conclude that the predictions made provide relevant values for the decision-making process of the company, as they indicate, with a certain degree of confidence, the forecast for the day when the next break in the network will occur at a specific point.

Based on the contributed results, CAGEPA can utilize the geolocation of the obtained predictions to carry out repairs on the network and preventive and/or predictive maintenance in advance (before the failure occurs), thus avoiding interruptions in the supply and inconvenience to the population. Additionally, they can optimize the mobilization of teams to carry out repairs in advance, including determining when each repair should take place and setting priorities for maintenance work.

Performing repairs after a failure occurrence can lead to significant disruptions and inconvenience for customers. By using predictive maintenance based on the obtained predictions, CAGEPA can proactively address potential issues before they escalate into full-blown failures, minimizing the impact on water supply and avoiding the need for emergency repairs. This proactive approach not only improves service reliability but also reduces operational costs and enhances overall customer satisfaction.

Planning and resource allocation are essential aspects of efficient operations for any utility company like CAGEPA. By proactively addressing potential failures based on predictive maintenance, the company can better manage its resources, optimize workforce scheduling, and ensure timely repairs without causing significant disruptions to water supply. This approach not only safeguards the company's reputation but also fosters enhanced customer satisfaction by minimizing inconvenience and ensuring the reliable delivery of services. Additionally, the ability to plan ahead and mitigate potential damages can lead to long-term cost savings and operational efficiency for the company.

6. Conclusions

Given the presented results, it is possible conclude that to predict in advance the occurrence of failures in water supply networks using data from the network itself, focusing on the historical record of past failures.

The results show that using the history of failures as input for predictive models allows CAGEPA to predict a significant percentage of failures in its network with considerable accuracy.

As future work, it is important to predict failures in network points that do not have a history of failures. However, the specific instantiation of the predictive model employed, along with the selection of relevant variables, are likely to necessitate differentiation. Additionally, this work can be expanded to be applied in other cities with the same company or in cities with different companies. Moreover, the utilization of flow simulation within an EPANET environment presents a highly intriguing approach worthy of consideration. Although this methodology is currently under consideration for future research endeavors, it is noteworthy that such information has not yet been made available by the company.

The model accuracy can be enhanced by incorporating new variables and utilizing different predictive models that may enrich the results in terms of precision. Incorporating additional variables and expanding the dataset holds promise for enhancing the model's accuracy. This potential arises from the prospect of leveraging a larger number of records during the model's training and validation phases. It is pertinent to note, however, that the company has not yet supplied any additional data. Furthermore, as a contribution to science, this work can be extrapolated and applied to other domains beyond water supply networks, such as energy supply networks, oil pipelines, communication networks, and so on.

Author Contributions: Conceptualization, V.d.S.M. and M.D.d.S.; methodology, V.d.S.M. and M.D.d.S.; software, V.d.S.M.; validation, M.D.d.S. and A.V.B.; formal analysis, V.d.S.M.; investigation, V.d.S.M.; resources, M.D.d.S.; data curation, M.D.d.S.; writing—original draft preparation, V.d.S.M.; writing—review and editing, V.d.S.M., M.D.d.S. and A.V.B.; visualization, V.d.S.M., M.D.d.S. and A.V.B.; supervision, V.d.S.M. and A.V.B.; project administration, V.d.S.M. and A.V.B.; funding acquisition, M.D.d.S. and A.V.B. All authors have read and agreed to the published version of the manuscript.

Funding: This work was supported by the National Council for Scientific and Technological Development (CNPq).

Data Availability Statement: Data are contained within the article.

Acknowledgments: First of all, I thank God, the Federal University of Paraíba, CAGEPA, my supervising professors, and everyone who collaborated in some way.

Conflicts of Interest: The authors declare no conflicts of interest.

References

1. Chatzigeorgakidis, G.; Karagiorgou, S.; Athanasiou, S.; Skiadopoulos, S. A MapReduce based k-NN joins probabilistic classifier. In Proceedings of the 2015 IEEE International Conference on Big Data (Big Data), Santa Clara, CA, USA, 29 October–1 November 2015; pp. 952–957. [CrossRef]
2. Luo, S.; Chu, V.W.; Zhou, J.; Chen, F.; Wong, R.K.; Huang, W. A multivariate clustering approach for infrastructure failure predictions. In Proceedings of the 2017 IEEE International Congress on Big Data (BigData Congress), Honolulu, HI, USA, 25–30 June 2017; pp. 274–281. [CrossRef]
3. Vališ, D.; Hasilová, K.; Forbelská, M.; Vintr, Z. Reliability modelling and analysis of water distribution network based on backpropagation recursive processes with real field data. *Measurement* **2020**, *149*, 107026. [CrossRef]
4. Robles-Velasco, A.; Muñuzuri, J.; Onieva, L.; Cortés, P. An evolutionary fuzzy system to support the replacement policy in water supply networks: The ranking of pipes according to their failure risk. *Appl. Soft Comput.* **2021**, *111*, 107731. [CrossRef]
5. Kabir, G.; Tesfamariam, S.; Loeppky, J.; Sadiq, R. Predicting water main failures: A Bayesian model updating approach. *Knowl.-Based Syst.* **2016**, *110*, 144–156. [CrossRef]
6. Giraldo-González, M.M.; Rodríguez, J.P. Comparison of Statistical and Machine Learning Models for Pipe Failure Modeling in Water Distribution Networks. *Water* **2020**, *12*, 1153. [CrossRef]
7. Pham, T.M.L.; Pham, H.H.; Do, N.A.T.; Le, D.H. Proposed probabilistic models of pipe failure in water distribution system. *MATEC Web Conf.* **2018**, *193*, 02002. [CrossRef]
8. Sistema Nacional de Informações sobre Saneamento SNIS. Diagnóstico Temático Serviços de Água e Esgoto. Gestão Técnica de Água. Technical Report, Secretaria Nacional de Saneamento Ambiental—Ministério das Cidades, 2023. Available online: https://www.gov.br/cidades/pt-br/acesso-a-informacao/acoes-e-programas/saneamento/snis/produtos-do-snis/diagnosticos/DIAGNOSTICO_TEMATICO_VISAO_GERAL_AE_SNIS_2023.pdf (accessed on 21 December 2023).
9. Zhang, C.; Ye, Z. Water pipe failure prediction using AutoML. *Facilities* **2021**, *39*, 36–49. [CrossRef]
10. Gorenstein, A.; Kalech, M.; Hanusch, D.F.; Hassid, S. Pipe fault prediction for water transmission mains. *Water* **2020**, *12*, 2861. [CrossRef]
11. Velasco, A.R.; Muñuzuri, J.; Onieva, L.; Palero, M.R. Trends and applications of machine learning in water supply networks management. *J. Ind. Eng. Manag.* **2021**, *14*, 45–54. [CrossRef]
12. De Clercq, D.; Smith, K.; Chou, B.; Gonzalez, A.; Kothapalle, R.; Li, C.; Dong, X.; Liu, S.; Wen, Z. Identification of urban drinking water supply patterns across 627 cities in China based on supervised and unsupervised statistical learning. *J. Environ. Manag.* **2018**, *223*, 658–667. [CrossRef]
13. Barton, N.A.; Farewell, T.S.; Hallett, S.H.; Acland, T.F. Improving pipe failure predictions: Factors affecting pipe failure in drinking water networks. *Water Res.* **2019**, *164*, 114926. [CrossRef]
14. Fan, X.; Wang, X.; Zhang, X.; Xiong (Bill), Y. Machine learning based water pipe failure prediction: The effects of engineering, geology, climate and socio-economic factors. *Reliab. Eng. Syst. Saf.* **2022**, *219*, 108185. [CrossRef]
15. Rifaai, T.M.; Abokifa, A.A.; Sela, L. Integrated approach for pipe failure prediction and condition scoring in water infrastructure systems. *Reliab. Eng. Syst. Saf.* **2022**, *220*, 108271. [CrossRef]

16. Ramírez, R.; Cobacho, R.; Torres, D.; López-Jiménez, P. Implementación de un modelo de predicción de fallos orientado a la gestión y estrategias de mantenimiento en redes de distribución de agua potable. *Ingeniería del Agua* **2019**, *23*, 247–258. [CrossRef]
17. Wols, B.; Thienen, P.V. Impact of climate on pipe failure: Predictions of failures for drinking water distribution systems. *Eur. J. Transp. Infrastruct. Res.* **2016**, *16*. [CrossRef]
18. Zhang, C.; Wu, H.; Bie, R.; Mehmood, R.; Kos, A. Dynamic Modeling of failure events in preventative pipe maintenance. *IEEE Access* **2018**, *6*, 12539–12550. [CrossRef]
19. Robles-Velasco, A.; Cortés, P.; Muñuzuri, J.; Onieva, L. Prediction of pipe failures in water supply networks using logistic regression and support vector classification. *Reliab. Eng. Syst. Saf.* **2020**, *196*, 106754. [CrossRef]
20. Kutyłowska, M. Neural network approach for failure rate prediction. *Eng. Fail. Anal.* **2015**, *47*, 41–48. [CrossRef]
21. Kabir, G.; Demissie, G.; Sadiq, R.; Tesfamariam, S. Integrating failure prediction models for water mains: Bayesian belief network based data fusion. *Knowl.-Based Syst.* **2015**, *85*, 159–169. [CrossRef]
22. Tang, K.; Parsons, D.J.; Jude, S. Comparison of automatic and guided learning for Bayesian networks to analyse pipe failures in the water distribution system. *Reliab. Eng. Syst. Saf.* **2019**, *186*, 24–36. [CrossRef]
23. Studziński, A.; Pietrucha-Urbanik, K. Failure Risk Analysis of Water Distributions Systems Using Hydraulic Models on Real Field Data. *Econ. Environ.* **2019**, *68*, 14. Available online: https://ekonomiaisrodowisko.pl/journal/article/view/106 (accessed on 18 February 2024).
24. Pietrucha-Urbanik, K.; Studziński, A. Qualitative analysis of the failure risk of water pipes in terms of water supply safety. *Eng. Fail. Anal.* **2019**, *95*, 371–378. [CrossRef]
25. Marquesone, R. *Big Data. Técnicas e Tecnologias para Extração de Valor dos Dados;* Casa do Código: São Paulo, Brazil, 2016.
26. Faceli, K. *Inteligência Artificial: Uma Abordagem de Aprendizado de Máquina*, 2nd ed.; LTC: Rio de Janeiro, Brazil, 2021; ISBN 978-85-216-3749-3.
27. Google. Visão Geral da API Elevation. 2024. Available online: https://mapsplatform.google.com/intl/pt-BR/products/#elevation (accessed on 18 February 2024).
28. Pandas. Pandas API Reference. 2024. Available online: https://pandas.pydata.org/docs/reference/api/pandas.DataFrame.corr.html (accessed on 20 February 2024).
29. Scikitlearn, D. Scikit-Learn: Machine Learning in Python. 2024. Available online: https://scikit-learn.org/stable/index.html (accessed on 20 April 2024).
30. Tensorflow. Keras | TensorFlow Core. 2020. Available online: https://www.tensorflow.org/guide/keras?hl=pt-br (accessed on 18 December 2023).
31. Scikitlearn, D. 3.3. Metrics and Scoring: Quantifying the Quality of Predictions—Scikit-Learn.org. 2023. Available online: https://scikit-learn.org/1.3/modules/model_evaluation.html#regression-metrics (accessed on 31 October 2023).

water

Article

A Novel IoT-Based Performance Testing Method and System for Fire Pumps

Shangcong Zhang [1,2], Yongfang Li [3], Xuefei Chen [4], Ruyi Zhou [3], Ziran Wu [1,2,*] and Taha Zarhmouti [1]

1 Engineering Research Center of Low-Voltage Apparatus of Zhejiang Province, Wenzhou University, Wenzhou 325035, China; zsc@wzu.edu.cn (S.Z.); 22521901003@stu.wzu.edu.cn (T.Z.)
2 Technology Institute of Wenzhou University in Yueqing, Wenzhou 325099, China
3 C-lin Electrical Co., Ltd., Yueqing 325600, China; pgb@xinling.com (Y.L.); xl01@xinling.com (R.Z.)
4 Shanghai TYIoT Technology Co., Ltd., Shanghai 200120, China; 18516240317@163.com
* Correspondence: naturex@wzu.edu.cn; Tel.: +86-18857745619

Abstract: Fire pumps are the key components of water supply in a firefighting system. At present, there is a lack of fire water pump testing methods that intelligently detect faulty states. Existing testing approaches require manual operation, which leads to low efficiency and accuracy. To solve the issue, this paper presents an automatic and smart testing approach that acquires measurements of the flow, pressure, shaft power and efficiency from smart sensors via an IoT network, so that performance curves are obtained in the testing processes. An IoT platform is developed for data conversion, transmission and storage. The Discrete Fréchet Distance is applied to evaluate the similarities between the acquired performance curves and metric performance curves, to determine the working condition of the fire pump. The weights of the measurement dimensions for distance computation are optimized by the Genetic Algorithm to improve the distinction between normal and faulty performance curves. Finally, the experimental results show that the proposed method can completely detect faulty states and prove its high practicality for real firefighting systems.

Keywords: fire pumps; performance testing; performance curves; Fréchet Distance

1. Introduction

Firefighting systems are of great importance in the safety and protection of buildings from fire accidents [1]. Traditional fire water supply systems rely on manual operation and inspection, which leads to slow responses and low reliability. Recently, smart firefighting systems have been proposed and developed due to their capability of enhancing the efficiency and effectiveness of firefighting operations. Reinforced by the Internet of Things (IoT) technologies [2,3], smart firefighting systems are able to perform intelligent and automated fire detection and suppression. IoT-based smart devices, such as smoke detectors, heat sensors, water sprinklers, etc., as well as advanced technologies such as wireless networking, smart sensors, machine learning and automation, are applied to provide intelligent monitoring, early detection and a rapid response to fire incidents.

Fire pumps are key components of a firefighting system [4]. The operating state of the fire pump determines whether the firefighting system can successfully suppress the fire. Fire pump failures and faults lead to low water supply performance, obstructing the firefighting processes, which is a critical issue that commonly occurs. Therefore, monitoring and inspecting the operating condition of fire pumps is crucial in the maintenance of firefighting systems. However, at present, most "smart" or "IoT" firefighting solutions only focus on fire detection hardware [5] and methodologies [6,7], as well as operational automation [8,9]. There is a lack of smart approaches to fire pump testing. Currently, the manual inspection of fire pumps is the conventional solution applied in fire water supply in most buildings. However, manual inspections are commonly performed annually or quarterly, due to the relatively high complexity and cost. Meanwhile, the accuracy cannot be fully guaranteed due to the variations in the skills of the inspectors. Therefore, at present, firefighting systems are at the risk of creating an unreliable fire water supply due to the limited efficiency and accuracy of manual inspection. A smart system that monitors and inspects fire pumps and intelligently detects faulty states needs to be investigated and developed.

There are a number of faults to lead to the complete failure or performance degradation of fire pumps. The issues that commonly cause fire pump faults are listed and described in Table 1.

Table 1. Frequently occurring issues in fire pumps.

No.	Issue	Description
1	Overheating	Insufficient cooling or motor cooling system faults cause overheating that leads to motor shutdown.
2	Impeller or shell damage	Destructive objects in the pump can damage the impeller or pump shell, reducing pump efficiency.
3	Seal leakage	Leakage in pump seals causes water loss and reduces pressure reduction.
4	Bearing failures	Failed bearings lead to friction increments, overheating and pump failures.
5	Pressure switch failures	Failed pressure switches lead to a failure in starting or stopping the pump.
6	Air or cavitation	Performance reductions and potential damage are caused air or cavitation in the pump.
7	Obstructions	Water flow and pump efficiency are reduced by blockages or obstructions in the suction or discharge pipelines.
8	Insufficient water supply	Inadequate water sources or low water pressure cause a reduction in flow and pressure.
9	Freezing	Freezing temperatures cause pump damage.
10	Dirt and debris	Accumulation of dirt, debris or contaminants causes pump performance degradation.
11	Misalignment	Poor alignment of the components in the pump system causes premature wear and potential failure.

The challenges in constructing a smart fire pump testing system include the following.

(1) Practicality: the testing system must be applied in practical firefighting applications. A testing structure and process that execute performance tests without interference in normal operation are required.
(2) Data transmission: a data transmission platform compatible with sensing devices using different communication modes and protocols is required.
(3) Evaluation methodology: an algorithm that accurately and efficiently detects faulty states is required.

Hence, this paper proposes a smart fire pump testing system that measures the flow, pressure, shaft power and efficiency by a set of IoT sensors and remotely inspects the condition of the fire pump in accordance with the measurements. An additional pipeline structure especially designed for testing is proposed. An analysis method distinguishing the normal and faulty states of the pump system by evaluating the acquired performance curves is presented. Compared to manual inspection, the testing system can automatically perform fire pump tests in a significantly higher frequency, and the tests can be evaluated by a unified and proven metric, so that the testing efficiency and accuracy can be improved. Integrating this testing system will ensure the water supply reliability of firefighting systems.

2. Related Work

Fire pump testing is part of the field of smart water supply, in which sensors are installed for monitoring and inspection. A smart water supply is determined by the integration of a set of products and informatics algorithms that are able to remotely and continuously monitor and diagnose problems, prioritize and manage maintenance issues and optimize all aspects of the water distribution network [10,11]. This requires limiting the energy costs, reducing water losses and improving the reliability [12,13]. The high importance of diagnostics in water supply systems and of informing operators about emergencies is emphasized. The implementation of these tasks requires the integration of a real-time remote monitoring mechanism, analysis of the collected measurement data and informatics tools identifying abnormal states [14,15]. For example, Ye et al. [13] designed an IoT-based smart water gird architecture that performed water information sensing, risk evaluation and management optimization intelligently in real time. Siew et al. [16] presented an extension of a hydraulic simulator to incorporate pressure-dependent demands. The utilization of a continuous nodal pressure-flow function coupled with a line search and backtracking procedure was applied for the improvement of the convergence rate and robustness of the algorithm. Simulations of practical water networks consisting of multiple sources, pipes, valves and pumps were successfully executed. Both normal and pressure-deficient conditions of water distribution systems can be modeled and analyzed.

Recently, IoT-based smart water supply has been integrated into Supervisory Control And Data Acquisition (SCADA) systems. Diagnostic methods are usually implemented in stages. The leakage detection methods are implemented first. Nevertheless, new methods of such detection are still being explored. Zahab et al. [17] presented a quasi-intelligent diagnostic method, integrating a SCADA system and a numerical model of the water supply system. The paper presented the effects of their implementation within the first three months of operation. Discussions of the advantages, as well as the identified limitations, recommended amendments and further developments, were also presented. Considering that the delivered water must be provided in adequate amounts and under appropriate pressure, the measurement of key attributes, such as the pressure, flow, energy, etc., of water supply systems to build diagnostic models has been recently widely investigated. These requirements necessitate the constant development of monitoring systems. Carriço et al. [18] proposed to apply four performance indices to assess energy efficiency based on the hydraulic energy balance. The implementation of this method demonstrated robustness and practicality, and the authors performed a comparison between different measures for the improvement of the energy efficiency. Stańczyk et al. [19] proposed to perform an empirical exponent analysis, determined on the basis of flow rate and pressure measurements for District Metered Areas (DMAs). Both supervised and unsupervised machine learning strategies were employed to establish an empirical exponent, and the accuracy of the operating condition qualification reached up to 90%. Li et al. [20] established a micro-smart water test bed that uses a flow sensor with an Arduino UNO and observation data model (ODM) database. Meanwhile, R-studio was employed to evaluate the performance of the micro-smart water test bed by comparing the relative errors between the experimental measurements and modeling outputs.

It can be observed that IoT-based smart testing approaches for water supply have been widely investigated and developed, both in terms of hardware and methodologies. Nevertheless, the research and development of smart fire pump testing have not been promoted thus far. Although manufacturers have already proposed fire pumps with IoT sensors so that the conditions can be remotely monitored, a method that can intelligently detect faulty states in fire pumps is still absent. There are only a small number of related studies. A similar study by Li et al. [21] investigated a fault diagnosis method for submersible pumps, but only electrical parameters were monitored and there was an absence of machine-learning-based modeling considering the comprehensive factors of faults. Du et al. [22] analyzed the effects caused by the failure of fire equipment by classifying the failure levels, but only a simple evaluation of functionalities was performed. Sobral et al. [23] proposed a periodic inspection strategy to estimate the availability of fire pumps, but the method still required multiple steps of manual analysis. Therefore, this paper proposes a smart fire pump testing system that contains both hardware for remote sensing and a method of fault detection, which is of high practicality in real implementations.

3. System Design

3.1. System Overview

The proposed system includes a fire pump with a set of smart sensors, an electric flow-control valve, a specially designed testing pipeline and an IoT network, as well as a cloud server. The architecture of the system is shown in Figure 1. The advantage of using the electric valve is to achieve automatic and remote flow control. It alters the water flow in accordance with the instructions from the cloud server via the IoT network. The testing pipeline is used only for fire pump tests so that interference in the main fire pipeline can be prevented. The IoT-based sensors measure the flow, water pressure, shaft power and power efficiency, and they upload all data to the cloud server via the IoT network. The cloud server, which remotely controls the tests, collects and analyzes testing data to evaluate the performance of the fire pump.

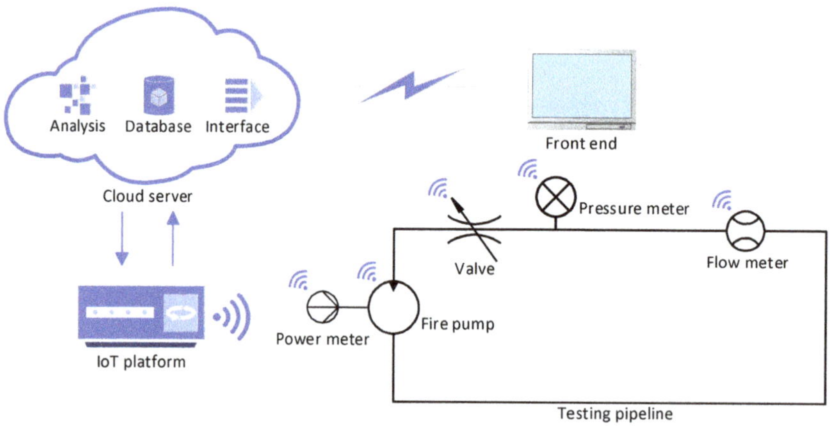

Figure 1. Architecture of the smart fire pump testing system.

The IoT network employs multiple communication modes, including Ethernet, Wi-Fi, cellular, LoRa, etc., in accordance with the practical communication conditions of the sensors. An IoT platform that converts and organizes data packages of different modes/protocols is designed and implemented to achieve transparent transmission between the server and the sensors. Figure 2 illustrates the architecture of the IoT platform. The platform consists of a section for general purposes of data reception, transmission, conversion and storage, as well as a section that is developed for customized computation and analysis. The platform connects to the IoT modules of the sensors directly or via edge-cloud gateways, integrating data from multiple sensors. The platform can also communicate with third-party IoT platforms.

In our case, a fire pump monitoring platform on the cloud server is constructed. On the platform, a back end for data transmission, storage, processing and analysis, as well as a front end for visualization and interaction, are developed. Figure 3 shows the web-based front end that displays the measurements from the sensors and the condition of the fire pump in real time.

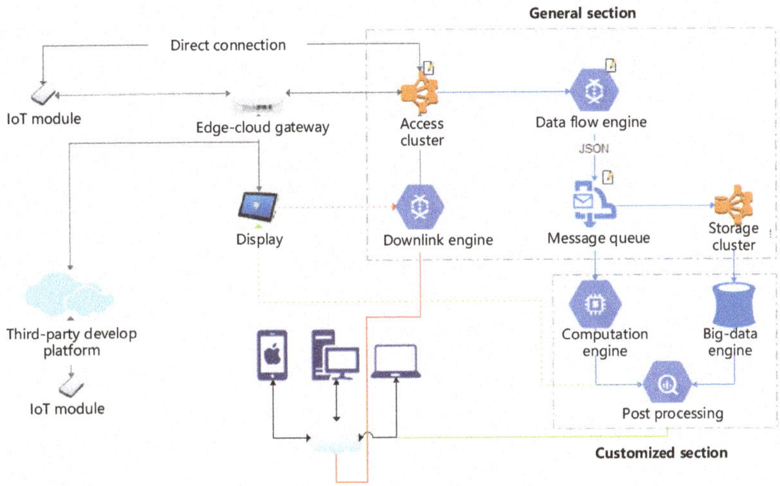

Figure 2. Architecture of the IoT platform.

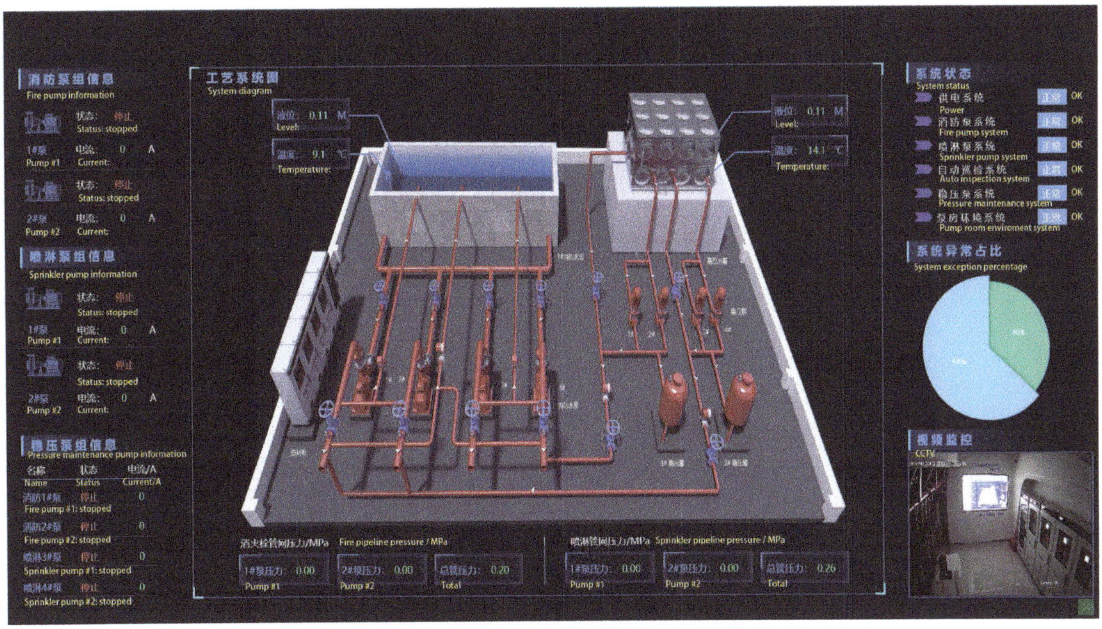

Figure 3. Web-based front end of the fire pump monitoring platform.

3.2. Fire Pump Testing Process

The fire pump in the system is automatically tested online. The performance testing process is described as follows.

(1) The cloud server transmits the flow control coefficient γ to the electric valve.
(2) The valve adjusts the flow to reach $q = \gamma Q$, where Q is the nominal flow of the pump.
(3) The water pressure p, the pump power P and the pump efficiency η are measured by sensors and transmitted to the cloud server.

(4) The system repeatedly performs steps 1 to 3 by altering γ from 0.0 to 1.5 in accordance with China Standard GB 50974-2014 [24], and all coefficients and measurements are saved in the cloud server.
(5) The cloud server forms a four-dimensional flow–pressure–power–efficiency curve that represents the performance of the fire pump at different flow conditions.
(6) A weighted multi-dimensional distance computation is performed to achieve similarity between the tested performance curve and the metric performance curve, so that the degradation of the fire pump can be evaluated.

The metric performance curve is also obtained by steps 1 to 5, with a well-conditioned fire pump. Therefore, the performance evaluation can be illustrated. In addition, the sequence of γ is defined by the user according to the application circumstances. Higher γ values refer to higher modeling precision but also a higher computational cost. In practice, a γ sequence containing 10–15 values is commonly used in the product certification of fire pumps.

4. Performance Evaluation

To evaluate the performance degradation of a fire pump, the similarity between the acquired performance curve (APC) and the metric performance curve (MPC) is assessed. The MPC can be achieved by computing the mean of a set of normal performance curves (NPCs). Therefore, faulty performance curves (FPCs) and NPCs need to be clustered separately so that they can be clearly distinguished. To verify the condition of an APC, the distance from the APC to the MPC should not be longer than the distances from any of the NPCs to the MPC.

Figure 4 illustrates the examples of three classes of performance curves in the space, where the flow, pressure and shaft power are represented by the axes, and the efficiency is represented by a colormap. The NPC is obviously closer to the MPC than the FPC. It is illustrated that there is almost no performance degradation in the NPC, and the small differences can be considered as measurement errors. Meanwhile, compared to the MPC, the FPC reflects significant degradation, i.e., the pressure and efficiency fall remarkably as the flow increases, indicating that potential faults could occur.

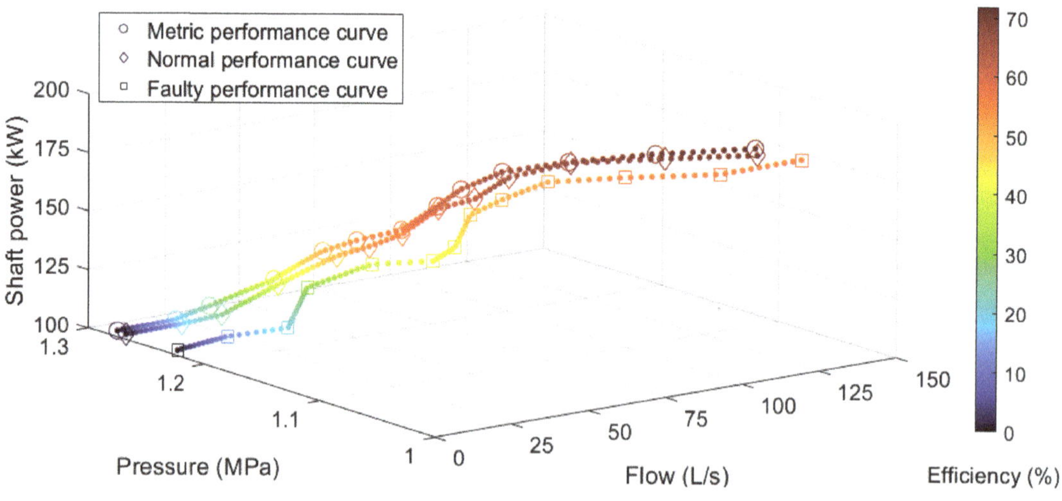

Figure 4. Examples of performance curves.

4.1. Distances between Performance Curves

To quantitively estimate the similarity between curves, the Fréchet Distance [25] in the four-dimensional flow–pressure–power–efficiency space is applied. According to its definition, the Fréchet Distance reduces as the similarity of two curves rises. In other words, NPCs have smaller Fréchet Distances to the MPC than those of FPCs. In this case, the implementation employs the Discrete Fréchet Distance [26,27], which can be presented as

$$\begin{aligned} F(x,y) &= \lim_{n\to\infty,\max|t_k-t_{k-1}|\to 0}\widetilde{F}^n(x,y) \\ &= \lim_{n\to\infty,\max|t_k-t_{k-1}|\to 0}\inf_{\alpha,\beta}\max_{t\in\{t_k\}}\{d(x(\alpha(t)),y(\beta(t)))\} \end{aligned} \tag{1}$$

where x and y are the two curves in a normalized n-dimensional space, and $\alpha(t)$ and $\beta(t)$ are two points sampled on x and y, respectively at time t. In the case of this paper, points on the curves are extracted from the measurements of the sensors. Since each curve contains the same number of measurements, the requirement in computing the Discrete Fréchet Distance can be met. We use the flow control coefficient dimension to replace the time dimension, so we can obtain

$$\begin{aligned} F(x,y) &= \lim_{n\to\infty,\max|\gamma_k-\gamma_{k-1}|\to 0}\widetilde{F}^n(x,y) \\ &= \lim_{n\to\infty,\max|\gamma_k-\gamma_{k-1}|\to 0}\inf_{\alpha,\beta}\max_{\gamma\in\{t_k\}}\{d(x(\alpha(\gamma)),y(\beta(\gamma)))\} \end{aligned} \tag{2}$$

Meanwhile, considering that measurements by different sensors should be weighted differently, we assign each dimension in the space a separate weight coefficient. Therefore, the weighted Euclidean distance between two points $U = (u_1, u_2, \ldots, u_n)$ and $V = (u_1, u_2, \ldots, u_n)$ in the space is defined as

$$d(U,V) = \sqrt{\sum_{i=1}^{n} w_i(u_i - v_i)^2} \tag{3}$$

where $W = (w_1, w_2, \ldots, w_n)$ are the corresponding weights of the dimensions. The weights are achieved by an optimization process with measurement data acquired in practical operations. In the case of this paper, every point on the curve has four dimensions, denoted as (γ, p, P, η), so the dimensionality $n = 4$.

4.2. Optimization of the Weights

The target of the optimization process is to minimize the mean distance within the MPC cluster (denoted as δ) and maximize the mean distance between the two clusters (denoted as σ). Therefore, we propose the ratio of δ to σ as the optimization objective. Denote the collection of all curve clusters as $\mathbf{C} = \{c_1, c_2, \ldots, c_m\}$, where c_1 is the MPC cluster. Hence, the following presents the optimization objective O:

$$\delta = \frac{1}{2k_1(k_1-1)}\sum_{x,y\in c_1} F(x,y) \tag{4}$$

$$\sigma = \frac{1}{2m(m-1)}\sum_{c_i,c_j\in\mathbf{C},i\neq j}\frac{1}{k_i\cdot k_j}\sum_{x\in c_i,y\in c_j} F(x,y) \tag{5}$$

$$\min O = \frac{\delta}{\sigma} \tag{6}$$

where k_i is the number of curves of class c_i. The Genetic Algorithm [28] is applied to find the best set of weights that can clearly distinguish the NPCs and FPCs, so that an APC can be correctly classified.

The process of the performance evaluation is illustrated in Figure 5.

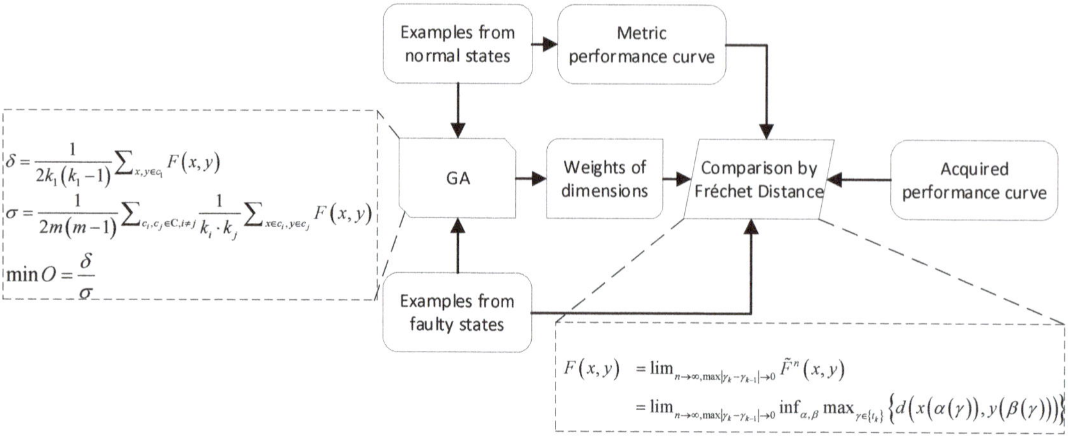

Figure 5. Complete performance evaluation process.

5. Experiments, Results and Discussion

5.1. Experimental Configuration

We construct an experimental platform for the proposed system using existing fire-fighting equipment produced by C-lin Electrical Co., Ltd. (Yueqing, China), one of the institutes associated with this research. Figure 6 shows the fire pump as well as its IoT-based control cabinet applied in the experiment. The specifications of the equipment are shown in Table 2.

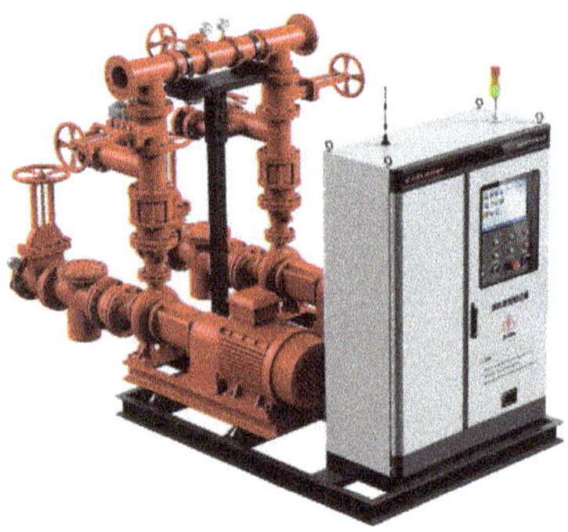

Figure 6. Fire pump device employed in the experiments.

Table 2. Specifications of the experimental platform.

Name	Value
Suction height	1 m
Rated pump rotational speed	2980 rpm
Rated pressure	1.2 MPa
Rated flow	85 L/s

Table 3 shows a set of measurements in the normal working state. The flow coefficient values, increasing from 0.00 to 1.50, are configured by the cloud server and transmitted to the pump. The flow, pressure, shaft power and efficiency values are the measurements acquired from the sensors. With an increase in the shaft power, the pressure remains higher than the metric (1.2 MPa) when the flow is under the rated value, and it drops slightly when the flow is overloaded, which meets the requirement of China Standard GB 50974-2014 [24]. Meanwhile, the shaft power and efficiency continue rising as the flow increases, proving that the pump system is in a good condition.

Table 3. Measurements in the normal working condition.

Flow Coefficient	Flow (L/s)	Pressure (MPa)	Shaft Power (kW)	Efficiency (%)
0.00	0.00	1.28	102.74	0.00
0.15	13.87	1.26	105.88	17.00
0.30	24.14	1.26	110.48	26.30
0.45	39.97	1.24	120.17	39.70
0.60	52.39	1.24	129.69	49.90
0.75	62.35	1.23	134.32	53.90
0.80	68.59	1.21	140.41	56.80
0.90	77.67	1.20	148.86	61.30
1.00	85.47	1.20	154.76	64.00
1.10	94.64	1.19	161.89	67.20
1.20	103.19	1.16	168.75	69.30
1.35	115.52	1.12	176.52	70.90
1.50	127.80	1.06	184.34	71.80

To include examples of high variety in the dataset, we change the state of the system by replacing the components in different conditions. The components are used and aged, but still work normally. Therefore, a relatively high variety of NPCs can be achieved. We also obtain a set of measurement data with faults, which are described in Table 1. The faults are replicated on the experimental platform by giving corresponding conditions, e.g., installing broken impellers, blocking the testing pipeline, breaking the seal, etc. Table 4 shows how the issues are replicated in the experiments.

Table 4. Replicated issues of fire pumps.

No.	Issue	Replication
1	Overheating	Overloading the motor without cooling.
2	Impeller or shell damage	Applying damaged impellers and shells.
3	Seal leakage	Applying failure seals.
4	Bearing failures	Applying damaged bearings.
5	Pressure switch failures	Applying damaged pressure switches.
6	Air or cavitation	Injecting air into the fire pump.
7	Obstructions	Placing obstructive items in the suction and discharge pipelines.
8	Insufficient water supply	Reducing the pressure of the water source.
9	Freezing	Running the fire pump system in a freezing temperature.
10	Dirt and Debris	Applying a fire pump with heavy dirt, debris and contaminants.
11	Misalignment	Misaligning the components of the fire pump.

Hence, an FPC set is also achieved. Finally, we obtain an NPC set of 60 examples, as well as an FPC set of another 60 examples. To verify the proposed method, we randomly divide the samples into training and validation sub-datasets by 3:1, i.e., 90 training examples and 30 validation ones.

5.2. Results and Discussion

We alter the key parameters of GA, including the population, crossover probability and mutation probability, to achieve an optimal solution of the weights. The following are the considerations made in determining the key parameters.

(1) Population: a larger population expands the search space of GA, but also increases the computational cost. Hence, under the premise of achieving high accuracy, it is expected to select a relatively small population value.
(2) Crossover probability and mutation probability: higher values of these two parameters also expand the search range of GA, but possibly cause a local optimum, which reduces the accuracy. The two parameters are selected in accordance with the test results.
(3) Testing values of the parameters: the testing values of the population (30, 50, 100), crossover probability (0.90, 0.93, 0.96) and mutation probability (0.10, 0.05, 0.01) are all determined by common GA parameter ranges, where optimal parameters can be found in most circumstances.

The weights are trained by the training sub-dataset and applied to the validation sub-dataset to determine the best solution. Every set of GA parameters is tested 3 times, in which different random divisions of the training and validation sub-datasets are used. Finally, we obtain the results as shown in Table 5.

Table 5. Test results varied by GA configuration.

Population	Crossover Probability	Mutation Probability	$O = \delta/\sigma$			Validation Accuracy		
			1	2	3	1	2	3
30	0.90	0.10	0.48	0.48	0.44	0.87	0.83	0.87
		0.05	0.43	0.48	0.46	0.90	0.87	0.90
		0.01	0.48	0.49	0.46	0.93	0.90	0.93
	0.93	0.10	0.41	0.39	0.39	0.93	0.80	0.97
		0.05	0.39	0.34	0.42	0.93	0.93	0.90
		0.01	0.38	0.34	0.37	0.93	0.90	1.00
	0.96	0.10	0.36	0.36	0.38	0.90	0.87	0.97
		0.05	0.30	0.32	0.35	1.00	0.90	1.00
		0.01	0.33	0.31	0.34	0.90	0.93	1.00
50	0.90	0.10	0.39	0.33	0.36	0.97	1.00	0.97
		0.05	0.33	0.34	0.40	1.00	0.97	0.93
		0.01	0.43	0.38	0.39	0.90	0.97	0.97
	0.93	0.10	0.32	0.30	0.27	1.00	0.97	1.00
		0.05	0.27	0.29	0.25	1.00	1.00	1.00
		0.01	0.28	0.31	0.25	1.00	0.97	1.00
	0.96	0.10	0.30	0.25	0.26	1.00	0.97	0.97
		0.05	0.28	0.24	**0.23**	1.00	1.00	**1.00**
		0.01	0.34	0.41	0.35	0.97	0.90	1.00
100	0.90	0.10	0.32	0.26	0.30	1.00	1.00	1.00
		0.05	0.30	0.28	0.24	0.97	0.97	1.00
		0.01	0.29	0.31	0.27	1.00	1.00	1.00
	0.93	0.10	0.23	0.26	0.23	1.00	1.00	1.00
		0.05	0.24	0.25	0.25	1.00	1.00	1.00
		0.01	0.30	0.29	0.23	1.00	0.97	1.00
	0.96	0.10	0.34	0.26	0.28	1.00	1.00	1.00
		0.05	0.30	0.28	0.24	1.00	1.00	1.00
		0.01	0.33	0.27	0.31	0.97	0.97	1.00
Unweighted Euclidian Distance			0.57	0.58	0.62	0.83	0.77	0.83

In Table 5, the optimization objective O represents the lowest ratio of δ to σ that indicates the highest distinction between the NPCs and FPCs. The validation accuracy shows its success in classifying APCs within the validation sub-dataset as normal or faulty states. We use both the optimization objective and the validation accuracy as metrics. It can be observed that with populations of 50 and 100, the accuracy in most conditions reaches 100%, which means that all tested FPCs can be recognized. Therefore, finally, we select a configuration with the lowest objective (0.23) and a relatively low cost (population = 50) as the best solution, as marked by the gray shading and bold text in Table 5. In this context, the crossover probability is 0.96 and the mutation probability is 0.05. Using this set of parameters, 100% detection of faulty states is achieved in all three tests with different dataset divisions, which proves the robustness of the proposed method. For real-world applications, the robustness of fault detection is related to the collection of faulty state examples. A more complete collection of faulty states can be used to construct a model with higher robustness and generalization abilities.

In comparison, we apply a Euclidian distance computation without optimized weights in Equation (2), i.e., each dimension is weighted equally in the Fréchet Distance computation, for APC classification. As shown in the bottom row of Table 5, in the three tests, the accuracies are 0.83, 0.77 and 0.83, respectively. Hence, it is proven that the proposed weight optimization method can significantly improve the accuracy in detecting faulty states in fire pumps. The effectiveness of the proposed method is verified.

6. Conclusions and Future Work

This paper proposes an IoT-based smart fire pump testing method and system that acquire measurements from sensors to form performance curves for degradation evaluation. The IoT network and platform is designed for data transmission. A method that identifies normal and faulty states using the acquired performance curves is proposed. Experiments prove that the method and system can achieve high detection accuracy in detecting faulty states.

In practice, for a specific firefighting system, the proposed fire pump testing system can be applied as follows:

(1) Sufficient measurement data under various normal and faulty states are collected to form a large dataset of NPCs and FPCs;
(2) The system obtains weights for the Fréchet Distance by the proposed method using GA, as well as computing the MPC;
(3) The system inspects the fire pump in a timely manner to obtain new APCs; the distance from every APC to the MPC and FPC is computed, to determine the corresponding state of the fire pump.

Since the entire testing process is fully automatic, tests can be executed weekly or even daily, to frequently inspect the pump state. Compared to conventional manual inspection, the proposed method achieves both high efficiency and accuracy in fire pump testing, which improves the reliability of the fire water supply. Since the hardware of the proposed system is designed to seamlessly replace traditional fire pumps in most circumstances, and the IoT sensors can all be wirelessly connected, there is low complexity in deploying such a testing system.

Nevertheless, at present, the fault detection models need to be trained separately for different types of fire pump systems. Hence, in future work, we plan to expand the database of examples in realistic working conditions via the proposed IoT platform, in order to build a general fault detection model for multiple types of fire pump systems. Meanwhile, the improvement of the modeling algorithm needs to be accomplished. The distribution of the curves in the performance space can be complicated, so different algorithms with a higher capability in distinguishing NPCs and FPCs need to be investigated.

In addition, the methodology of the proposed system is scalable to be utilized in other applications in the firefighting or water supply fields, e.g., end-water testing, leakage detection, etc. The proposed testing system can be integrated into existing firefighting systems with the functionalities of smart fire detection, automatic sprinkling and smoke exhaust, and a comprehensive firefighting framework can be constructed.

Finally, there are inevitable limitations in the proposed fire pump testing method. The implementation of the proposed testing system is limited by several factors: the requirement of IoT infrastructure leads to high costs in some circumstances; the acquisition of modeling data from users is a potential issue due to concerns about data security; and the pricing of the proposed system is higher than that of traditional fire pumps, which could raise marketing difficulties.

Author Contributions: Conceptualization, Y.L.; methodology, Z.W.; software, R.Z.; validation, Y.L. and X.C.; formal analysis, S.Z. and X.C.; investigation, S.Z.; resources, T.Z. and R.Z.; data curation, R.Z.; writing—original draft preparation, S.Z. and T.Z.; writing—review and editing, Z.W.; visualization, T.Z.; supervision, Z.W.; project administration, Z.W.; funding acquisition, Y.L. All authors have read and agreed to the published version of the manuscript.

Funding: This research was funded by the Yueqing Science and Technology Bureau, grant number 2023001. The APC was funded by C-lin Electrical Co., Ltd.

Data Availability Statement: Data sharing is not available due to commercial confidentiality.

Conflicts of Interest: Author Yongfang Li and Ruyi Zhou were employed by the company C-lin Electrical Co., Ltd. Author Xuefei Chen was employed by the company Shanghai TYIoT Technology Co., Ltd. The remaining authors declare that the research was conducted in the absence of any commercial or financial relationships that could be construed as a potential conflict of interest.

References

1. Till, R.C.; Coon, J.W. Unwanted Fire and Fire Growth. In *Fire Protection: Detection, Notification, and Suppression*; Till, R.C., Coon, J.W., Eds.; Springer International Publishing: Cham, Switzerland, 2019; pp. 1–10. ISBN 978-3-319-90844-1.
2. Chui, K.T.; Gupta, B.B.; Liu, J.; Arya, V.; Nedjah, N.; Almomani, A.; Chaurasia, P. A Survey of Internet of Things and Cyber-Physical Systems: Standards, Algorithms, Applications, Security, Challenges, and Future Directions. *Information* **2023**, *14*, 388. [CrossRef]
3. He, W. IoT System for Intelligent Firefighting in the Electric Power Industry. *J. Shanghai Jiaotong Univ. Sci.* **2021**, *26*, 686–689. [CrossRef]
4. Till, R.C.; Coon, J.W. Fire Pumps and Water Supplies. In *Fire Protection: Detection, Notification, and Suppression*; Till, R.C., Coon, J.W., Eds.; Springer International Publishing: Cham, Switzerland, 2019; pp. 45–61. ISBN 978-3-319-90844-1.
5. Khalaf, O.; Abdulsaheb, G. IOT Fire Detection System Using Sensor with Arduino_http. *AUS* **2019**, *26*, 74–78. [CrossRef]
6. Komarasamy, D.; Gokuldhev, M.; Hermina, J.J.; Gokulapriya, M.; Manju, M. Review for Detecting Smoke and Fire in Forest Using Different Technologies. *IOP Conf. Ser. Mater. Sci. Eng.* **2020**, *993*, 012056. [CrossRef]
7. Bu, F.; Gharajeh, M.S. Intelligent and Vision-Based Fire Detection Systems: A Survey. *Image Vis. Comput.* **2019**, *91*, 103803. [CrossRef]
8. Jeevanandham, A.T.; Sivamurgan, P. IoT Based Automatic Fire Alarm System. *Bull. Sci. Res.* **2020**, *2*, 29–34. [CrossRef]
9. Huang, L.-C.; Chang, H.-C.; Chen, C.-C.; Kuo, C.-C. A ZigBee-Based Monitoring and Protection System for Building Electrical Safety. *Energy Build.* **2011**, *43*, 1418–1426. [CrossRef]
10. Tao, T.; Li, J.; Xin, K.; Liu, P.; Xiong, X. Division Method for Water Distribution Networks in Hilly Areas. *Water Supply* **2015**, *16*, 727–736. [CrossRef]
11. Lee, S.W.; Sarp, S.; Jeon, D.J.; Kim, J.H. Smart Water Grid: The Future Water Management Platform. *Desalin. Water Treat.* **2015**, *55*, 339–346. [CrossRef]
12. Lenzi, C.; Bragalli, C.; Bolognesi, A.; Artina, S. From Energy Balance to Energy Efficiency Indicators Including Water Losses. *Water Supply* **2013**, *13*, 889–895. [CrossRef]
13. Shirzad, A.; Tabesh, M. New Indices for Reliability Assessment of Water Distribution Networks. *J. Water Supply Res. Technol. Aqua* **2016**, *65*, 384–395. [CrossRef]
14. Albino, V.; Berardi, U.; Dangelico, R.M. Smart Cities: Definitions, Dimensions, Performance, and Initiatives. *J. Urban Technol.* **2015**, *22*, 3–21. [CrossRef]
15. Allen, M.; Preis, A.; Iqbal, M.; Whitttle, A. Case Study: A Smart Water Grid in Singapore. *Water Pract. Technol.* **2012**, *7*, wpt2012089. [CrossRef]
16. Siew, C.; Tanyimboh, T.T. Pressure-Dependent EPANET Extension. *Water Resour. Manag.* **2012**, *26*, 1477–1498. [CrossRef]
17. El-Zahab, S.; Zayed, T. Leak Detection in Water Distribution Networks: An Introductory Overview. *Smart Water* **2019**, *4*, 5. [CrossRef]
18. Carriço, N.; Covas, D.; Alegre, H.; do Céu Almeida, M. How to Assess the Effectiveness of Energy Management Processes in Water Supply Systems. *J. Water Supply Res. Technol. Aqua* **2014**, *63*, 342–349. [CrossRef]

19. Stańczyk, J.; Burszta-Adamiak, E. The Analysis of Water Supply Operating Conditions Systems by Means of Empirical Exponents. *Water* **2019**, *11*, 2452. [CrossRef]
20. Li, J.; Lee, S.; Shin, S.; Burian, S. Using a Micro-Test-Bed Water Network to Investigate Smart Meter Data Connections to Hydraulic Models. In Proceedings of the World Environmental and Water Resources Congress 2018, Minneapolis, MN, USA, 3–7 June 2018; pp. 342–350. [CrossRef]
21. Li, L.; Hua, C.; Xu, X. Condition Monitoring and Fault Diagnosis of Electric Submersible Pump Based on Wellhead Electrical Parameters and Production Parameters. *Syst. Sci. Control Eng.* **2018**, *6*, 253–261. [CrossRef]
22. Du, Y.; Mao, X.; Xu, D.; Ren, F. Analyzing the Effects of Failure on Fire Equipment in Building by FAST. *Procedia Eng.* **2012**, *45*, 655–662. [CrossRef]
23. Sobral, J.; Ferreira, L.A. Availability of Fire Pumping Systems under Periodic Inspection. *J. Build. Eng.* **2016**, *8*, 285–291. [CrossRef]
24. *GB 50974-2014*; Technical Code for Fire Protection Water Supply and Hydrant Systems. Ministry of Housing and Urban-Rural Development of the People Republic of China: Beijing, China, 2014.
25. Fréchet, M.M. Sur quelques points du calcul fonctionnel. *Rend. Circ. Mat. Palermo* **1906**, *22*, 1–72. [CrossRef]
26. Eiter, T.; Mannila, H. *Computing Discrete Frechet Distance*; Technische Universität Wien: Vienna, Austria, 1994.
27. Alt, H.; Godau, M. Computing the Fréchet Distance between Two Polygonal Curves. *Int. J. Comput. Geom. Appl.* **1995**, *5*, 75–91. [CrossRef]
28. Suratno, T.; Rarasati, N.; Gusmanely, Z. Optimization of Genetic Algorithm for Implementation Designing and Modeling in Academic Scheduling. *EKSAKTA Berk. Ilm. Bid. MIPA* **2019**, *20*, 17–24. [CrossRef] [PubMed]

water

MDPI

Article

Evaluating the Effectiveness of Coagulation–Flocculation Treatment Using Aluminum Sulfate on a Polluted Surface Water Source: A Year-Long Study

Hichem Tahraoui [1,2], Selma Toumi [3], Meriem Boudoukhani [4], Nabil Touzout [3], Asma Nour El Houda Sid [5], Abdeltif Amrane [6,*], Abd-Elmouneïm Belhadj [2], Mohamed Hadjadj [7], Yacine Laichi [7], Mohamed Aboumustapha [7], Mohammed Kebir [8], Abdellah Bouguettoucha [1], Derradji Chebli [1], Aymen Amin Assadi [9] and Jie Zhang [10,*]

1 Laboratoire de Génie des Procédés Chimiques, Department of Process Engineering, University of Ferhat Abbas, Setif 42000, Algeria; hichemm.tahraouii@gmail.com (H.T.); abdallah.bouguettoucha@univ-setif.dz (A.B.); derradji_chebli@yahoo.fr (D.C.)

2 Laboratory of Biomaterials and Transport Phenomena (LBMTP), University Yahia Fares, Médéa 26000, Algeria; belhadj_1@yahoo.fr

3 Faculty of Sciences, University of Medea, Nouveau Pole Urbain, Medea University, Medea 26000, Algeria; toumiselma24@gmail.com (S.T.); nabil12marin.eco@gmail.com (N.T.)

4 Material and Environmental Laboratory (LME), University of Medea, Ain D'heb, Medea 26001, Algeria; boudoukhani.meriemlme@gmail.com

5 Chemical Engineering Department, Process Engineering Faculty, University of Salah Boubnider Costantine 3, Constantine 25000, Algeria; asma.sid@univ-constantine3.dz

6 Ecole Nationale Supérieure de Chimie de Rennes, University of Rennes, CNRS, ISCR—UMR6226, 35000 Rennes, France

7 Center for Scientific and Technical Research in Physico-Chemical Analysis CRAPC, BP 384, Bou-Ismail 42000, Algeria; mohamed.hadjadj@gmail.com (M.H.); laichi@gmail.com (Y.L.); aboumustaphamohamed@gmail.com (M.A.)

8 Research Unit on Analysis and Technological Development in Environment (URADTE-CRAPC), BP 384, Bou-Ismail 42000, Algeria; medkebir@yahoo.fr

9 College of Engineering, Imam Mohammad Ibn Saud Islamic University, IMSIU, Riyadh 11432, Saudi Arabia; aaassadi@imamu.edu.sa

10 School of Engineering, Merz Court, Newcastle University, Newcastle upon Tyne NE1 7RU, UK

* Correspondence: abdeltif.amrane@univ-rennes1.fr (A.A.); jie.zhang@newcastle.ac.uk (J.Z.)

Abstract: Safeguarding drinking water is a major public health and environmental concern because it is essential to human life but may contain pollutants that can cause illness or harm the environment. Therefore, continuous research is necessary to improve water treatment methods and guarantee its quality. As part of this study, the effectiveness of coagulation–flocculation treatment using aluminum sulfate ($Al_2(SO_4)_3$) was evaluated on a very polluted site. Samplings were taken almost every day for a month from the polluted site, and the samples were characterized by several physicochemical properties, such as hydrogen potential (pH), electrical conductivity, turbidity, organic matter, ammonium (NH^+_4), phosphate (PO_4^{3-}), nitrate (NO_3^-), nitrite (NO_2^-), calcium (Ca^{2+}), magnesium (Mg^{2+}), total hardness (TH), chloride (Cl^-), bicarbonate (HCO_3^-), sulfate (SO_4^{2-}), iron (Fe^{3+}), manganese (Mn^{2+}), aluminum (Al^{3+}), potassium (K^+), sodium (Na^+), complete alkalimetric titration (TAC), and dry residue (DR). Then, these samples were treated with $Al_2(SO_4)_3$ using the jar test method, which is a common method to determine the optimal amount of coagulant to add to the water based on its physicochemical characteristics. A mathematical model had been previously created using the support vector machine method to predict the dose of coagulant according to the parameters of temperature, pH, TAC, conductivity, and turbidity. This $Al_2(SO_4)_3$ treatment step was repeated at the end of each month for a year, and a second characterization of the physicochemical parameters was carried out in order to compare them with those of the raw water. The results showed a very effective elimination of the various pollutions, with a very high rate, thus demonstrating the effectiveness of the $Al_2(SO_4)_3$. The physicochemical parameters measured after the treatment showed a significant reduction in the majority of the physicochemical parameters. These results demonstrated that the coagulation–flocculation treatment with $Al_2(SO_4)_3$ was very effective in eliminating the various pollutions present in the raw water. They also stress the importance of continued research in the field of water treatment to improve the quality of drinking water and protect public health and the environment.

Keywords: water quality; coagulation–flocculation; aluminum sulfate; jar test; physicochemical parameters

1. Introduction

Water quality is a significant indicator of the health of our environment and the safety of our drinking water supply [1,2]. Water quality can be influenced by a variety of factors, including pollution, human activities, climate changes, and water resource management [3,4]. Monitoring water quality is crucial for detecting contaminants and pathogens in the water in order to take measures to protect public health and the environment [1]. Water quality testing may involve chemical, microbiological, and physical analyses to assess the presence and quantity of substances such as nitrate (NO_3^-), pesticides, heavy metals, pathogenic microorganisms, and organic matter [5]. To improve water quality and eliminate pollution, there are several techniques available depending on the source and type of pollution [6]. Wastewater from households and industries can be treated using technologies such as filtration, sedimentation, disinfection, and reverse osmosis before being discharged into rivers, lakes, or oceans [7].

Recent strides in water treatment signify remarkable progress in addressing micropollutants and substances resistant to conventional processes [8–13]. Within this context, a meticulous exploration of specific coagulants and their optimal dosage becomes indispensable [14]. The safeguarding of environmental sustainability and human health hinges crucially on ensuring the non-toxicity of water treatment agents. In this pivotal context, aluminum sulfate ($Al_2(SO_4)_3$) emerges as a distinguished contender, not only celebrated for its efficacy but also lauded for its commendable non-toxic nature [15]. This attribute underscores its significance in contributing to water treatment strategies that prioritize both environmental conservation and the well-being of human populations [15].

In the realm of water treatment, the deployment of aluminum sulfate has surfaced as a pivotal strategy for eliminating suspended particles, organic matter, metals, and other impurities from drinking water and wastewater [16]. Noted for its effectiveness in forming flocs—particle aggregates—by reacting with impurities in water, aluminum sulfate coagulation offers distinctive advantages [17]. These flocs, being denser than water, readily separate through sedimentation or filtration [18,19]. Beyond its efficacy in removing contaminants such as suspended matter, heavy metals, and specific organic pollutants, aluminum sulfate coagulation is acknowledged for its relative simplicity and cost-effectiveness [20,21]. The dosage of aluminum sulfate assumes particular significance as a critical parameter in achieving optimal treatment efficiency [22]. Determining the appropriate coagulant dosage is pivotal in the design and operation of a water treatment system [23]. Inadequate dosage may lead to treatment inefficiency and compromised water quality, while excessive dosage can result in additional costs and environmental concerns associated with excess coagulant discharge [24]. The multifaceted nature of $Al_2(SO_4)_3$ coagulation allows for its integration with other water treatment methods, ensuring maximum efficiency in reducing water contamination levels [19,25,26]. It can be employed as a pretreatment step before filtration, disinfection, or reverse osmosis, enhancing the overall effectiveness of the water treatment process [27,28].

Optimizing the optimal coagulant dosage can be achieved through laboratory methods such as the jar test [29]. The jar test is a laboratory procedure used to ascertain the optimal amount of coagulant–flocculant to add to water for impurity removal [30]. This test is typically conducted prior to the implementation of a large-scale water treatment system [30]. The test involves taking a sample of raw water and dividing it into six or eight equal-volume samples in glass containers. Each container is stirred with a different amount of coagulant–flocculant, and the formation of flocs is observed [31]. The containers are then left to settle for a specified period, usually an hour, allowing for the flocs to form and settle at the bottom of the container [32]. After the settling period, water is sampled from each container at different heights and analyzed to determine the quantity of suspended particles, organic matter, and other contaminants [31]. The test results help determine the optimal coagulant–flocculant dosage for achieving the best treated water quality.

Other methods for optimizing the optimal coagulant dosage include using mathematical models to predict the optimal dosage based on characteristics of the raw water, such as hydrogen potential (pH), hardness, and organic content [33]. These models can be employed to determine the optimal coagulant dosage for specific water conditions [33]. It is also important to consider other factors in coagulant dosage optimization, such as water temperature, coagulant concentration, mixing time, and settling time [34]. These factors can affect treatment efficiency and must be taken into account when optimizing the coagulant dosage [34]. Lastly, regular monitoring of treated water quality is essential to ensure that the coagulant dosage used remains optimal and to adjust the dosage if necessary based on changes in raw water characteristics [35]. As there are several environmental factors that influence the coagulation process and can vary significantly depending on the source of raw water, it is important to take these factors into consideration. Mathematical models can help predict the optimal coagulant dosage for specific water conditions, but it is crucial to account for environmental factors and conduct regular water quality monitoring to ensure consistent conditions and effective coagulation [36]. Furthermore, the rising levels of contaminants in raw water due to climate change may necessitate ongoing adaptation of water treatment processes to maintain safe drinking water quality [37]. Thus, water quality is influenced by numerous environmental parameters such as temperature, pH, electrical conductivity, turbidity, organic matter, ammonium ($NH^+{}_4$), phosphate ($PO_4{}^{3-}$), $NO_3{}^-$, nitrite ($NO_2{}^-$), calcium (Ca^{2+}), magnesium (Mg^{2+}), total hardness (TH), chloride (Cl^-), bicarbonate ($HCO_3{}^-$), sulfate ($SO_4{}^{2-}$), iron (Fe^{3+}), manganese (Mn^{2+}), aluminum (Al^{3+}), potassium (K^+), sodium (Na^+), complete alkalimetric titration (TAC), and dry residue (DR) [38–40].

These parameters can originate from various sources including human activities, weather conditions, soil geological characteristics, and local ecology [40]. Regular monitoring of these parameters is crucial to ensure good water quality and safe use for both humans and the environment. This innovative study represents a significant advancement in the understanding and application of $Al_2(SO_4)_3$ as a coagulant in severely polluted environments [40].

The central objective of this research was to comprehensively evaluate the performance and effectiveness of a specific coagulant in treating highly contaminated water over an extended 12-month period. To carry out this detailed assessment, a meticulous methodology was employed. Sample collection occurred nearly every day throughout each month, allowing us to capture potential variations in environmental conditions. These samples underwent a carefully controlled storage and mixing process at the end of each month, producing data representative of the key physicochemical properties found in the collected and mixed contaminated water. A pivotal aspect of this research involved the utilization of a previously validated mathematical model based on the support vector machine method [41,42]. This model was specifically designed to accurately predict the optimal dosage of $Al_2(SO_4)_3$ required for the coagulation process, taking into account seasonal variations and fluctuations in environmental parameters such as temperature, pH, TAC, conductivity, and turbidity. This $Al_2(SO_4)_3$ treatment step was systematically repeated at the end of each month throughout the entire study year. After each treatment, we conducted a second comprehensive characterization of the physicochemical parameters to compare the results with those of the original raw water.

Our study stands out due to its unique context and innovative approach in evaluating the efficacy of coagulation–flocculation treatment using $Al_2(SO_4)_3$ for surface water purification, especially within an exceptionally polluted site. Although $Al_2(SO_4)_3$ is a well-established water treatment method, our research significantly enriches existing knowledge by focusing on several key aspects:

- Unprecedented Environmental Challenge: Our study addresses the pressing need to assess water treatment under severe pollution levels, providing invaluable insights into the adaptability of coagulation–flocculation treatment in environments where pollutant concentrations surpass controlled settings.
- High-Frequency Sampling Regime: By adopting an innovative daily sampling regimen over a month, our study unveils dynamic pollutant removal trends, showcasing the potential nuances missed with less frequent sampling intervals.
- Advanced Predictive Modeling: The incorporation of a state-of-the-art mathematical model, based on the support vector machine method, advances the practicality of coagulant dosage determination, offering a potentially more efficient and adaptable approach.
- Longitudinal Analysis: Spanning a year with monthly $Al_2(SO_4)_3$ treatments, our longitudinal approach accounts for seasonal fluctuations, providing insights into the treatment method's performance over extended periods, a unique feature in water treatment research.
- Comprehensive Parameter Characterization: The extensive characterization of a wide array of physicochemical parameters offers an all-encompassing view of water quality before and after treatment, enhancing our understanding of treatment impacts on various facets of water chemistry.
- Diverse Parameter Coverage: Our study encompasses an extensive range of water quality parameters, including turbidity, organic matter, PO_4^{3-}, Fe^{3+}, Mg^{2+}, SO_4^{2-}, NH_4^+, Ca^{2+}, Mg^{2+}, TH, Cl^-, and HCO_3^-, among others, providing a holistic perspective on $Al_2(SO_4)_3$ efficacy.
- Indirect Parameter Reduction Mechanisms: Demonstrating the intricate interactions of $Al_2(SO_4)_3$ with water quality parameters, such as NH_4^+, Ca^{2+}, Mg^{2+}, TH, Cl^-, and HCO_3^-, among others, via mechanisms such as complexation and pH changes, elucidates novel pathways for pollutant removal.

This comprehensive investigation underscores the significance of $Al_2(SO_4)_3$ treatment in pollutant removal and water quality improvement, particularly in severely contaminated contexts. Furthermore, it underscores the perpetual need for ongoing research in water treatment, offering broader implications for the preservation of public health and the environment. Importantly, this research breaks new ground as the first of its kind to delve into $Al_2(SO_4)_3$ utilization as a coagulant in high-pollution scenarios, potentially reshaping perspectives on environmental management and water resource protection.

2. Methods and Materials

2.1. Mode of Collection and Preservation of Water Samples

Over a period of 12 months, a rigorous sampling campaign was conducted daily at the highly polluted surface water site of Oued Chiffa, situated between the regions of Médéa and Blida. Each day, one liter of water was collected from various points within the contaminated area to ensure the representativeness of the samples. At the end of each month, a composite sample of 30 L was created from these daily collections. The specific choice of this site was motivated by its elevated levels of pollution, aiming to assess the effectiveness of coagulation processes in an intensely contaminated environment. It is noteworthy that, during the initial phase of the study, samples were taken from different locations each day. However, for the remainder of the months, the samples were collected from the same locations, in accordance with the indications provided in Figure 1. The study year was 2019, and samples were systematically collected from January to December 2019, providing a comprehensive dataset for analysis.

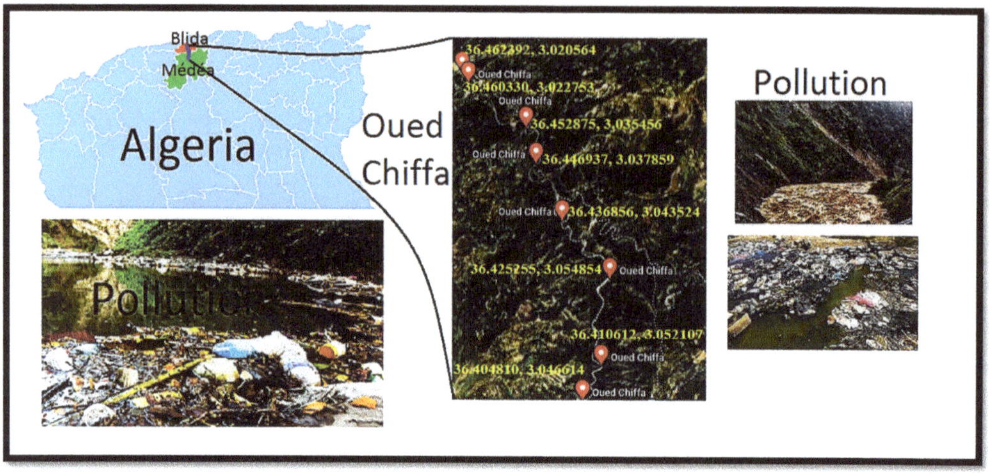

Figure 1. Mapping of sampling points and visualizations of polluted areas during the study.

During the sampling process, meticulous attention was paid to various critical factors to ensure the accuracy and integrity of the collected samples. Specifically, clean glass bottles were used to preserve the samples, and they were collected from a depth of 15 to 30 cm below the water surface to minimize potential surface layer influences. The pH and temperature of each daily water sample were meticulously measured on-site before initiating the acidification process and subsequent preservation in a refrigerator set at a constant temperature of 4 °C. These critical steps played a central role in ensuring the accuracy of the dataset, allowing for real-time adjustments to align the mixture with actual conditions. By calculating monthly average values before the treatment phase, the aim was to capture a nuanced understanding of temporal variations in water quality, enriching the depth of analytical insights.

To maintain chemical stability and prevent contamination, the samples were acidified with sulfuric acid at a purity of 97% (2 mL/L) [43–45]. The pre-analysis acidification of raw water samples with sulfuric acid is a crucial step, aiming to establish a known and constant pH value to ensure reproducibility and accuracy in subsequent measurements. This process serves various purposes, including fixing the pH at an acidic value (typically lower than 2) to stabilize the measured parameters over time and preventing the precipitation of ions and compounds as insoluble salts at higher pH levels, thereby maintaining solubility and accurate measurements [45]. Additionally, acidification contributes to sample preservation by inhibiting microbial growth and minimizing undesirable chemical reactions that could alter the sample's composition over time. It also facilitates control over the chemical activity of ions in water by maintaining a constant, low pH, ensuring consistent conditions for each analyzed sample. Moreover, optimizing specific chemical reactions under acidic conditions enhances the reactivity of reagents and ensures the accuracy of the overall analysis [43,45].

Additionally, great care was taken to sterilize the bottles, which was achieved through thorough rinsing with distilled water followed by autoclaving, thereby eliminating any risks of cross-contamination [45]. To protect the samples from the influence of light, they were stored in opaque containers. Immediately after collection, the samples were promptly cooled by placing them in a cooler maintained at 4 °C, ensuring optimal preservation by slowing degradation processes [43,45]. Measures were taken to prevent freeze-related alterations when temperatures dropped below freezing. Prior to each analysis, a meticulous homogenization process was systematically carried out with the use of impeccable gloves and avoidance of any contaminating contact [45]. Samples were hermetically sealed to preserve their purity, prevent air contact, and minimize the risk of evaporation. Furthermore, consistent temperature stability was maintained throughout the process, thanks to the use of insulated containers during transport. Continuous monitoring of storage temperature and sample condition was undertaken meticulously, with each step rigorously documented, including collection dates, storage conditions, and any relevant events associated with each sample.

2.2. Jar Test

In this study, $Al_2(SO_4)_3$ in octadecahydrate form ($Al_2(SO4)_3 \bullet 18H_2O$) was chosen for its crucial role in water treatment, leveraging its unique properties. With a molecular weight of approximately 666.429 g/mol, this form readily dissolves in water, promoting uniform distribution and efficient interaction with contaminants, crucial for removing impurities from water samples. Prior to the jar test, a comprehensive analysis, including pH evaluations during daily sampling (without acidifications), was conducted. The pH of the mixture was adjusted to the monthly average value (at the end of the month) by adding NaOH (1 M) on the treatment day, reflecting the actual pH values after acidification operations and preserving the sample integrity. Additionally, monthly average temperature values were calculated for water samples, adjusting each sample to this temperature before treatment with $Al_2(SO_4)_3$ for realistic conditions.

To enhance operational efficiency and reduce chemical waste, a support vector machine (SVM) mathematical model was employed to predict the optimal coagulant dosage. This model integrates pH, electrical conductivity, temperature, turbidity, and TAC values from combined samples (the mixture) [41,42]. This technique, validated and previously published in our earlier article [41,42], adapts to changing conditions, aligning with goals of profitability and environmental sustainability to minimize operational costs and chemical waste and addressing complexities related to climate change and fluctuating parameters [41,42].

Following the determination of the optimal dosage, the study proceeded to the jar test. Over a period of more than 12 months, jar tests (Floculateur Fisher 1198) were carried out with only three stirrers out of the six available and one-liter beakers containing 500 mL of mixed monthly samples. Controlled agitation at 180 rpm for 5 min [43,46], followed by intentional reduction to 40 rpm over the next 30 min, facilitated optimal interaction and floc development [42]. After a settling period of 45 min, the supernatant was extracted and filtered through specialized paper, enabling a comprehensive comparison between treated and untreated water samples [41–43]. This systematic approach allowed for a thorough evaluation and precise quantification of the effectiveness of $Al_2(SO_4)_3$.

2.3. PhysicoChemical Analysis

After daily samplings and meticulous adherence to precautions, a monthly physicochemical analysis of combined samples was conducted in two stages: an initial assessment followed by a second analysis after coagulation–flocculation treatment [41,45]. This comprehensive analysis measures 21 key parameters, as previously mentioned. This process is repeated at the end of each month for a year, strictly following established protocols and guidelines [47]. Detailed protocols for each parameter are available in the supplementary data.

2.4. Statistical Analysis

The measurements were conducted three times for each data point, and the results of the experimental data were presented as the average along with the standard deviation (SD). The statistical assessment (performed using ANOVA) was carried out using OriginPro 8 software with a 95% confidence level (where $p < 0.05$). Tukey's one-way comparison test was applied for post hoc analysis.

Table 1, which compiles the minimum ('min'), median ('mean'), maximum ('max'), and standard deviation ('STD') values of the data, provides a comprehensive statistical analysis. Additionally, these minimum, median, and maximum values for each parameter in polluted water (raw water) were derived from a thorough analysis conducted throughout the year. Its inclusion aims to highlight the elevated level of pollution at the specified site, thereby underscoring its failure to meet global standards for treated water.

Table 1. Comprehensive statistical analysis of minimum, median, maximum, and standard deviation values for polluted water parameters throughout the year.

Key Parameters	Unit	Min	Mean	Max	STD
Potential hydrogen	-	7.65	7.79	8.19	0.14
Electrical conductivity	$\mu S\ cm^{-1}$	2870.00	3315.00	3410.00	207.82
Turbidity	NTU	3.01	4.94	9.26	2.03
Organic matter	mg/L	20.50	25.30	29.2	2.55
Ammonium	mg/L	0.64	0.79	0.88	0.08
Phosphate	mg/L	0.53	0.75	0.98	0.13
Nitrate	mg/L	55.00	66.05	96.977	10.81
Nitrite	mg/L	0.24	0.32	0.38	0.03
Calcium	mg/L	240.28	254.30	276.35	10.73
Magnesium	mg/L	168.04	197.20	202.10	10.35
Total hardness	°F	130.08	145.74	153.29	6.99
Chloride	mg/L	510.00	554.70	575.10	19.96
Bicarbonate	mg/L	400.00	448.27	496.42	32.39
Sulfate	mg/L	540.00	652	900	119.50
Iron	mg/L	Negligible	Negligible	Negligible	/
Manganese	mg/L	Negligible	Negligible	Negligible	/
Aluminum	mg/L	Negligible	Negligible	Negligible	/
Potassium	mg/L	21.50	23.20	25.20	1.41
Sodium	mg/L	2150.00	270.00	380.00	55.03
Complete alkalimetric titration	mg/L of $CaCO_3$	313.54	415.47	428.75	29.82
Dry residue	mg/L	2100.00	2494.00	2980.00	297.69

3. Results and Discussion

As previously highlighted, the SVM model developed and published in our previous article was employed in this study to predict the coagulant doses required for treating contaminated water [41,42]. This prediction was achieved by incorporating the pH, electrical conductivity, temperature, turbidity, and TAC parameters of the raw water [41,42]. The coagulant doses obtained through this method are documented in Table 2. This predictive approach based on SVM demonstrated significant accuracy, thereby validating the utility of this model in optimizing the coagulation–flocculation process.

Table 2. Coagulant–flocculant doses (mg/L) by month.

January	February	March	April	May	June	July	August	September	October	November	December
427.94	414.64	427.35	427.03	427.62	427.69	426.87	427.19	427.14	427.08	427.25	427.91

Table 2 illustrates the variation in coagulant–flocculant doses used each month in the treatment of contaminated water. The values exhibit slight fluctuations from one month to another, suggesting that the quantity of coagulant–flocculant required may depend on environmental conditions or the characteristics of the raw water throughout the months. These data are crucial for ensuring effective water treatment throughout the year. Indeed, the coagulant–flocculant dose values exhibit minor month-to-month variations due to the similarity in environmental conditions and characteristics of the raw water throughout the year. In other words, the minimal variations in coagulant–flocculant doses reflect the fact that the 21 physicochemical parameters of the raw water remained relatively constant and highly polluted throughout the year (Table 1). This indicates that the raw water characteristics did not change significantly from one month to another, which explains the close proximity of the coagulant–flocculant doses.

To initiate the analysis, histograms for each physicochemical parameter in raw and treated water were plotted for easy evaluation of the effectiveness of $Al_2(SO_4)_3$. It thus contributed to highlighting the effectiveness of this specific process.

3.1. Potential Hydrogen Activity (pH) of Water

Before adding $Al_2(SO_4)_3$ to raw water samples, a meticulous readjustment of their initial pH was carried out using a NaOH solution. This step was necessary as the samples had been initially acidified for preservation purposes. As convincingly highlighted by Figure 2 (raw waters), the pH values ranged from 7.65 to 8.19, demonstrating the strict adherence to the standards set for drinking water by the Official Journal of the Algerian Republic. As perspicuously illustrated by Figure 2, depicting the characteristics of raw waters, the pH values exhibited a varied range, ranging between 7.65 and 8.19. This meticulous approach ensures the reliability of the results and underscores the commitment to practices in compliance with rigorous standards in force.

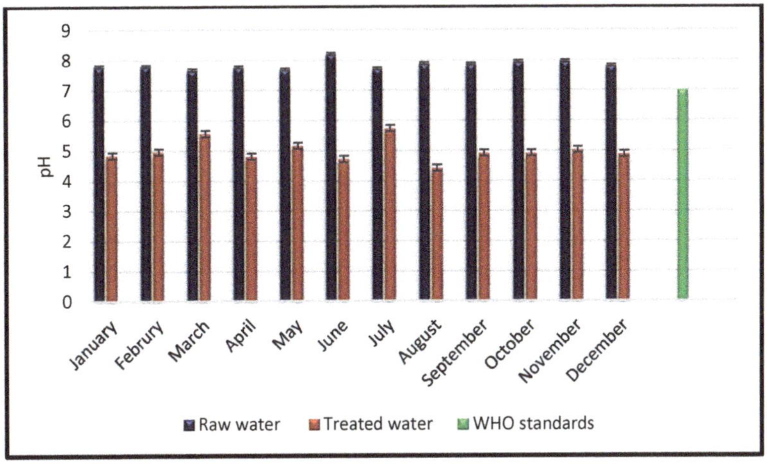

Figure 2. Variation in monthly means of pH as a function of time (months) of raw and treated waters.

The introduction of $Al_2(SO_4)_3$ resulted in a careful transformation of the raw water samples, leading to a slight decrease in pH. The pH values converged towards slightly acidic ranges, ranging from pH 4.8 to 5.7 (as depicted in Figure 2). This pH adjustment can be attributed to the coagulation process, during which $Al_2(SO_4)_3$ neutralized negative charges within the liquid, generating hydroxide ions (OH^-) and, consequently, causing a modest decrease in pH. It is essential to note that even though the pH of the water reached as low as pH 4.8 following $Al_2(SO_4)_3$ treatment, it is generally accepted that this variation does not result in substantial issues.

Furthermore, it is worth highlighting that in conventional drinking water treatment facilities pH correction is typically performed at the end of the treatment process, just before distribution. It is also pertinent to mention that there is a documented decrease in the effectiveness of chlorine against microbial contamination beyond a pH value of 8.19 [47–49]. Nevertheless, the pH adjustments resulting from $Al_2(SO_4)_3$ treatment in this study remained well within acceptable ranges, thereby ensuring that water quality met the relevant standards without compromising safety. This detailed analysis enhances our understanding of the pH reaction when using the coagulant $Al_2(SO_4)_3$.

3.2. Electrical Conductivity

Variation in the electrical conductivity of water samples has been considered a critical indicator of mineralization and dissolved ion content. In this section, we have examined in detail the chemical mechanisms that were responsible for the observed reduction in electrical conductivity following the introduction of $Al_2(SO_4)_3$ as a treatment.

As depicted significantly in Figure 3, the electrical conductivity values exhibited noteworthy variations between the raw and treated water samples. Raw water samples showed a wide range of conductivity, ranging from 2870 to 3410 µS/cm each month (Figure 3). It should be emphasized that these values exceeded the maximum limit set by the World Health Organization (WHO) standards, which was established at 2800 µS/cm. However, a detailed examination of the data in Figure 3 reveals a distinct tendency of obtained values for treated water to remain below those of raw water, especially concerning the maximum value, which slightly exceeds 1000 µS/cm. The action of $Al_2(SO_4)_3$ has thus significantly contributed to the reduction in electrical conductivity levels for treated water. In harmony with the findings of Rodier et al., it is established that a conductivity exceeding 600 µS/cm indicates substantial mineralization of waters, a feature clearly evident in the case of treated water [47]. Guidelines set by USEPA highlight that electrical conductivity tends to maintain relative stability within a distribution network, provided the balance with pipe materials is upheld. Substantial conductivity variations, as clearly illustrated by the comprehensive comparison between raw water and treated water conductivity levels (Figure 3), could potentially signal issues related to corrosion [41,50].

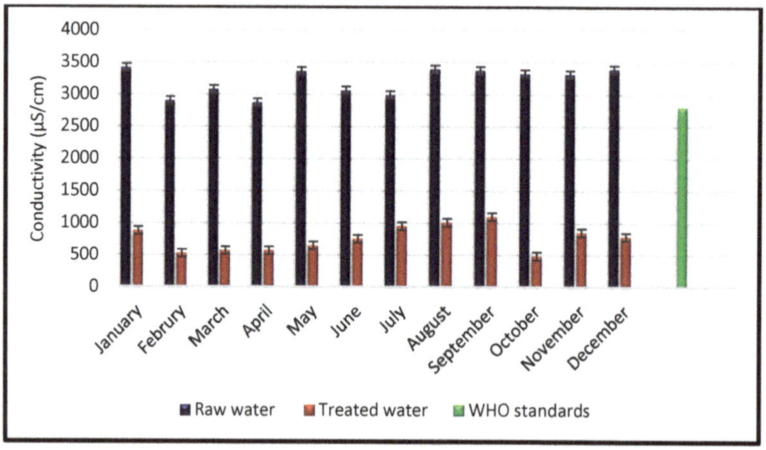

Figure 3. Variation in monthly mean conductivity (µS/cm) over time (months) for raw and treated waters.

It is important to note that an 85.60% reduction in electrical conductivity was achieved through the coagulation process. Indeed, $Al_2(SO_4)_3$ is commonly used in water treatment for the coagulation and flocculation of suspended particles and impurities. Its primary role is to group these particles into larger flocs, making it easier to remove them from water through settling or filtration. This can lead to improved water clarity and a reduction in turbidity. However, concerning water conductivity, the addition of $Al_2(SO_4)_3$ generally does not directly reduce conductivity. Water conductivity is mainly influenced by the presence of dissociated ions such as Na^+, K^+, Ca^{2+}, Mg^{2+}, Cl^-, $SO_4{}^{2-}$, $HCO_3{}^-$, and others [51]. $Al_2(SO_4)_3$ itself dissolves into Al^{3+} ions and $SO_4{}^{2-}$ ions in water, but this does not significantly alter water conductivity. However, $Al_2(SO_4)_3$ can indirectly affect water conductivity through other parameters, such as [47,52–54]:

(1) pH: The water's pH can influence the effectiveness of $Al_2(SO_4)_3$ in reducing conductivity. Generally, $Al_2(SO_4)_3$ is more effective as a coagulant at slightly acidic or neutral pH levels, typically in the pH range of 5 to 8. When the water's pH is maintained within this range, $Al_2(SO_4)_3$ can form aluminum hydroxide ($Al(OH)_3$) precipitates that adsorb suspended particles and impurities. These precipitates contribute to

the formation of larger and heavier flocs, which are easier to remove from water. However, if the water's pH is too low (acidic, pH $\leq$ 2) or too high (alkaline, pH $\geq$ 12), the formation of $Al(OH)_3$ precipitates can be hindered, thus reducing coagulation efficiency. Additionally, water conductivity can be influenced by the presence of dissolved ions, including OH^- and hydronium ions (H_3O^+), which depend on the water's pH. A significant pH change can indirectly affect conductivity by altering the concentration of these ions in the water.

(2) Complexation with Other Physicochemical Parameters: $Al_2(SO_4)_3$ complexation with other physicochemical parameters can potentially contribute to reducing water conductivity, especially if these parameters are associated with suspended particles. For example, if certain physicochemical parameters (such as dissolved ions) are bound to suspended particles, $Al_2(SO_4)_3$ can play a role in their coagulation and flocculation, leading to their removal from water. This could result in a reduction in conductivity, as conductive dissolved ions would be removed along with the suspended particles. In general, $Al_2(SO_4)_3$ is commonly used in water treatment to remove various parameters and impurities. The main parameters that $Al_2(SO_4)_3$ can easily eliminate include suspended particles, turbidity, organic matter, $PO_4{}^{3-}$, heavy metals, and certain inorganic impurities. However, these parameters (such as suspended particles, turbidity, organic matter, $PO_4{}^{3-}$, and heavy metals) can influence water conductivity in different ways. Here is how these parameters interact with conductivity.

- Suspended Particles and Turbidity: Suspended particles, responsible for water turbidity, may contain adsorbed dissolved ions on their surface. When $Al_2(SO_4)_3$ is used to coagulate and flocculate these particles, it can lead to the aggregation of particles into larger flocs. If these suspended particles contain conductive ions, their removal through coagulation and sedimentation can contribute to a reduction in water conductivity. However, the impact on conductivity will depend on the relative contribution of these particles to the total conductivity of the water.
- Organic Matter: Dissolved organic matter, containing various organic ions, plays a significant role in conductivity. The use of $Al_2(SO_4)_3$ for organic matter removal can result in a decrease in conductivity by eliminating specific dissolved organic ions. The extent of this removal is contingent upon the characteristics of the organic matter. Below are examples of organic ions:
    - Carboxylate ions: Derived from carboxylic acids featuring the -COOH functional group. Notable examples include the acetate ion (CH_3COO^-) and formate ion ($HCOO^-$).
    - Sulfonate ions: Originating from sulfonic acids, these ions exhibit the -SO_3H functional group. An illustrative example is the methylsulfonate ion ($CH_3SO_3{}^-$).
    - Phenolate ions: Arising from phenols with an aromatic ring structure, a representative example is the phenolate ion ($C_6H_5O^-$).
    - Amino acids: The fundamental building blocks of proteins, certain amino acids possess ionizable groups. Glycine, for instance, can exist as a zwitterion with both positive and negative charges.
    - Quaternary ammonium ions: Positively charged nitrogen atoms are commonly found in organic compounds like surfactants. The tetramethylammonium ion ($N(CH_3)_4{}^+$) serves as an example.
    - Phosphonate ions: Derived from phosphonic acids containing phosphorus, the methylphosphonate ion ($CH_3PO_3{}^{2-}$) is an illustrative instance.
    - Sulfate and Sulfonate Esters: Formed through the esterification of sulfuric or sulfonic acids, these ions include the ethyl sulfate ion ($C_2H_5SO_4{}^-$) and methylsulfonate ester ($CH_3SO_3{}^-$).

It is crucial to recognize that the specific organic ions present in water are contingent upon the source of organic matter and the composition of the organic compounds. When $Al_2(SO_4)_3$ is employed for organic matter removal, it engages with these organic ions, leading to the formation of insoluble aluminum complexes and subsequently contributing to a reduction in conductivity. The efficacy of this process is intricately linked to the characteristics of the organic matter present in the water.

- Phosphates: Dissolved PO_4^{3-} does not significantly contribute to conductivity since it has charged ions. $Al_2(SO_4)_3$ is often used to precipitate PO_4^{3-} by forming insoluble precipitates, which helps reduce the PO_4^{3-} concentration in water. While the reduction in PO_4^{3-} does not have a direct impact on conductivity, it can influence overall water quality.
- Heavy Metals: Dissolved heavy metals such as Fe^{3+} and Mn^{2+} can contribute to water conductivity as charged ions. $Al_2(SO_4)_3$ can promote the precipitation of these metals as insoluble precipitates, removing them from water. This can reduce the concentration of dissolved metal ions and potentially reduce conductivity.

(3) Interactions between Water Quality Parameters (Suspended Particles, Turbidity, Organic Matter, PO_4^{3-}, Heavy Metals) and Ions Directly Impacting Conductivity (Na^+, Ca^{2+}, Mg^{2+}, HCO_3^-, Cl^-): These interactions can be complex and depend on the chemical composition of the water as well as environmental conditions. Here is how these parameters can interact:

- Suspended Particles and Turbidity: Suspended particles often cause water turbidity. Ions such as Ca^{2+} and Mg^{2+} can bind to suspended particles, contributing to the formation of precipitates. Coagulation with coagulants like $Al_2(SO_4)_3$ can help group these particles into larger flocs, facilitating their removal and reducing turbidity.
- Organic Matter: Dissolved organic matter in water can complex with metal ions like Fe^{3+} and Mn^{2+}, forming organometallic complexes. These complexes can contribute to water color and other quality issues. $Al_2(SO_4)_3$ can promote the precipitation of these complexes, aiding in the removal of organic matter.
- Phosphates: PO_4^{3-} can react with Ca^{2+} and Mg^{2+} ions to form insoluble precipitates such as Ca^{2+}, Mg^{2+} and PO_4^{3-}. However, under certain conditions, they can also react with heavy metals to form soluble complexes. Water treatment may aim to reduce PO_4^{3-} concentration by promoting their precipitation.
- Heavy Metals: Heavy metals like Fe^{3+} and Mn^{2+} can precipitate in the presence of Ca^{2+} and Mg^{2+} ions, forming insoluble precipitates. $Al_2(SO_4)_3$ is commonly used to promote this precipitation and remove these metals from water.
- Na^+, HCO_3^-, Cl^- Ions: In general, Na^+, HCO_3^-, and Cl^- ions do not directly react with the aforementioned water quality parameters. However, they can influence the overall chemical balance of water, pH, and other properties, which can, in turn, affect the solubility of compounds and chemical reactions in water.

3.3. *Turbidity*

Turbidity curves have been meticulously crafted for both the raw and treated water samples, offering a detailed and comprehensive representation in Figure 4. A thorough analysis of Figure 4 has revealed a notable and extensive variation in turbidity values within the raw water, ranging from 1.59 to 9.26 NTU. It is worth noting that these measurements have significantly exceeded the turbidity standards set by the WHO, which has established a maximum limit of 5 NTU. Conversely, when we shift our focus to the treated water samples, as illustrated in Figure 4, it becomes evident that the recorded turbidity values have consistently remained well below 2 NTU. These results have not only complied with but also significantly surpassed the defined tolerance threshold of <5 NTU, as documented in references [41,55]. The turbidity reduction rate has reached an impressive level, achieving a remarkable reduction of 98.05%. This accomplishment is a clear testament to the exceptional efficiency of the treatment process, demonstrating its ability to eliminate the vast majority of suspended particles responsible for the initial turbidity in raw water. This substantial reduction unequivocally reaffirms the effectiveness of employing $Al_2(SO_4)_3$ in the clarification process, underlining its pivotal role in enhancing water quality.

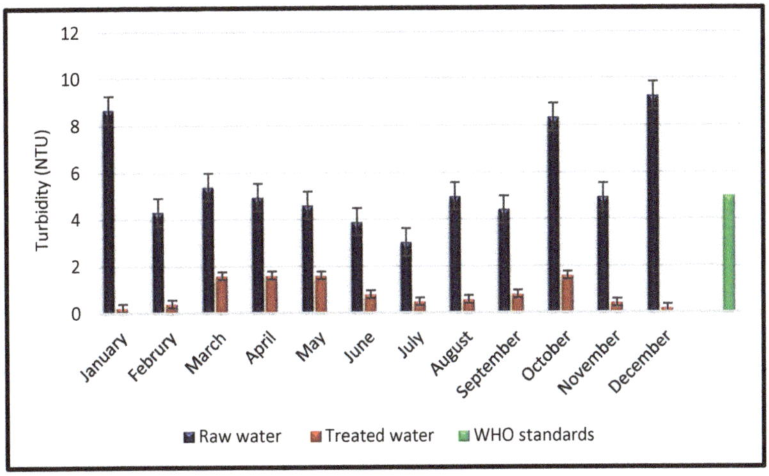

Figure 4. Variation in mean turbidity (NTU) over time (months) for raw and treated waters.

The process of reducing turbidity through $Al_2(SO_4)_3$ centers on promoting the formation of flocs, which act as agents for capturing and removing suspended particles. This mechanism involves two fundamental steps [56]: coagulation and flocculation, which work synergistically to purify water by aggregating impurities into larger, heavier structures, facilitating their removal.

Coagulation is the initial stage of this process, where $Al_2(SO_4)_3$ is introduced into the water. Upon contact with the water, $Al_2(SO_4)_3$ dissociates into Al^{3+} ions and SO_4^{2-} ions [57]. These Al^{3+} ions play a crucial role in destabilizing the suspended particles present in the water. They neutralize the electrostatic charges on the surface of these particles, which typically repel each other due to their like charges. By neutralizing these charges, the Al^{3+} ions allow for the suspended particles to come closer together, overcoming their natural electrostatic repulsion [57].

Flocculation follows coagulation and is the stage where the formation of flocs occurs. During this step, gentle mixing or stirring is applied to the water, which encourages the previously destabilized particles to collide and stick together [57]. This collision and adhesion process leads to the formation of larger, heavier aggregates, known as flocs. The flocs continue to grow as they capture more suspended particles and impurities in the water [57].

The key to the effectiveness of this process lies in the size and weight of the resulting flocs. Because they are larger and denser than individual particles, these flocs settle more rapidly in the water, aided by gravity. This settling process is known as sedimentation. As the flocs settle to the bottom of the water container or treatment tank, they entrap and encapsulate suspended particles and impurities within their structure [57].

Once the flocs have settled, they can be easily separated from the clear water through physical methods such as sedimentation basins or filtration systems [57]. The clear water is then collected for distribution, while the captured impurities remain trapped in the settled flocs [58].

It is essential to emphasize that turbidity, as observed, originates from suspended particles in water, including various components such as organic debris like clay and microscopic organisms [41,59]. Additionally, it is noteworthy that turbidity values often experience a significant increase during and after periods of precipitation. Furthermore, high turbidity can influence water temperature, leading to an elevation, while concurrently affecting the concentration of dissolved oxygen, which tends to decrease [41,56]. The significance of these parameters should not be underestimated, as their substantial influence can have far-reaching implications, impacting not only overall water quality but also the equilibrium and stability of the surrounding ecosystem [41,42,60].

3.4. Pollution Parameters

3.4.1. Organic Matter

The graphical representations depicting organic matter concentrations in both untreated and treated water samples have been meticulously crafted, unveiling a wealth of intricate details within Figure 5. Expanding upon the insights gleaned from the analysis conducted in Figure 5, it becomes increasingly apparent that organic matter values in the untreated water samples exhibit significantly elevated levels, with a striking range spanning from 20.5 to 29.2 mg/L. This range of measurements naturally beckons further investigation into whether these values align with global regulatory standards. Nevertheless, the most pivotal observation emerges from the treated water samples (Figure 5), where the rigorous implementation of the $Al_2(SO_4)_3$ coagulation–flocculation process [41,61] culminates in a remarkable reduction in organic matter levels. The results unequivocally demonstrate that the coagulation–flocculation process with $Al_2(SO_4)_3$ has achieved an impressive reduction in organic matter, reaching an outstanding rate of 99.62%. This figure undeniably confirms the unquestionable effectiveness of this method in diminishing organic matter levels in treated water.

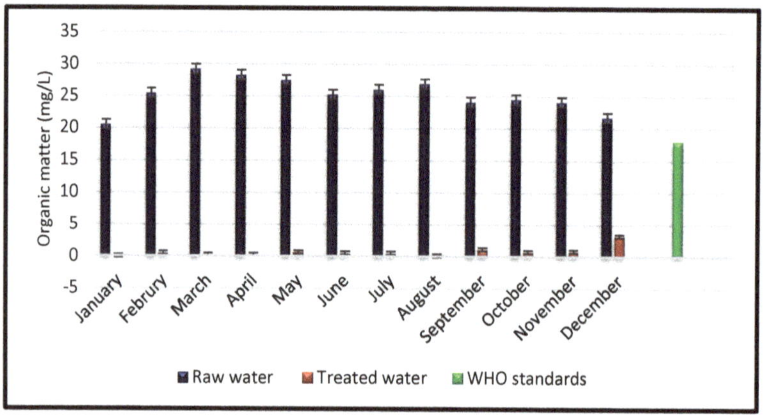

Figure 5. Variation in monthly mean organic matter (mg/L) over time (months) for raw and treated waters.

The specific types and concentrations of dissolved organic carbon (DOC) in river or stream water can vary widely depending on the characteristics of the water source, surrounding land use, geological features, and human activities in the area. Dissolved organic carbon refers to the fraction of organic carbon in water that is in a dissolved form. It includes a variety of organic compounds that have leached into the water from the surrounding environment.

The types of dissolved organic carbon in river or stream water can include:

- Humic Substances: Humic substances are complex organic compounds formed during the decay of plant and animal matter. They include fulvic acids, humic acids, and hymatomelanic acids. These substances contribute to the brown color often seen in water bodies.
- Amino Acids and Proteins: Organic nitrogen compounds, including amino acids and proteins, can be present in dissolved form in water. These compounds may come from decaying organic matter, animal waste, or anthropogenic sources.

- Carbohydrates: Sugars and other carbohydrates can be dissolved in water, contributing to the overall dissolved organic carbon content. These compounds may come from plant material or runoff from agricultural areas.
- Organic Acids: Various organic acids, such as fulvic acid and acetic acid, can be found in water. These acids contribute to the acidity of the water and can play a role in complexation reactions with metals.
- Dissolved Organic Nitrogen (DON): In addition to dissolved organic carbon, nitrogen-containing organic compounds may also be present. These include amines, amides, and other nitrogen-containing organic molecules.
- Microbial Metabolites: The activities of microorganisms in the water can lead to the production of organic compounds as metabolic byproducts. This can include substances like exopolysaccharides and extracellular enzymes.

The specific composition and concentrations of these dissolved organic carbon components will vary based on the specific conditions of the river or stream, including factors such as land use, vegetation, climate, and anthropogenic activities in the watershed.

$Al_2(SO_4)_3$ plays a pivotal role in water treatment by establishing a foundation for effectively reducing the presence of organic matter. When introduced into water, $Al_2(SO_4)_3$ undergoes a chemical reaction, resulting in the formation of $Al(OH)_3$ precipitates, often referred to as "flocs". These flocs carry a positive charge, and their presence in the water augments the overall capacitance and enhances electrostatic interactions within the aquatic environment [15,47,62]. The adsorption process involving these positively charged flocs plays a crucial role in significantly reducing the concentration of dissolved organic matter in the water. In fact, this reduction can be quite remarkable, with reduction rates reaching as high as 99.62%. This substantial decrease in dissolved organic matter is of paramount importance in enhancing water quality. The removal of unwanted organics from the water has far-reaching implications. First and foremost, it contributes to the safety of the water supply. High levels of dissolved organic matter can introduce contaminants and impurities, potentially jeopardizing the health of those who consume or use the water. Additionally, it aids in ensuring that the water complies with stringent water quality standards, which are set to safeguard public health and the environment. By employing $Al_2(SO_4)_3$ to generate flocs that efficiently adsorb and eliminate dissolved organic matter, water treatment processes achieve a notable improvement in water quality. This, in turn, fosters a safer and more reliable water supply, meeting the expectations and requirements of both regulatory bodies and the communities they serve. The reduction in dissolved organic matter through the action of $Al_2(SO_4)_3$ thus underscores its vital role in the broader context of water treatment and the provision of clean, safe, and compliant drinking water.

In concordance with the discoveries made by Harrat and Achour, it is pertinent to underscore the potential formation of chlorination byproducts, including chloroform, resulting from the interaction between organic matter in surface waters and chlorine. These byproducts carry potential long-term toxicity risks, necessitating careful consideration.

This cautionary note underscores the paramount importance of meticulous deliberation in the decision-making processes related to water management [41,60,63]. It serves as a poignant reminder of the pivotal role played by $Al_2(SO_4)_3$ in the reduction in organic matter, ultimately contributing to the improvement of water quality and adherence to international regulatory standards.

3.4.2. Ammonium

The curves depicting the concentrations of NH_4^+ in both the raw water and treated water samples have been meticulously constructed, showcasing an exceptional level of precision, as exemplified in Figure 6. An analysis of these graphs reveals a striking conclusion: the levels of NH_4^+ in raw water exceed established drinking water standards (greater than 0.5 mg/L) (Figure 6), while treated water samples convincingly meet potability criteria (Figure 6), demonstrating a reduction rate of 100% (especially as the initial values were very low). This marked disparity underscores the undeniable efficiency of the treatment process using $Al_2(SO_4)_3$.

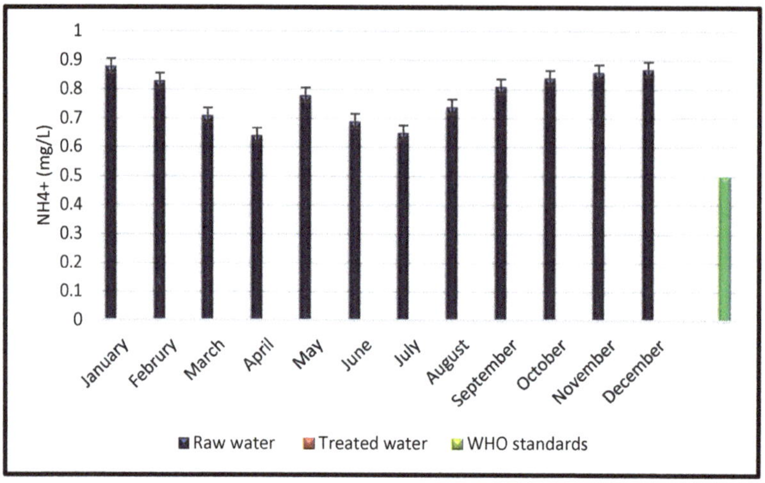

Figure 6. Variation in monthly averages of ammonium (mg/L) over time (months) for raw and treated waters.

Indeed, the reduction in NH_4^+ in water can be indirectly influenced by factors such as conductivity, organic matter, turbidity, and pH. Understanding these interactions is crucial for optimizing water treatment processes and ensuring water quality in various contexts, from drinking water supply to the preservation of aquatic ecosystems. Here is how these parameters can interact [34,47,64,65]:

- Conductivity: Water conductivity is essentially a measure of its ability to conduct electrical current and is primarily influenced by the presence of dissolved ions. NH_4^+ is one of these ions, and its concentration in water contributes to the total conductivity. Thus, water containing a significant amount of NH_4^+ typically has higher conductivity. By reducing water conductivity (as found), you can potentially reduce the concentration of NH_4^+ (especially when the initial values are very low), although this also depends on the contribution of other ions to conductivity.
- Organic Matter: Dissolved organic matter in water can be a key factor in reducing NH_4^+. Some organic compounds can complex with NH_4^+, forming complexes that may be less soluble. This can potentially promote the precipitation or adsorption of NH_4^+, contributing to its reduction.
- Turbidity: While turbidity itself is not directly related to NH_4^+, it can have an indirect effect. Suspended particles in water can adsorb or trap NH_4^+, altering its availability in the water. Reducing turbidity can, therefore, help maintain a more stable NH_4^+ concentration in the water.
- pH: The pH of water is a crucial parameter for understanding the chemistry of NH_4^+. At higher pH values (such as those mentioned between 7.65 and 8.19), a greater proportion of NH_4^+ can convert to ammonia (NH_3), which is a gaseous and less ionic form. This conversion can be desirable in some situations to reduce the toxicity of NH_4^+.

The presence of NH_3 nitrogen in surface waters can be attributed to a complex array of sources. These sources encompass the existence of plant materials along watercourses, the decomposition of organic matter of animal or human origin (humans excrete between 15 and 30 g of urea daily), industrial discharges, and fertilizer inputs, among others [41,47]. According to the comprehensive findings of Rodier et al., the application of chlorine-based oxidation treatment proves effective in removing dissolved NH_3 nitrogen from water. In this regard, it should be noted that biological oxidation of NH_3 could lead to the formation of anaerobic zones within certain parts of distribution networks, thereby generating unpleasant tastes and potentially contributing to pipe corrosion [47]. These findings emphasize the critical importance of maintaining adequate levels of NH_4^+ in water resources intended for human consumption, not only to ensure public health but also to preserve the integrity and long-term sustainability of the hydraulic infrastructure.

3.4.3. Phosphate

The curves representing the concentrations of PO_4^{3-} in both the raw water and treated water samples are displayed in Figure 7. A closer examination of Figure 7 reveals that the recorded levels exceed the limit of 0.5 mg/L in raw water sources. This observation emphasizes the natural presence of phosphorus in water sources, a characteristic influenced by the properties of the terrain and the decomposition of organic matter. It is worth noting that concentrations above 0.5 mg/L can be interpreted as potential indicators of pollution [41,47]. However, a striking contrast is observed in treated water samples, where complete removal of PO_4^{3-} is evident (Figure 7), demonstrating a reduction rate of 100% (Especially the initial values were very low). This finding unequivocally highlights the efficiency of the treatment process using $Al_2(SO_4)_3$. The residual presence of PO_4^{3-} in treated water is practically nonexistent, showcasing the exceptional performance of this chemical compound in contaminant removal.

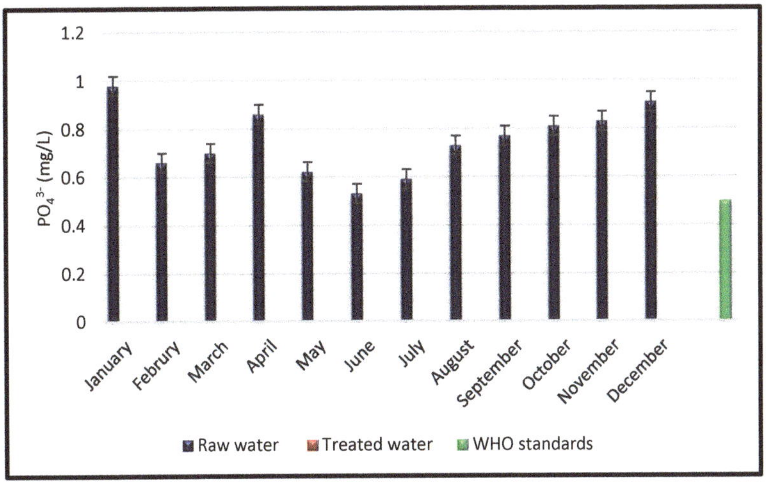

Figure 7. Variation in monthly averages of phosphate (mg/L) over time (months) for raw and treated waters.

Indeed, the chemical reaction between $Al_2(SO_4)_3$ and PO_4^{3-} ions plays a pivotal role in water management and purification. When these two compounds come into contact, they react to form insoluble aluminum phosphate precipitates, creating a solid complex [66,67]. This precipitation process is of paramount importance as it effectively removes unwanted PO_4^{3-} ions from water. Once formed, these precipitates can be removed through separation methods such as sedimentation or filtration, significantly contributing to the reduction in PO_4^{3-} concentrations in water [68]. This chemical reaction thus plays an essential role in preserving water quality and protecting the environment by limiting algal proliferation and preventing eutrophication of aquatic ecosystems [69].

These remarkable results convincingly confirm the significant role of $Al_2(SO_4)_3$, which has proven to be a crucial factor in achieving a substantial reduction in PO_4^{3-} levels in treated water. This conclusion underscores the positive impact of $Al_2(SO_4)_3$ usage on water quality enhancement and the preservation and balance of the surrounding aquatic ecosystem.

3.4.4. Nitrate

Research conducted by Rodier et al. demonstrates that NO_3^- results from natural decomposition processes, primarily orchestrated by nitrogenous microorganisms present in plant and animal proteins and originating from excrement [45]. It is also noteworthy that precipitation can introduce NO_3^- from nitrogen oxides and NH_3 present in the atmosphere. Furthermore, bacterial oxidation of NH_3 can play a significant role in elevating NO_3^- levels [41,70].

It is essential to consider the health implications of high NO_3^- concentrations in water. These adverse effects, such as mucous membrane irritation, nausea, vomiting, bloody stools, and even hypertension risks [41,71], underscore the crucial importance of monitoring and reducing NO_3^- levels in water resources intended for human consumption. This approach aims to ensure the health and well-being of communities dependent on these water sources.

Carefully plotted curves that visualize NO_3^- concentrations within the raw water and treated water samples are detailed in Figure 8. A thorough analysis of NO_3^- concentration variations reveals significant differences, with values ranging from 96.977 mg/L to 55 mg/L in raw water (Figure 8). These measurements manifestly exceed the current regulatory standards for waters, thereby highlighting a contamination issue to be considered. In contrast, the recorded values for treated water demonstrate a notable reduction and fall below those of raw water (Figure 8).

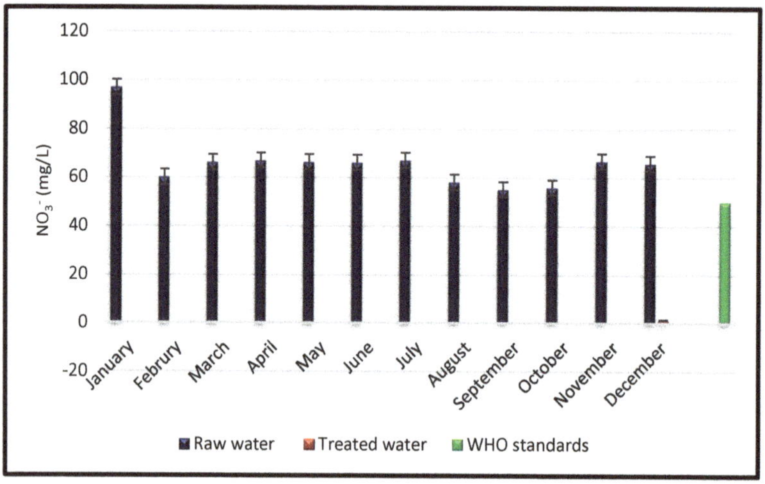

Figure 8. Variation in monthly averages of nitrate (mg/L) over time (months) for raw and treated waters.

It is imperative to emphasize that the WHO recommends a limit of 50 mg/L (NO_3^-) for waters intended for human consumption [47]. After introducing $Al_2(SO_4)_3$, a reduction rate of 100% was recorded. This finding indisputably highlights the intrinsic efficiency of the treatment process using $Al_2(SO_4)_3$ in the drastic reduction in NO_3^- levels, thus contributing to compliance with water quality standards.

The chemical reaction between $Al_2(SO_4)_3$ and NO_3^- ions does not naturally lead to the formation of insoluble precipitates [72]. However, it is conceivable that this reduction involves the use of specific pH conditions, particular reagent concentrations, or other reaction parameters that promote NO_3^- reduction or precipitation [72]. These processes can be tailored to specific water treatment or effluent treatment needs.

In our study, the significant reduction in NO_3^- within a pH range of 7.65 to 8.19 suggests that these pH conditions were favorable for NO_3^- reduction when using $Al_2(SO_4)_3$ as a treatment agent. Several mechanisms may explain this phenomenon [47,73–76]:

1. Precipitate Formation: In this pH range, $Al_2(SO_4)_3$ can react with NO_3^- to form aluminum nitrate precipitates or other insoluble reaction products. These precipitates can then be removed through sedimentation or filtration, resulting in NO_3^- reduction in the water.
2. Adsorption: $Al_2(SO_4)_3$ can also adsorb $NO^-{}_3$ ions onto the surface of its particles or formed precipitates, thereby removing them from the aqueous solution.
3. Direct Chemical Reaction: Under specific pH conditions, it is possible that $Al_2(SO_4)_3$ directly reacts with $NO^-{}_3$ to convert it into a non-reactive form or precipitate it:

- Effects on Physicochemical Parameters: Furthermore, the efficient removal of physicochemical parameters using $Al_2(SO_4)_3$ can have a positive impact on NO_2^- and NO_3^- concentrations in the water. Here are some possible reasons.
- Reduction in Organic Load: The removal of organic matter by $Al_2(SO_4)_3$ can reduce the availability of organic compounds that promote the growth of nitrifying bacteria, contributing to the reduction in NO_2^- and NO_3^-.
- Reduction in Turbidity and Suspended Matter: The elimination of turbidity and suspended matter can contribute to water clarification, thereby improving the efficiency of biological processes responsible for converting NO_2^- into NO_3^- and subsequently reducing them.
- Impact on pH and Reaction Conditions: The use of $Al_2(SO_4)_3$ can influence the pH of the water, creating specific reaction conditions that favor the reduction in $NO^-{}_2$ and $NO^-{}_3$.

4. Removal of other Ions:

- The precipitates formed by the reaction between $Al_2(SO_4)_3$ and certain ions, such as PO_4^{3-}, can trap other ions in solution. This precipitation can potentially sequester NO_2^- and NO_3^-, removing them from the aqueous phase and contributing to their reduction.
- Reduction in Chemical Oxygen Demand (COD): The removal of organic matter and other impurities by $Al_2(SO_4)_3$ can reduce the chemical oxygen demand (COD) of the water. This can improve water quality in terms of chemical reactivity and promote conditions that reduce the presence of NO_2^- and NO_3^-.

5. Not to forget, there chemical reactions involving NO_2^- and NO_3^- ions in relation to other physicochemical parameters such as pH, electrical conductivity, turbidity, organic matter, and metal ions.

- Reactions between NO_3^- and NO_2^-: NO_3^- and NO_2^- ions are forms of oxidized nitrogen in water. They can be involved in conversion reactions, especially in natural environments and water treatment systems. Under acidic conditions (low pH) and in the presence of reducing microorganisms, NO_3^- can be reduced to NO_2^- in a process called denitrification. This reaction involves electron transfer:

$$NO_3^- + 2e^- + 2H^+ \rightarrow NO_2^- + H_2O \tag{1}$$

Conversely, under more alkaline conditions (high pH), NO_2^- can be oxidized to NO_3^-. This reaction can be catalyzed by nitrifying bacteria:

$$2NO_2^- + O_2 \rightarrow 2NO_3^- \tag{2}$$

These reactions are essential for the nitrogen cycle in aquatic ecosystems.

- Reactions of NO_2^- Decomposition: NO_2^- ions can undergo decomposition under acidic conditions:

$$2NO_2^- + 8H^+ + 6e^- \rightarrow N_2 + 4H_2O \quad (3)$$

This reaction produces nitrogen gas and water.

- Oxidation and Reduction Reactions with Other Ions: Oxidation and reduction reactions involving NO_3^- and NO_2^- ions can be influenced by the presence of other metal ions. For example, Fe^{3+} can act as an oxidant for NO_2^- ions, forming ferrous Fe^{3+}. This can be important in denitrification processes.
- Influence of pH and HCO_3^- Ions: The pH of the solution is a crucial factor in reactions involving NO_3^- and NO_2^- ions. Reduction reactions are generally favored in acidic conditions, while oxidation reactions are favored in alkaline conditions. The presence of HCO_3^- ions can also influence these reactions as they can act as buffers and modify the solution's pH.

3.4.5. Nitrite

Meticulously plotted curves representing NO_2^- concentrations within the raw water and treated water samples are detailed in Figure 9. A comprehensive examination of the data reveals a reduced presence of NO_2^-, as clearly highlighted in Figure 9. Consistent with Kurnaz's observations (2004), NO_2^- is typically found at modest levels in natural waters. However, it is noteworthy that even at these levels, the measured concentrations exceed the WHO regulatory standards (maximum value of 0.1 mg/L). Nevertheless, a striking contrast emerges for treated water samples, with a significant observation: NO_2^- values have been reduced to zero (100% reduction rate, especially as the initial values were very low) (Figure 9), thus underscoring the complete elimination of this harmful substance. This undeniable discovery highlights the intrinsic efficiency of the treatment process using $Al_2(SO_4)_3$, ensuring a drastic reduction in NO_2^- levels and thereby ensuring compliance with rigorous water quality standards.

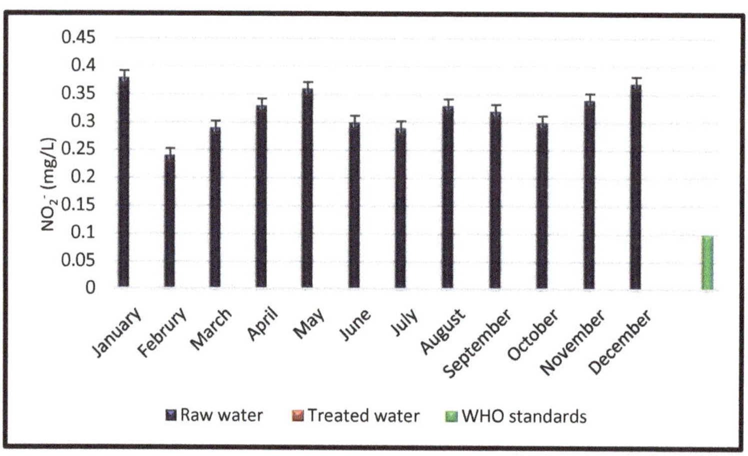

Figure 9. Variation in monthly averages of nitrite (mg/L) over time (months) for raw and treated waters.

The origins of NO_2^- in raw waters are diverse and can be attributed to several factors, including [41]:

- Partial oxidation of NH_3 nitrogen, leading to NO_2^- formation under the influence of nitrifying bacteria [41].
- Reduction of NO_3^-, primarily of agricultural origin, into NO_2^- through the action of denitrifying bacteria [77].

Research by Rodier et al. (2009) also mentions the possibility of NO_2^- formation as a byproduct during the water chlorination process. This precaution is crucial to maintaining adequate chlorination while ensuring an optimal level of safety for consumers [47].

Regarding water intended for human consumption, it is important to note that the WHO recommends a provisional guideline value of 3 mg/L (NO_2^-) [47], thus highlighting the fundamental importance of monitoring and maintaining NO_2^- levels in accordance with health guidelines to ensure the purity and safety of drinking water for the health and well-being of the population.

Our study has illuminated several relevant aspects related to the removal of $NO^-{}_2$ from water using $Al_2(SO_4)_3$ as a treatment agent. The results reveal a complex yet effective mechanism that relies on several key elements: [15,47,74,78].

- Decomposition of NO_2^-: NO_2^- present in water can undergo decomposition under acidic conditions, typically at a low pH. This chemical reaction leads to the production of gaseous nitrogen (N_2) and water (H_2O). This NO_2^- decomposition is an essential element of the elimination process and significantly contributes to its reduction in water.
- Oxidation and reduction reactions with other ions: Oxidation and reduction reactions involving NO_3^- and NO_2^- ions can be modulated by the presence of other metal ions, notably Fe^{3+}. These reactions can play a crucial role in water treatment systems, promoting the conversion of NO_2^- into NO_3^- or their complete elimination.
- Influence of pH and HCO_3^- ions: The pH of the solution plays a crucial role in reactions between NO_2^- and other ions. Acidic conditions generally favor reduction reactions, while alkaline conditions favor oxidation reactions. Additionally, the presence of HCO_3^- ions can act as a buffer, altering the solution's pH and, consequently, influencing the progression of reactions.

3.5. *Overall Mineral*

3.5.1. Calcium, Magnesium, and Total Hardness

The meticulously crafted concentration curves depicting the levels of Ca^{2+}, Mg^{2+}, and TH in both untreated and treated water samples are intricately detailed in Figures 10–12. A comprehensive examination of Figure 10 reveals a notable variation in Ca^{2+} levels in untreated water, spanning from 240.3 to 276.35 mg/L. Similarly, a meticulous analysis of monthly average Mg^{2+} concentrations, as presented in Figure 11, unveils a range of fluctuation from 168.04 to 202.1 mg/L in untreated water. It is of utmost importance to underscore that both of these parameters surpass the prescribed WHO standards for water quality.

In consonance with insights offered by Rodier et al., the productivity of freshwater ecosystems has been intricately correlated with the presence of Ca^{2+} and Mg^{2+} salts, reaching its zenith with a total Ca^{2+} alkalinity of 25 mg/L [47]. This association underscores the pivotal role of these elements in the chemical composition and inherent quality of natural water bodies.

Water hardness, arising from the collective concentrations of Ca^{2+} and Mg^{2+}, assumes a central position in the assessment of water potability. The remarkably elevated levels documented for these two elements, as depicted in Figure 12, unequivocally classify the water as "very hard".

It is important to emphasize that water hardness not only affects human health but also significantly influences pipe corrosion. In line with Rodier's research, hard water, characterized by high concentrations of Ca^{2+} and Mg^{2+}, generally exhibits lower corrosivity compared to soft water. This discrepancy can be attributed to the formation of a protective layer of carbonates that precipitate onto the inner surfaces of pipes, effectively mitigating the risk of corrosion [41,79,80]. These intricate interactions between water parameters and their ramifications on the quality and stability of water distribution infrastructure accentuate the critical necessity for judicious and balanced management of these essential elements to ensure continuous access to high-quality potable water.

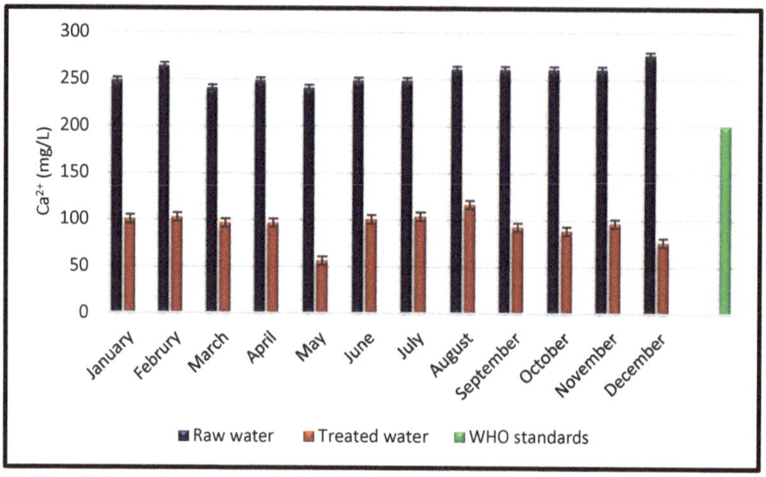

Figure 10. Variation in monthly averages of calcium (mg/L) over time (months) for raw and treated waters.

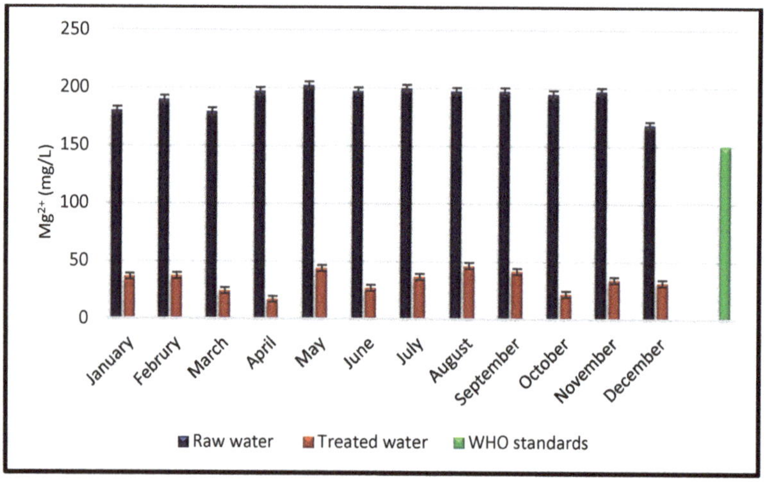

Figure 11. Variation in monthly averages of magnesium (mg/L) over time (months) for raw and treated waters.

Following the introduction of $Al_2(SO_4)_3$ into the contaminated water, concentrations of Ca^{2+} and Mg^{2+} in treated water samples underwent a reduction attributable to $Al_2(SO_4)_3$ treatment, albeit remaining slightly elevated (Figures 10 and 11). Nonetheless, they remain within the limits recommended by the WHO. Our findings have unequivocally demonstrated that the prudent application of $Al_2(SO_4)_3$ resulted in substantial reductions, ranging from 55.35% to 76.66% for Ca^{2+} ions and from 76.58% to 91.37% for Mg^{2+} ions. The reduction in total water hardness ranged from 67.40% to 78.69%. These results underscore the effectiveness of $Al_2(SO_4)_3$ as a coagulant for mitigating water hardness under specific conditions, thereby facilitating the precipitation of Ca^{2+} and Mg^{2+} ions as insoluble salts.

As indicated in the conductivity reduction mechanism, Ca^{2+} and Mg^{2+} ions are among the key parameters that influence conductivity. Therefore, by reducing conductivity (as found), Ca^{2+} and Mg^{2+} ions will be automatically reduced by one of the mechanisms mentioned in the conductivity reduction process, notably pH [81].

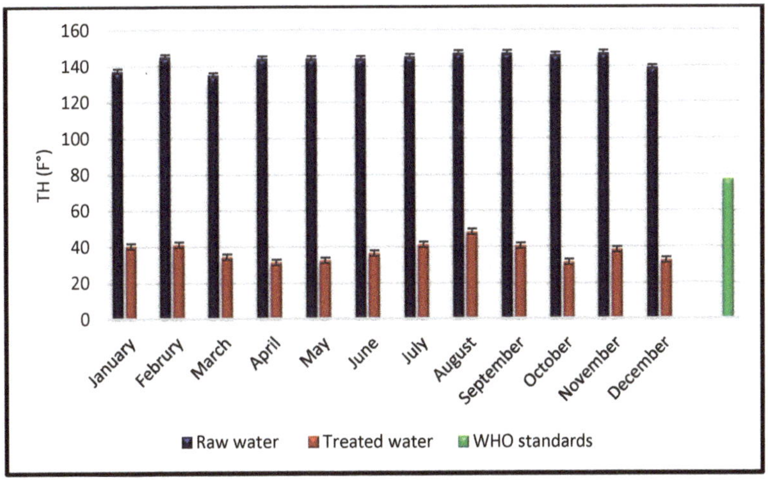

Figure 12. Variation in monthly averages of total hardness (mg/L) over time (months) for raw and treated waters.

Indeed, when $Al_2(SO_4)_3$ is introduced into water, it reacts with the Ca^{2+} and Mg^{2+} ions present in the solution. The mechanism of this reaction is based on the formation of insoluble precipitates, primarily calcium hydroxide ($Ca(OH)_2$) and magnesium hydroxide ($Mg(OH)_2$). This process is influenced by the pH of the water, with the pH range of 7.65 to 8.19 being particularly favorable. In this pH range, OH^- are relatively abundant, thereby facilitating the formation of precipitates. The chemical equation summarizing this mechanism is as follows:

Formation of hydroxide precipitates:

$$Al^{3+} + 3H_2O \rightarrow Al(OH)_3 + 3H^+ \tag{4}$$

Precipitation reactions for Ca^{2+} ions and Mg^{2+} ions, respectively:

$$Ca^{2+} + 2OH^- \rightarrow Ca(OH)_2 \text{ (precipitate)} \tag{5}$$

$$Mg^{2+} + 2OH^- \rightarrow Mg(OH)_2 \text{ (precipitate)} \tag{6}$$

Subsequently, these insoluble precipitates are removed from the system through sedimentation or filtration processes, resulting in a substantial reduction in the concentrations of Ca^{2+} and Mg^{2+} ions in the water and, consequently, a reduction in water hardness.

The initial pH of the water between 7.65 and 8.19 promotes the formation of these precipitates, thereby optimizing the effectiveness of $Al_2(SO_4)_3$ as a water hardness reduction agent.

3.5.2. Chloride

The meticulously plotted curves representing Cl^- concentrations within the raw water and treated water samples are detailed in Figure 13. Careful observation of these curves reveals a significant variation in Cl^- levels in raw water, ranging between 510 mg/L and 562.52 mg/L (Figure 13). This concentration range notably exceeds the standards established by the WHO water quality guidelines. However, the treatment of water samples using the $Al_2(SO_4)_3$-based process has yielded remarkable results by reducing Cl^- concentrations to considerably lower levels, with most falling below 90 mg/L (Figure 13). It is worth noting that a slight exception was observed in November, when a value of 124 mg/L was recorded. Nevertheless, even this value, remaining below the limits set by the rigorous Official Journal of the Algerian Republic standards (2011) [41,55], demonstrates the unequivocal effectiveness of the treatment process.

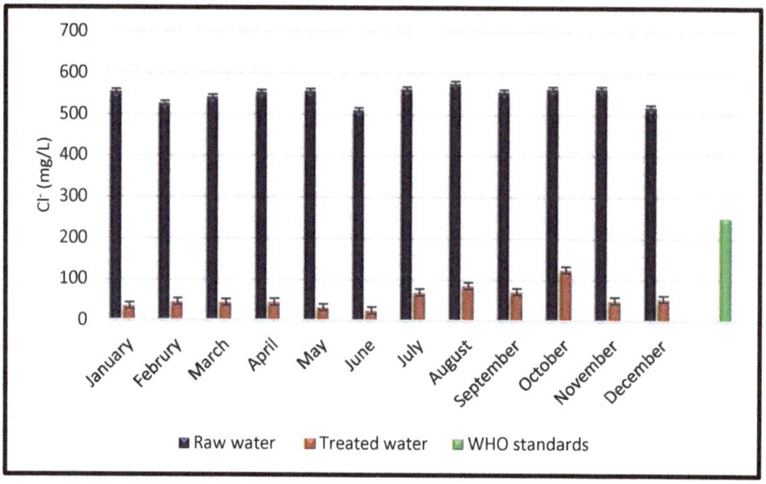

Figure 13. Variation in monthly averages of chloride (mg/L) over time (months) for raw and treated waters.

In addition to its impact on water quality compliance, Cl^- has broader implications. Its effect on the taste of water is an important aspect to consider, especially when concentrations reach or exceed 250 mg/L, which can alter the perceived quality of drinking water. Interestingly, the presence of Ca^{2+} and Mg^{2+} can mitigate this unpleasant taste. However, beyond the taste aspect, Cl^- can also play a role in the corrosion of water distribution infrastructure, particularly in the case of stainless steel materials. The concentration of 50 mg/L marks a critical threshold beyond which the risk of corrosion increases significantly [47]. These factors underscore the need for constant monitoring of Cl^- levels to ensure high-quality drinking water and the long-term sustainability of distribution networks. In conclusion, a comprehensive analysis of Cl^- concentrations in the raw water and treated water samples, as detailed in Figure 13, highlights the exceptional efficiency of the $Al_2(SO_4)_3$-based treatment in reducing Cl^- levels. This approach ensures compliance with rigorous water quality standards while preserving public health and the integrity of water distribution infrastructure. The water treatment process involving the use of $Al_2(SO_4)_3$ in coagulation–flocculation provides an effective method for reducing the concentration of Cl^- ions in water. This treatment relies on complex chemical reactions that occur in a specifically controlled environment. $Al_2(SO_4)_3$ is introduced into the water to be treated, generating Al^{3+} ions and ${SO_4}^{2-}$ ions. Maintaining the pH of the water within an optimal range, typically between 7.65 and 8.19 in this case, is crucial to promote the coagulation process. Subsequently, the formation of $Al(OH)_3$ occurs under the influence of the appropriate pH, where Al^{3+} ions react with water molecules to create $Al(OH)_3$ according to the equation [15]:

$$2Al^{3+} + 6H_2O \rightarrow 2Al(OH)_3 + 6H^+ \quad (7)$$

Following this, Cl^- ions present in the water interact with $Al(OH)_3$, leading to the formation of insoluble alumina (Al_2O_3). The reaction is described as follows:

$$2Al(OH)_3 + 6Cl^- + 3/2O_2 \rightarrow Al_2O_3 + 6OH^- + 3Cl_2 \quad (8)$$

In this reaction, chlorine undergoes oxidation by losing electrons, and it is oxidized from a lower oxidation state (in chloride ions, Cl^-) to a higher oxidation state in chlorine gas (Cl_2). The oxidation of chlorine is accompanied by the reduction of aluminum ions to form insoluble alumina (Al_2O_3) and hydroxide ions (OH^-) as byproducts.

This reaction results in the formation of insoluble alumina, chlorine gas (Cl_2), and OH^- as byproducts.

The formed alumina appears in the form of solid, insoluble flocs. These flocs have the ability to capture various impurities present in the water, such as suspended particles, organic matter, and excess ions. Due to their greater weight compared to water, the precipitated alumina flocs settle at the bottom of the treatment tank. The precipitated alumina flocs are then removed from the treatment tank, resulting in a substantial reduction in the concentration of Cl^- ions, as well as other impurities, in the treated water [47].

This mechanism explains how $Al_2(SO_4)_3$, under appropriate pH conditions and with an adequate dosage, can effectively reduce the concentration of Cl^- ions in water by promoting the formation of insoluble alumina. It is worth noting that the efficiency of this process depends on various factors and must be tailored to the specific water treatment conditions [82]. The reduction rate of 95.13% found in this study demonstrates the success of this process in this particular context.

3.5.3. Bicarbonate

The meticulously plotted curves representing HCO_3^- concentrations within the raw water and treated water samples are detailed in Figure 14. Upon careful scrutiny of Figure 14, it is evident that HCO_3^- levels in raw water exceed the defined limits for potable water. In contrast, treated water samples adhere to stringent potability standards, unmistakably illustrating the efficiency of the treatment process using $Al_2(SO_4)_3$ (Figure 14).

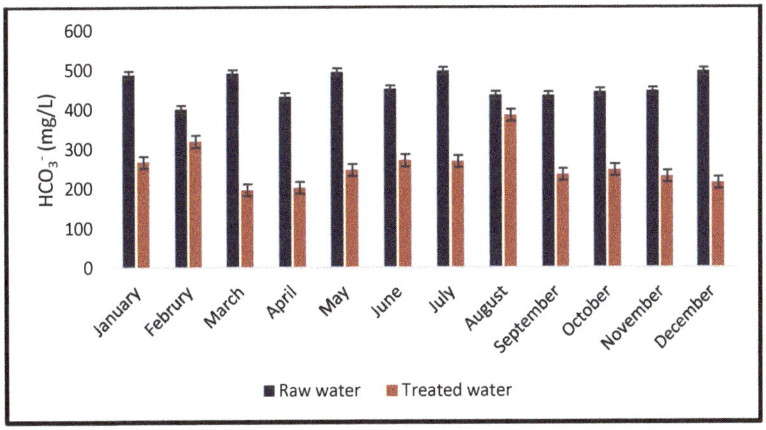

Figure 14. Variation in monthly averages of bicarbonate (mg/L) over time (months) for raw and treated waters.

Rodier et al. astutely highlighted that the HCO_3^- concentration in water is closely correlated with pH and temperature, two key parameters that exert significant influence on the chemical composition of water [47]. Natural levels of total alkalinity, which can reach up to 400 mg/L of $CaCO_3$, although considered relatively moderate, should not be overlooked due to their potential to induce corrosion issues in plumbing systems, underscoring the necessity for cautious and balanced management of these concentrations [41,83].

However, the relevance of maintaining adequate HCO_3^- levels extends beyond mere normative compliance. In reality, these concentrations play a crucial role in preserving water quality as well as the long-term integrity of water distribution infrastructure [41]. The innovative treatment process using $Al_2(SO_4)_3$ that is examined in this study not only ensures the sanitary safety of water but also helps mitigate potential corrosion-related risks, thus preserving the durability and integrity of distribution networks. This holistic approach underscores the importance of comprehensive chemical parameter management to ensure a supply of high-quality water while safeguarding public health and the sustainability of water infrastructure [84].

HCO_3^- ions are present in water and contribute to its alkalinity. To reduce the HCO_3^- concentration in water, it is generally necessary to use specific treatment techniques, such as the addition of acids (such as sulfuric acid) to neutralize the HCO_3^- and reduce alkalinity [85]. These acids react with HCO_3^- to produce water, carbon dioxide (CO_2), and the corresponding salts:

$$\begin{gathered} HCO_3^- + H^+ \rightarrow H_2O + CO_2 + \text{salts} \\ \text{Example: } 2HCO_3^- + H_2SO_4 \rightarrow 2H_2O + 2CO_2 + SO_4^{2-} \end{gathered} \quad (9)$$

This reaction decreases the HCO_3^- concentration and lowers the alkalinity of water.

As previously observed for the pH parameter, a decrease in the water's pH was caused by the introduction of $Al_2(SO_4)_3$ into the water, which can explain the reduction in HCO_3^- in our case [85]. $Al_2(SO_4)_3$ can chemically react with the HCO_3^- in the water, forming precipitates that are removed from the system during the coagulation–flocculation process. The reduction in the pH from 8.19 to 4.5 is likely due to this chemical reaction between $Al_2(SO_4)_3$ and HCO_3^-, which can result in the precipitation of HCO_3^- as solid. This process can be effective in removing some of the HCO_3^- from the water.

Furthermore, a significant decrease in pH from 8.19 to 4.5 indicates acidification of the water, which can effectively lead to a decrease in the HCO_3^- concentration in the water [82]. When the pH is significantly lowered, HCO_3^- can transform into gaseous carbon dioxide (CO_2) and water (H_2O). This can also explain the reduction in HCO_3^- concentration that was observed.

All of these explanations demonstrate a reduction rate of 60%, which was achieved and highlights the effectiveness of $Al_2(SO_4)_3$ in this process.

3.5.4. Sulfate

Meticulously plotted curves representing SO_4^{2-} concentrations within the raw water and treated water samples were meticulously crafted and are presented in detail in Figure 15. A thorough analysis of this figure brings to light a significant characteristic: SO_4^{2-} levels in raw water markedly exceed established thresholds, reaching values even beyond 540 mg/L (Figure 15). This observation clearly underscores that SO_4^{2-} concentrations in raw water do not adhere to the standards prescribed by the WHO water quality norms (Figure 15). However, the efficacy of the treatment process, employing $Al_2(SO_4)_3$, has succeeded in reducing these SO_4^{2-} concentrations in treated water, though a few values remain slightly above the limit. Importantly, these values still fall within the strict limits recommended by the Official Journal of the Algerian Republic (2011) (<400 mg/L), irrefutably affirming the effectiveness of the treatment process.

Research by Rodier et al. has already highlighted that the use of $Al_2(SO_4)_3$ in water coagulation processes can lead to an increase in SO_4^{2-} levels [47]. However, the addition of $Al_2(SO_4)_3$ to water can sometimes lead to a reduction in the SO_4^{2-} concentration in the solution. This observation may seem counterintuitive. This reduction can be explained by various mechanisms, including reactions with other ions present in the water, pH changes induced by $Al_2(SO_4)_3$, or specific reactions related to the particular application of water treatment. These findings emphasize the importance of understanding, in detail, water chemistry and the reactions that occur when chemicals are added for water treatment.

It is also important to note that excessive SO_4^{2-} concentrations can alter the taste of water, imparting an undesirable bitterness and potentially causing laxative effects in individuals unaccustomed to such concentrations [47]. This observation underscores the vital imperative of maintaining appropriate SO_4^{2-} levels in drinking water to ensure both consumer health safety and satisfaction.

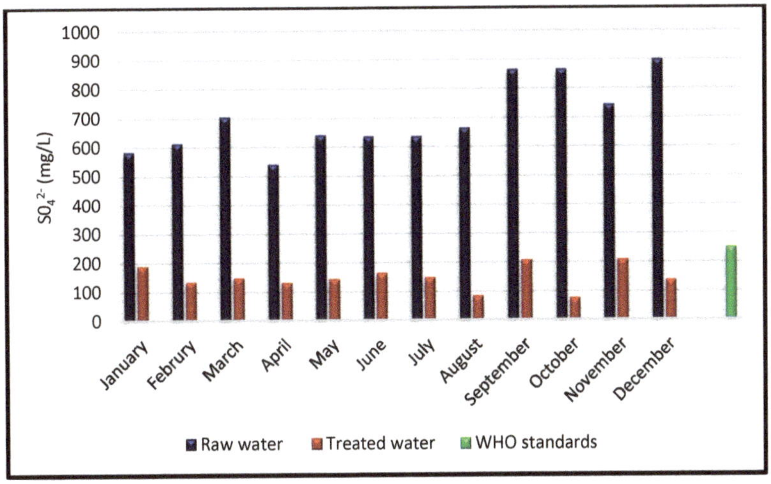

Figure 15. Variation in monthly averages of sulfate (mg/L) over time (months) for raw and treated waters.

The rigorous examination of SO_4^{2-} concentrations, coupled with the evident success of $Al_2(SO_4)_3$ treatment, eloquently illustrates the critical necessity to diligently monitor and manage chemical water parameters at all stages of purification [41,86,87]. This comprehensive approach not only ensures compliance with quality standards but also helps preserve taste and consumer safety, thereby reinforcing confidence in the consistent provision of reliable and wholesome drinking water.

3.6. *Undesirable Parameters*

3.6.1. Iron

Carefully plotted curves representing Fe^{3+} concentrations within the raw water and treated water samples reflect the result of meticulous investigation, and are detailed in Figure 16. A thorough analysis of these curves reveals an intriguing feature: concentrations of Fe^{3+} in both the raw and treated water samples remain remarkably low, maintaining an almost negligible presence, below 0.3 mg/L [41]. Notably, it is remarkable to observe that Fe^{3+} concentration values for raw water samples have been brought down to undetectable levels following treatment (total reduction, especially the initial values were very low), unequivocally demonstrating the undisputed effectiveness of the treatment process in eliminating Fe^{3+} presence from the water (Figure 16). Indeed, $Al_2(SO_4)_3$ reacts with Fe^{3+} ions in the water, resulting in the formation of precipitates and contributing to the removal of Fe^{3+}. The chemical reactions can be articulated as follows.

The detailed mechanistic depiction of the reaction between aluminum sulfate and iron(III) ions, forming iron(III) hydroxide, includes the following simplified steps:

- Dissociation of aluminum sulfate:

$$Al_2(SO_4)_3 \rightarrow 2Al^{3+} + 3SO_4^{2-} \quad (10)$$

- Reaction with hydroxide ions:

$$2Al^{3+} + 6OH^{-} \rightarrow 2Al(OH)_3 \quad (11)$$

- Formation of the aluminum–iron(III) complex:

$$Fe^{3+} + Al(OH)_3 \rightarrow Fe(OH)_3 + Al^{3+} \quad (12)$$

- Precipitation of the complex:

$$Fe(OH)_3 \rightarrow Fe(OH)_3\downarrow \quad (13)$$

In this mechanistic sequence, Al^{3+} ions resulting from the dissociation of aluminum sulfate react with OH^- ions to produce $Al(OH)_3$. Subsequently, Fe^{3+} irons interact with aluminum hydroxide, forming a complex that precipitates as iron(III) hydroxide ($Fe(OH)_3$).

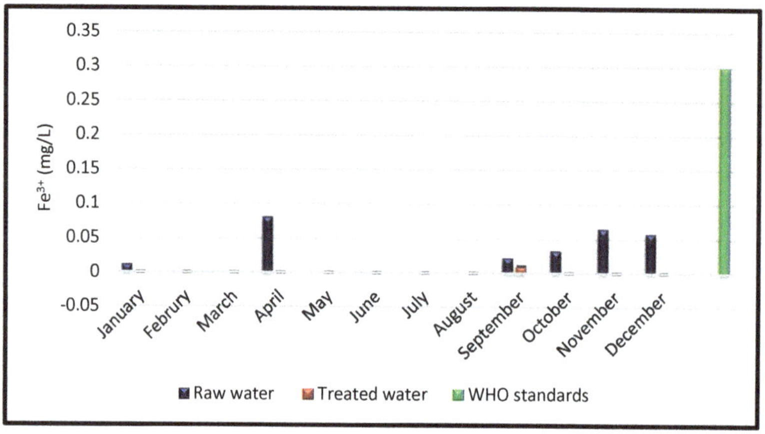

Figure 16. Variation in monthly averages of iron (mg/L) over time (months) for raw and treated waters.

The precipitate formed is $Fe(OH)_3$, which is insoluble in water and precipitates as a solid. Fe^{3+} is thus removed from the solution in the form of a precipitate. This method is commonly used in water treatment to eliminate Fe^{3+}, which is crucial for improving the quality of drinking water or meeting environmental requirements for wastewater discharge. Once the precipitate is formed, it can be separated from the water using various water treatment methods.

On the other hand, in-depth research by Rodier et al. provides a captivating perspective by highlighting that even at concentrations as minimal as 0.05 mg/L, Fe^{3+} can impart an unpleasant taste to water, potentially affecting consumer perception [47]. This observation underscores the crucial importance of maintaining adequate Fe^{3+} levels to ensure a pleasant and discomfort-free consumption experience.

Furthermore, extensive studies by Friedman et al. have established a substantial link between pipeline corrosion and the release of contaminants like Fe^{3+} into the water distribution network. This interconnection also raises questions regarding the potential impact of cast Fe^{3+} pipelines, which could contribute to Fe^{3+} introduction in the system [41]. Additionally, the accumulation of Fe^{3+} within this system can result from intricate oxidation and settling processes, emphasizing the vital importance of implementing preventive measures to mitigate these undesirable phenomena.

The vigilance given to monitoring and careful management of Fe^{3+} concentrations, combined with tangible proof of treatment efficacy, signifies a rigorous approach to ensuring the quality and safety of drinking water [41]. This proactive approach goes beyond mere compliance with stringent quality standards; it aims to guarantee ongoing consumer satisfaction and reinforce confidence in the reliability of their drinking water supply.

3.6.2. Manganese

Meticulously traced curves representing Mn^{2+} concentrations within raw water and treated water samples are the result of a meticulous and detailed scientific approach, and are revealed in depth in Figure 17. Scrutinizing Figure 17 closely reveals a captivating revelation: Mn^{2+} concentrations in both the raw and treated water samples remain minute, maintaining an impressively low trace level, below 0.05 mg/L (Figure 17). This remarkable finding unequivocally highlights the unwavering effectiveness of the treatment process deployed in eliminating Mn^{2+} presence from the water. Equally impressive is the fact that Mn^{2+} concentrations in raw water samples have been reduced to undetectable levels following treatment (total reduction, especially when the initial values were very low), offering tangible proof of the undeniable efficacy of this treatment process.

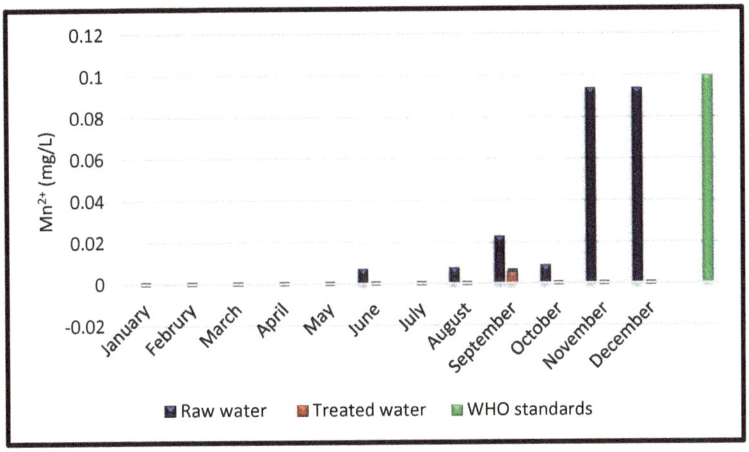

Figure 17. Variation in monthly averages of manganese (mg/L) over time (months) for raw and treated waters.

Indeed, the introduction of $Al_2(SO_4)_3$ into water containing Mn^{2+} ions initiates a reaction leading to the formation of insoluble precipitates—specifically manganese(II) hydroxide ($Mn(OH)_2$) [88]. The overarching chemical reaction governing the precipitation of Mn^{2+} in the presence of $Al_2(SO_4)_3$ is delineated as follows:

- Dissociation of aluminum sulfate:

$$Al_2(SO_4)_3 \rightarrow 2Al^{3+} + 3SO_4{}^{2-} \quad (14)$$

- Reaction with hydroxide ions:

$$2Al^{3+} + 6OH^- \rightarrow 2Al(OH)_3 \quad (15)$$

- Formation of the aluminum–manganese(III) complex:

$$Mn^{2+} + Al(OH)_3 \rightarrow Mn(OH)_2 + Al^{3+} + OH^- \quad (16)$$

- Precipitation of the complex:

$$Mn(OH)_2 \rightarrow Mn(OH)_2\downarrow \quad (17)$$

In this reaction, Mn^{2+} ions react with hydroxide ions, OH^-, derived from the decomposition of $Al_2(SO_4)_3$, thus forming insoluble $Mn(OH)_2$ precipitate in the water. This precipitate can then be removed by settling or filtration, thereby contributing to water purification by eliminating undesirable Mn^{2+} ions [89].

It is crucial to emphasize that the primary source of Mn^{2+} in surface waters largely stems from industrial activities and fertilizer plants, as confirmed by extensive research by Rodier et al. [47]. In these contexts, Mn^{2+} tends to appear in an oxidized and precipitated form, rendering it particularly amenable to removal through traditional clarification processes [41].

Beyond its impact on water taste quality, it is imperative to note that Mn^{2+} can impart an unpleasant taste to water, even at low concentrations (0.02 mg/L). Moreover, it can lead to the formation of a black layer inside pipelines, which may detach and alter the visual appearance of water, rendering it uninviting for consumption [47]. This finding underscores the crucial importance of carefully monitoring Mn^{2+} levels, even at minimal rates, to ensure a pleasant consumption experience free from taste or visual disturbances. In summary, the rigorous approach adopted to manage Mn^{2+} concentration significantly contributes to preserving the quality and acceptability of drinking water, thus enhancing consumer confidence and satisfaction with their water supply [41].

3.6.3. Aluminum

The carefully traced curves representing Al^{3+} concentrations within raw water and treated water samples reflect the outcome of a meticulous and rigorous scientific investigation, detailed comprehensively in Figure 18.

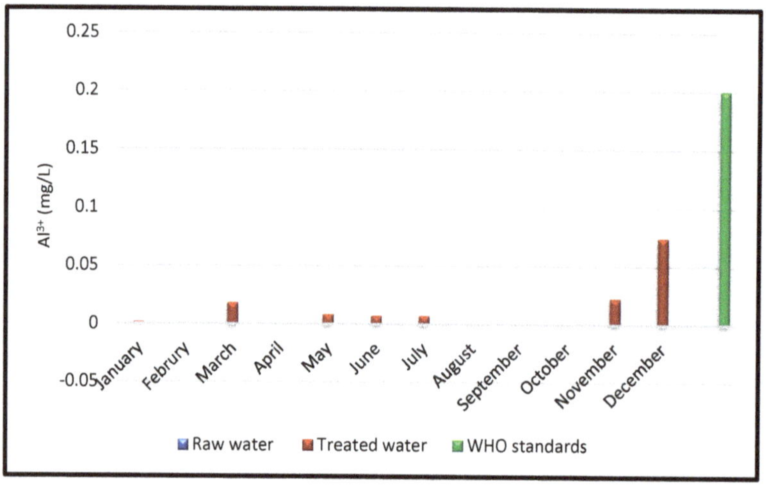

Figure 18. Variation in monthly averages of aluminum over time (months) for raw and treated waters.

A thorough analysis of data pertaining to raw water reveals both an intriguing and significant observation: Al^{3+} concentrations are virtually imperceptible. However, following the application of treatment involving the use of $Al_2(SO_4)_3$, a marginal presence of Al^{3+} emerges, characterized by negligible levels that align perfectly with standards governing treated waters.

It is essential to highlight the pivotal role played by $Al_2(SO_4)_3$ as a coagulant in water treatment processes. This substance, widely employed to facilitate coagulation and clarification, introduces Al^{3+} into the water as an inevitable byproduct [41,90]. However, cautious vigilance is warranted due to the observation that the use of $Al_2(SO_4)_3$ may lead to a 50% likelihood of increasing Al^{3+} concentrations in treated water compared to raw water [41,91].

This discovery underscores the need to closely monitor Al^{3+} levels in treated water, thereby reinforcing commitment to safety and quality standards. Furthermore, it raises pertinent concerns about potential effects on human health, prompting a judicious and prudent management approach toward the use of Al^{3+}-based coagulants in treatment processes. As a result, a proactive and balanced approach is imperative to maintaining drinking water quality while minimizing potential risks associated with Al^{3+} presence [41]. This health-focused approach enhances consumer confidence in the ongoing reliability and safety of their drinking water supply.

3.7. Alkaline Parameters

3.7.1. Potassium

The meticulously traced curves representing K^+ concentrations within raw water and treated water samples are the result of a rigorous scientific methodology, detailed extensively in Figure 19. A thorough analysis of Figure 19 reveals a striking characteristic: K^+ concentrations in raw water do not meet potability standards. In contrast, treated waters adhere perfectly to these stringent norms (Figure 19). It is widely known that $Al_2(SO_4)_3$ itself generally does not have a significant effect on the concentration of K^+ in water. The chemical reactions that occur when $Al_2(SO_4)_3$ is added to water are primarily related to the coagulation of suspended particles and the formation of $Al(OH)_3$ precipitates. K^+ is a distinct chemical element and does not directly react with $Al_2(SO_4)_3$ within this process [19]. However, it can indirectly react with $Al_2(SO_4)_3$ in certain situations, depending on the chemical composition of the water and specific conditions. For instance [34,76,81,92]:

- pH of the water: $Al_2(SO_4)_3$ is often used in water treatment to adjust pH. When added to water, it can react with OH^- present in the water to form $Al(OH)_3$ precipitates. This reaction can influence the water's pH by increasing it, which, in turn, can affect the solubility of K^+ salts in the water.
- Complexation: In some waters, K^+ ions can form complexes with OH^- ions or with $Al(OH)_3$ precipitates, depending on the pH and concentration. These complexes can impact the availability of K^+ in the water. Potassium exhibits versatile complexation behaviors across different chemical contexts. Here are illustrative instances of potassium complexes:
    - Potassium Hydroxide Complex (KOH): In alkaline solutions, potassium ions can form complexes with hydroxide ions, resulting in potassium hydroxide (KOH). The equilibrium reaction is:

$$K^+ + OH^- \rightleftharpoons KOH \quad (18)$$

    - Aluminum Potassium Complex: In waters containing aluminum ions, potassium ions can form complexes with aluminum hydroxide precipitates. The complexation reaction may involve the formation of species like $Al(OH)_4{}^-$ and can be represented as:

$$K^+ + (Al(OH)_4)^- \rightarrow K(Al(OH)_4) \quad (19)$$

    - Potassium Cyanide Complex (KCN): Potassium can form complexes with cyanide ions, resulting in potassium cyanide (KCN). The equilibrium can be expressed as:

$$K^+ + CN^- \rightleftharpoons KCN \quad (20)$$

    - Potassium Tetrafluoroborate Complex (KBF_4): Potassium can also form complexes with fluoroborate ions, as seen in the compound potassium tetrafluoroborate (KBF_4):

$$K^+ + BF_4{}^- \rightleftharpoons KBF_4 \quad (21)$$

    - Potassium Chloride Complex (KCl): While not a traditional complex, in certain contexts, the dissolution of potassium chloride in water can be considered as a form of complexation:

$$KCl \rightleftharpoons K^+ + Cl^- \quad (22)$$

These examples illustrate different scenarios where potassium ions interact with other ions to form complexes, and each complex has its unique properties and significance in various chemical systems.

As seen in our study, a significant reduction in the concentration of K^+ in the polluted water, approximately a 97.42% reduction, occurred after the addition of $Al_2(SO_4)_3$ in the treatment process. This notable reduction can be attributed to several interrelated factors. Firstly, the initial pH of the polluted water, ranging between 7.65 and 8.19, played a crucial role. At this slightly alkaline pH, $Al_2(SO_4)_3$ reacted with OH^- ions in the water to form $Al(OH)_3$ precipitates ($Al(OH)_3$). These precipitates acted as coagulation agents, facilitating the agglomeration of suspended particles. Additionally, the high pH influenced the solubility of K^+ salts, reducing their presence in the treated water. This reduction in K^+ could also be due to complex interactions between $Al_2(SO_4)_3$ and K^+ ions, forming complexes that decreased the concentration of soluble K^+.

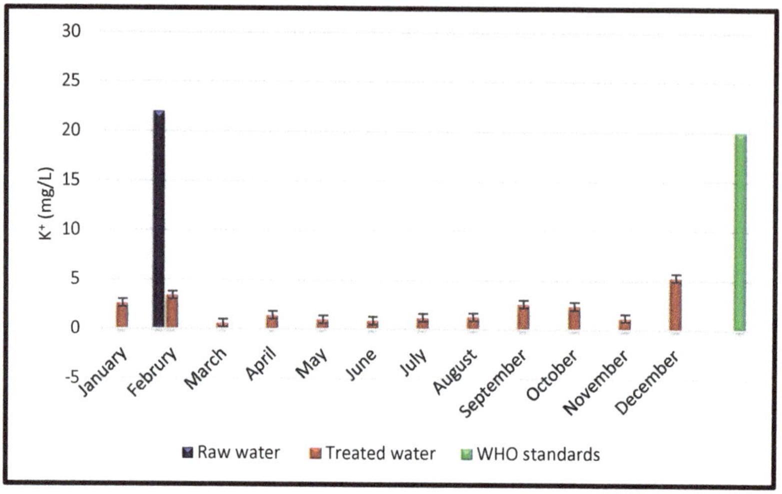

Figure 19. Variation in monthly averages of potassium (mg/L) over time (months) for raw and treated waters.

This proactive approach to treatment using $Al_2(SO_4)_3$ and monitoring of K^+ concentrations, coupled with the concrete demonstration of its effectiveness, testifies to responsible and thoughtful management of the drinking water supply. It enhances consumer confidence in the reliability and continuous quality of the water they consume daily, while highlighting sustained efforts to meet the most demanding standards of water safety and quality [41].

3.7.2. Sodium

The meticulously elaborated curves representing Na^+ concentrations within raw water and treated water samples reflect a rigorous scientific approach, the results of which are detailed in Figure 20. A thorough analysis of Figure 20 highlights a remarkable observation: Na^+ concentrations in raw water exceed the defined limits for drinking water, while treated waters maintain perfect compliance with these stringent standards (Figure 20). This undeniable finding attests convincingly to the exceptional effectiveness of the $Al_2(SO_4)_3$-based treatment in reducing Na^+ levels. As previously mentioned, $Al_2(SO_4)_3$ is commonly used to coagulate suspended particles in water, allowing them to clump together into larger flocs that can be easily removed through filtration. However, it does not have a direct effect on the Na^+ concentration. Nevertheless, it can have an indirect effect on the Na^+ concentration in water, although this effect is generally limited and depends on specific water treatment conditions. When $Al_2(SO_4)_3$ is used as a coagulant in water treatment, it promotes the formation of flocs by aggregating suspended particles. These flocs primarily contain solid impurities present in the water, such as organic matter, colloids, suspended particles, etc. In the coagulation–flocculation process, some soluble impurities, including Na^+ compounds, may also be trapped or adsorbed by the flocs formed by $Al_2(SO_4)_3$. This can result in a reduction in the concentration of Na^+ in the treated water. This demonstrates the 92.64% reduction that has been recorded.

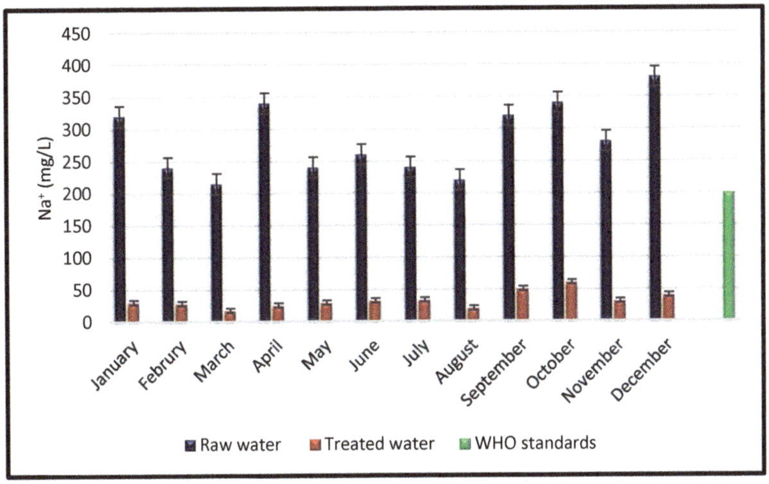

Figure 20. Variation in monthly averages of sodium (mg/L) over time (months) for raw and treated waters.

It is relevant to note that this achievement extends beyond merely satisfying regulatory requirements; it also holds deeper significance. By ensuring adequate Na^+ levels in drinking water, this treatment process actively contributes to preserving consumer health and well-being. Treated waters become a trusted source for the population, delivering quality water that can be consumed with confidence [41].

The positive implications also extend to the realms of infrastructure safety and environmental sustainability. By maintaining Na^+ concentrations at acceptable levels, potential risks of pipeline and equipment corrosion are reduced, which in turn can extend their useful life and minimize maintenance costs.

3.7.3. Total Alkalinity

The carefully traced curves representing TAC concentrations within raw water and treated water samples testify to a meticulous and systematic scientific approach, the results of which are minutely detailed in Figure 21. A meticulous analysis of Figure 21 reveals a striking characteristic: TAC values in raw water samples are remarkably high, far exceeding established standards for ensuring water potability. However, this situation takes a significant turn with treated water samples, where a substantial reduction in TAC levels is observed (Figure 21), bringing them into alignment with stringent potability criteria. This undeniable observation vividly highlights the exceptional effectiveness of the $Al_2(SO_4)_3$-based treatment process in significantly reducing TAC levels, thereby contributing to ensuring the quality of distributed water.

Unlike acidity, which reflects the concentration of hydrogen ions in water, alkalinity represents the presence of bases and salts of weak acids. In the context of natural waters, alkalinity primarily originates from HCO_3^-, carbonates, and hydroxides present. However, it should be noted that other salts of weak acids can also be detected, potentially introducing interferences in the analysis process, such as humic acids, PO_4^{3-}, citrates, and tartrates. Additionally, the presence of ionic silica can impact measurements, especially when the water pH exceeds the threshold of 8.5. This awareness underscores the importance of considering these factors when assessing total alkalinity in water [41,47]. Indeed, $Al_2(SO_4)_3$ itself does not significantly add or remove bases or acids from water. However, it can influence water hardness, which is a measure of the concentration of dissolved Ca^{2+} and Mg^{2+} cations in water. When $Al_2(SO_4)_3$ reacts with these cations, it can form insoluble precipitates, thus reducing water hardness [47]. This can sometimes lead to a decrease in the TAC of the water, although it depends on the composition of the original water.

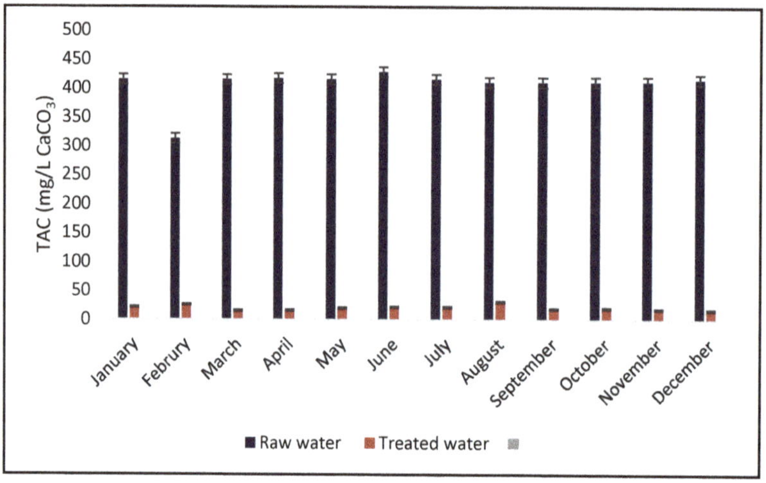

Figure 21. Variation in monthly averages of total alkalinity (mg/L of $CaCO_3$) over time (months) for raw and treated waters.

In our study, a significant removal was recorded for parameters such as electrical conductivity, turbidity, organic matter, NH_4^+, PO_4^{3-}, NO_3^-, NO_2^-, Ca^{2+}, Mg^{2+}, TH, Cl^-, HCO_3^-, SO_4^{2-}, Fe^{3+}, Mn^{2+}, Al^{3+}, K^+, and Na^+ in our water treatment system using $Al_2(SO_4)_3$. It is indeed possible that this has an indirect impact on TAC. TAC is largely determined by the presence of HCO_3^- and carbonate (CO_3^{2-}) ions in the water, among other alkaline species [47]. By reducing the concentration of these other ions and forming precipitates with other compounds present in the water, $Al_2(SO_4)_3$ can contribute to a decrease in TAC by removing alkaline chemical species. However, it is important to note that the impact on TAC may vary depending on specific treatment conditions and the initial composition of the water. Therefore, our finding of a reduction in TAC after the addition of $Al_2(SO_4)_3$ seems consistent with the results obtained for other parameters.

The use of $Al_2(SO_4)_3$ in water treatment has demonstrated significant results in modifying the TAC of water. When applied under initial pH conditions ranging from 7.65 to 8.19, this coagulant has shown exceptional effectiveness in reducing water TAC. This reduction, measured at 96.13%, attests to ability of $Al_2(SO_4)_3$ to interact with alkaline compounds in water, leading to the formation of insoluble precipitates. This process not only facilitated water clarification but also resulted in a decrease in pH. These findings highlight the crucial importance of $Al_2(SO_4)_3$ as an effective treatment agent for reducing alkalinity under specific water quality conditions.

Furthermore, it demonstrates that $Al_2(SO_4)_3$ also has an indirect impact on TAC by removing alkaline compounds, in addition to its contribution to the removal of other water contaminants. This further enriches our understanding of the efficacy of this coagulant in water treatment.

3.8. Dry Residue

The carefully plotted curves depicting the concentrations of DR within raw water and treated water samples bear witness to the methodical and rigorous application of a scientific approach, the detailed results of which are thoroughly expounded in Figure 22.

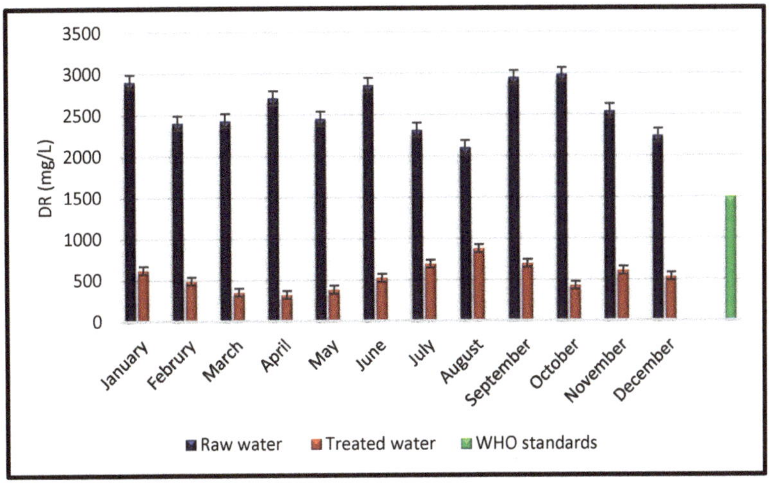

Figure 22. Variation in the monthly averages of the dry residue (mg/L) as a function of time (months) of the raw and treated waters.

The determination of DR in water holds fundamental importance in evaluating its chemical composition. This measurement estimates the total amount of non-volatile and suspended solids present in the water after thorough evaporation. It serves as a key indicator for assessing the quality and composition of dissolved elements in the analyzed water [41,47,93].

Figure 22 offers a vivid visualization of the spatial distribution of DR levels in the raw and treated water samples. A comprehensive analysis reveals that the values of DR in raw water significantly exceed the limits defined for drinking water, with some values even surpassing 2000 mg/L (Figure 22). Conversely, treated water samples exhibit levels that scrupulously adhere to strict potability standards (Figure 22). This tangible observation undeniably highlights the exceptional efficiency of the $Al_2(SO_4)_3$-based treatment process in substantially reducing DR levels, thereby significantly contributing to ensuring the superior quality of water intended for distribution. Indeed, the use of $Al_2(SO_4)_3$ in water treatment has resulted in a substantial reduction in various water quality parameters, encompassing electrical conductivity, turbidity, organic matter, NH_4^+, PO_4^{3-}, NO_3^-, NO_2^-, Ca^{2+}, Mg^{2+}, TH, Cl^-, HCO_3^-, SO_4^{2-}, Fe^{3+}, Mn^{2+}, Al^{3+}, K^+, and Na^+, as well as the total alkalinity content (TAC). Since all of these parameters have been reduced, including those that directly influence DR, it is evident that DR would be significantly reduced, where DR represents the cumulative amount of dissolved solids in the water, encompassing salts, minerals, and various other compounds. This reduction in DR was indeed remarkable, with an impressive reduction rate of 88.14% observed in the recorded data.

It should be emphasized that the result obtained for DR can be influenced by various parameters, such as temperature and the duration of the evaporation process. At temperatures between 100 and 105 °C, a portion or all of the interstitial water as well as the water of crystallization of certain compounds can evaporate. When the temperature exceeds 105 °C, hydrogen carbonates can transform into carbonates, resulting in the release of carbon dioxide. Beyond these temperatures, complex chemical reactions can occur, involving the loss of NH_3, hydrogen sulfide, and organic matter. Different analysis temperatures, such as 180 °C and 525 °C, thus enable the dissection of the different components of DR and a better understanding of its multifaceted composition [41,47,93].

In summary, the rigorous scientific approach adopted to evaluate DR in water, coupled with tangible proof of treatment effectiveness, attests to a committed stance in preserving the optimal quality of drinking water, considering the intricate aspects of its composition and characteristics.

4. Conclusions

This comprehensive study on the quality of raw and treated water across a series of key parameters has underscored the crucial importance of the $Al_2(SO_4)_3$-based treatment process in ensuring high-quality drinking water. The obtained results have enabled an in-depth analysis of concentrations in various elements such as NO_2^-, Ca^{2+}, Mg^{2+}, Cl^-, SO_4^{2-}, Fe^{3+}, K^+, Na^+, DR, and many others. The remarkable effectiveness of the $Al_2(SO_4)_3$-based treatment has been clearly demonstrated through a significant reduction in levels of various contaminants throughout the year, thereby ensuring compliance with national and international potability standards. Elevated concentrations of certain elements in raw water, such as Ca^{2+}, Mg^{2+}, and Cl^-, have been substantially mitigated after treatment, while other elements like Fe^{3+} have been virtually eliminated, attesting to the undeniable efficiency of the process. These findings also highlight the complexity and interdependence of different parameters influencing water quality, such as alkalinity, hardness, and the presence of dissolved substances. They underscore the need for a comprehensive and balanced approach to maintain the quality of distributed water, while considering public health and infrastructure corrosion considerations. Ultimately, this study demonstrates that continuous monitoring, in-depth scientific research, and the application of effective treatment methods are essential to ensure a safe and high-quality supply of drinking water. The collected data and drawn conclusions contribute not only to the understanding of water's chemical composition but also to the implementation of proactive strategies aimed at upholding the health and safety of the population, while preserving precious water resources.

Supplementary Materials: The following supporting information can be downloaded at: https://www.mdpi.com/article/10.3390/w16030400/s1, Table S1. Water Quality Analysis Throughout the Year: A Comprehensive Overview of Physicochemical Parameters in Raw and Treated Water; Table S2. WHO standards of water quality parameters.

Author Contributions: Conceptualization, H.T., S.T., M.B., N.T., A.N.E.H.S., A.A., A.-E.B., M.H., Y.L., M.A., M.K., A.B., D.C., A.A.A. and J.Z.; Methodology, H.T., S.T., M.B., N.T., A.N.E.H.S., A.A., A.-E.B., M.H., Y.L., M.A., M.K., A.B., D.C., A.A.A. and J.Z.; Software, H.T., S.T., N.T., A.N.E.H.S., A.A., A.-E.B., M.H., Y.L., M.A., M.K., A.B., D.C., A.A.A. and J.Z.; Validation, H.T., S.T., A.A., A.-E.B., M.H., Y.L., M.A., M.K., A.B., D.C., A.A.A. and J.Z.; Formal analysis, H.T., S.T., M.B., N.T., A.N.E.H.S., A.A., A.-E.B., M.H., Y.L., M.A., M.K., A.B., D.C., A.A.A. and J.Z.; Investigation, H.T., S.T., M.B., N.T., A.N.E.H.S., A.A., A.-E.B., M.H., Y.L., M.A., M.K., A.B., D.C., A.A.A. and J.Z.; Resources, H.T., S.T., M.B., A.A., A.-E.B., M.H., Y.L., M.A., M.K., A.B., D.C. and J.Z.; Data curation, H.T., S.T., M.B., N.T., A.A., M.H., Y.L., M.A., M.K., D.C. and J.Z.; Writing—original draft, H.T. and S.T.; Writing—review & editing, M.B., N.T., A.N.E.H.S., A.A., A.-E.B., M.K., A.B., D.C., A.A.A. and J.Z.; Visualization, H.T., S.T., N.T., A.N.E.H.S., A.A., A.-E.B., M.K., A.B., A.A.A. and J.Z.; Supervision, A.A., A.-E.B. and J.Z.; Project administration, H.T., A.A., A.A.A. and J.Z. All authors have read and agreed to the published version of the manuscript.

Funding: This research received no external funding.

Data Availability Statement: All relevant data is included within the article.

Conflicts of Interest: The authors declare no conflict of interest.

References

1. Wen, X.; Chen, F.; Lin, Y.; Zhu, H.; Yuan, F.; Kuang, D.; Jia, Z.; Yuan, Z. Microbial Indicators and Their Use for Monitoring Drinking Water Quality—A Review. *Sustainability* **2020**, 12, 2249. [CrossRef]
2. Mammeri, A.; Tiri, A.; Belkhiri, L.; Salhi, H.; Brella, D.; Lakouas, E.; Tahraoui, H.; Amrane, A.; Mouni, L. Assessment of Surface Water Quality Using Water Quality Index and Discriminant Analysis Method. *Water* **2023**, 15, 680. [CrossRef]
3. Armstrong, A.K.; Krasny, M.E.; Schuldt, J.P. *Communicating Climate Change: A Guide for Educators*; Cornell University Press: Ithaca, NY, USA, 2018; ISBN 1-5017-3080-0.

4. Bouchelkia, N.; Tahraoui, H.; Amrane, A.; Belkacemi, H.; Bollinger, J.-C.; Bouzaza, A.; Zoukel, A.; Zhang, J.; Mouni, L. Jujube Stones Based Highly Efficient Activated Carbon for Methylene Blue Adsorption: Kinetics and Isotherms Modeling, Thermodynamics and Mechanism Study, Optimization via Response Surface Methodology and Machine Learning Approaches. *Process Saf. Environ. Prot.* **2022**, *170*, 513–535. [CrossRef]
5. Shmeis, R.M.A. Water Chemistry and Microbiology. In *Comprehensive Analytical Chemistry*; Elsevier: Amsterdam, The Netherlands, 2018; Volume 81, pp. 1–56, ISBN 0166-526X.
6. Xiao, H.; Ji, W. Relating Landscape Characteristics to Non-Point Source Pollution in Mine Waste-Located Watersheds Using Geospatial Techniques. *J. Environ. Manag.* **2007**, *82*, 111–119. [CrossRef] [PubMed]
7. Saravanan, A.; Kumar, P.S.; Jeevanantham, S.; Karishma, S.; Tajsabreen, B.; Yaashikaa, P.R.; Reshma, B. Effective Water/Wastewater Treatment Methodologies for Toxic Pollutants Removal: Processes and Applications towards Sustainable Development. *Chemosphere* **2021**, *280*, 130595. [CrossRef] [PubMed]
8. Czech, B.; Zygmunt, P.; Kadirova, Z.C.; Yubuta, K.; Hojamberdiev, M. Effective Photocatalytic Removal of Selected Pharmaceuticals and Personal Care Products by Elsmoreite/Tungsten Oxide@ ZnS Photocatalyst. *J. Environ. Manag.* **2020**, *270*, 110870. [CrossRef] [PubMed]
9. Gallo-Cordova, A.; Castro, J.J.; Winkler, E.L.; Lima, E., Jr.; Zysler, R.D.; del Puerto Morales, M.; Ovejero, J.G.; Streitwieser, D.A. Improving Degradation of Real Wastewaters with Self-Heating Magnetic Nanocatalysts. *J. Clean. Prod.* **2021**, *308*, 127385. [CrossRef]
10. Iovino, P.; Chianese, S.; Canzano, S.; Prisciandaro, M.; Musmarra, D. Photodegradation of Diclofenac in Wastewaters. *Desalin. Water Treat.* **2017**, *61*, 293–297.
11. Erto, A.; Chianese, S.; Lancia, A.; Musmarra, D. On the Mechanism of Benzene and Toluene Adsorption in Single-Compound and Binary Systems: Energetic Interactions and Competitive Effects. *Desalin. Water Treat.* **2017**, *86*, 259–265. [CrossRef]
12. Salvestrini, S.; Fenti, A.; Chianese, S.; Iovino, P.; Musmarra, D. Electro-Oxidation of Humic Acids Using Platinum Electrodes: An Experimental Approach and Kinetic Modelling. *Water* **2020**, *12*, 2250. [CrossRef]
13. Liu, Z.; Hosseinzadeh, S.; Wardenier, N.; Verheust, Y.; Chys, M.; Hulle, S.V. Combining Ozone with UV and H_2O_2 for the Degradation of Micropollutants from Different Origins: Lab-Scale Analysis and Optimization. *Environ. Technol.* **2019**, *40*, 3773–3782. [CrossRef] [PubMed]
14. Rosenblum, J.S.; Sitterley, K.A.; Thurman, E.M.; Ferrer, I.; Linden, K.G. Hydraulic Fracturing Wastewater Treatment by Coagulation-Adsorption for Removal of Organic Compounds and Turbidity. *J. Environ. Chem. Eng.* **2016**, *4*, 1978–1984. [CrossRef]
15. Sigg, L.; Behra, P.; Stumm, W. *Chimie des Milieux Aquatiques*; Dunod: Paris, France, 2001; ISBN 2-10-004739-6.
16. Rui, L.M.; Daud, Z.; Latif, A.A.A. Treatment of Leachate by Coagulation-Flocculation Using Different Coagulants and Polymer: A Review. *Int. J. Adv. Sci. Eng. Inf. Technol.* **2012**, *2*, 1–4. [CrossRef]
17. Alansari, A.Y. A Comprehensive Study of Drinking Water Coagulation with Aluminum Sulfate. Ph.D. Thesis, The University of North Carolina at Charlotte, Charlotte, NC, USA, 2021.
18. Lu, C.; Xu, Z.; Zhang, Y.; Dong, B. Rapid and Clean Dewatering of Dredged Sediment via Coupling of a Two-Step Flocculation and Geotextile Filtration. *Chem. Eng. J.* **2023**, *480*, 147604. [CrossRef]
19. Jiang, J.-Q. The Role of Coagulation in Water Treatment. *Curr. Opin. Chem. Eng.* **2015**, *8*, 36–44. [CrossRef]
20. Owodunni, A.A.; Ismail, S. Revolutionary Technique for Sustainable Plant-Based Green Coagulants in Industrial Wastewater Treatment—A Review. *J. Water Process Eng.* **2021**, *42*, 102096. [CrossRef]
21. Iwuozor, K.O. Prospects and Challenges of Using Coagulation-Flocculation Method in the Treatment of Effluents. *Adv. J. Chem.-Sect. A* **2019**, *2*, 105–127. [CrossRef]
22. Prokopova, M.; Novotna, K.; Pivokonska, L.; Cermakova, L.; Cajthaml, T.; Pivokonsky, M. Coagulation of Polyvinyl Chloride Microplastics by Ferric and Aluminium Sulphate: Optimisation of Reaction Conditions and Removal Mechanisms. *J. Environ. Chem. Eng.* **2021**, *9*, 106465. [CrossRef]
23. Heddam, S.; Bermad, A.; Dechemi, N. Applications of Radial-Basis Function and Generalized Regression Neural Networks for Modeling of Coagulant Dosage in a Drinking Water-Treatment Plant: Comparative Study. *J. Environ. Eng.* **2011**, *137*, 1209–1214. [CrossRef]
24. Ernest, E.; Onyeka, O.; David, N.; Blessing, O. Effects of pH, Dosage, Temperature and Mixing Speed on the Efficiency of Water Melon Seed in Removing the Turbidity and Colour of Atabong River, Awka-Ibom State, Nigeria. *Int. J. Adv. Eng. Manag. Sci.* **2017**, *3*, 239833. [CrossRef]
25. Abebe, L.S.; Chen, X.; Sobsey, M.D. Chitosan Coagulation to Improve Microbial and Turbidity Removal by Ceramic Water Filtration for Household Drinking Water Treatment. *Int. J. Environ. Res. Public Health* **2016**, *13*, 269. [CrossRef] [PubMed]
26. Ukiwe, L.N.; Ibeneme, S.I.; Duru, C.E.; Okolue, B.N.; Onyedika, G.O.; Nweze, C.A. Chemical and Electro-Coagulation Techniques in Coagulation-Flocculation in Water and Wastewater Treatment—A Review. *J. Adv. Chem.* **2014**, *9*, 2321–2807.
27. Jiang, S.; Li, Y.; Ladewig, B.P. A Review of Reverse Osmosis Membrane Fouling and Control Strategies. *Sci. Total Environ.* **2017**, *595*, 567–583. [CrossRef] [PubMed]
28. Yang, Z.; Gao, B.; Yue, Q. Coagulation Performance and Residual Aluminum Speciation of $Al_2(SO_4)_3$ and Polyaluminum Chloride (PAC) in Yellow River Water Treatment. *Chem. Eng. J.* **2010**, *165*, 122–132. [CrossRef]

29. Andréasson, J. Efficient Coagulant Dosing for Sustainable Water Treatment. An Empirical Study on Reducing Chemical Consumption in Swedish Drinking Water Treatment Plants. Master's Thesis, Chalmers University of Technology, Gothenburg, Sweden, 2023.
30. Sher, F.; Hanif, K.; Rafey, A.; Khalid, U.; Zafar, A.; Ameen, M.; Lima, E.C. Removal of Micropollutants from Municipal Wastewater Using Different Types of Activated Carbons. *J. Environ. Manag.* **2021**, *278*, 111302. [CrossRef] [PubMed]
31. Jackson, G.E.; Letterman, R.D. Granular Media Filtration in Water and Wastewater Treatment: Part 2. *Crit. Rev. Environ. Sci. Technol.* **1980**, *11*, 1–36.
32. Johnson, R.M. Role of Sediment Type, Concentration, and Salinity on Clay Flocs. Master's Thesis, University of South Carolina, Columbia, SC, USA, 2023.
33. Achite, M.; Samadianfard, S.; Elshaboury, N.; Sharafi, M. Modeling and Optimization of Coagulant Dosage in Water Treatment Plants Using Hybridized Random Forest Model with Genetic Algorithm Optimization. *Environ. Dev. Sustain.* **2023**, *25*, 11189–11207. [CrossRef]
34. Shak, K.P.Y.; Wu, T.Y. Coagulation–Flocculation Treatment of High-Strength Agro-Industrial Wastewater Using Natural Cassia Obtusifolia Seed Gum: Treatment Efficiencies and Flocs Characterization. *Chem. Eng. J.* **2014**, *256*, 293–305. [CrossRef]
35. Sheng, D.P.W.; Bilad, M.R.; Shamsuddin, N. Assessment and Optimization of Coagulation Process in Water Treatment Plant: A Review. *ASEAN J. Sci. Eng.* **2023**, *3*, 79–100.
36. Wadkar, D.V.; Karale, R.S.; Wagh, M.P. Application of Cascade Feed Forward Neural Network to Predict Coagulant Dose. *J. Appl. Water Eng. Res.* **2022**, *10*, 87–100. [CrossRef]
37. Khan, S.J.; Deere, D.; Leusch, F.D.; Humpage, A.; Jenkins, M.; Cunliffe, D. Extreme Weather Events: Should Drinking Water Quality Management Systems Adapt to Changing Risk Profiles? *Water Res.* **2015**, *85*, 124–136. [CrossRef]
38. Saalidong, B.M.; Aram, S.A.; Otu, S.; Lartey, P.O. Examining the Dynamics of the Relationship between Water pH and Other Water Quality Parameters in Ground and Surface Water Systems. *PLoS ONE* **2022**, *17*, e0262117. [CrossRef] [PubMed]
39. Azhari, H.E.; Cherif, E.K.; Sarti, O.; Azzirgue, E.M.; Dakak, H.; Yachou, H.; Esteves da Silva, J.C.; Salmoun, F. Assessment of Surface Water Quality Using the Water Quality Index (IWQ), Multivariate Statistical Analysis (MSA) and Geographic Information System (GIS) in Oued Laou Mediterranean Watershed, Morocco. *Water* **2022**, *15*, 130. [CrossRef]
40. Wu, Z.; Wang, X.; Chen, Y.; Cai, Y.; Deng, J. Assessing River Water Quality Using Water Quality Index in Lake Taihu Basin, China. *Sci. Total Environ.* **2018**, *612*, 914–922. [CrossRef] [PubMed]
41. Tahraoui, H.; Belhadj, A.-E. Optimisation de l'Élimination des Micropolluants Organiques. Ph.D. Thesis, University Yahia Fares, Médéa, Algeria, 2021.
42. Tahraoui, H.; Belhadj, A.E.; Moula, N.; Bouranene, S.; Amrane, A. Optimisation and Prediction of the Coagulant Dose for the Elimination of Organic Micropollutants Based on Turbidity. *Kem. Ind.* **2021**, *70*, 675–691. [CrossRef]
43. Tahraoui, H.; Belhadj, A.-E.; Triki, Z.; Boudella, N.R.; Seder, S.; Amrane, A.; Zhang, J.; Moula, N.; Tifoura, A.; Ferhat, R.; et al. Mixed Coagulant-Flocculant Optimization for Pharmaceutical Effluent Pretreatment Using Response Surface Methodology and Gaussian Process Regression. *Process Saf. Environ. Prot.* **2022**, *169*, 909–927. [CrossRef]
44. Singh, B.; Kumar, P. Pre-Treatment of Petroleum Refinery Wastewater by Coagulation and Flocculation Using Mixed Coagulant: Optimization of Process Parameters Using Response Surface Methodology (RSM). *J. Water Process Eng.* **2020**, *36*, 101317. [CrossRef]
45. Rodier, J. *L'analyse de Léau: Eaux Naturelles, Eaux Résiduaires, Eaux de Mer (Chimie, Physico-Chimie, Bactériologique, Biologie)*; Bordas Editions: Paris, France, 2009.
46. Hadadi, A.; Imessaoudene, A.; Bollinger, J.-C.; Bouzaza, A.; Amrane, A.; Tahraoui, H.; Mouni, L. Aleppo Pine Seeds (*Pinus Halepensis* Mill.) as a Promising Novel Green Coagulant for the Removal of Congo Red Dye: Optimization via Machine Learning Algorithm. *J. Environ. Manag.* **2023**, *331*, 117286. [CrossRef]
47. Rodier, J.; Legube, B.; Merlet, N.; Brunet, R. *L'analyse de l'eau—9e éd.: Eaux Naturelles, Eaux Résiduaires, Eau de Mer*; Dunod: Paris, France, 2009; ISBN 978-2-10-054179-9.
48. Anayed, N.; Dhaouadi, L.; Tounsi, W.; Benhmidène, A.; Hamrouni, B.A.; Boukchina, R.; Belkadhi, M.S. Valorisation Des Eaux de Retour des Serres Agricoles Chauffée Par l'eau Géothermique Dans Le Sud Tunisien. *J. Oasis Agric. Sustain. Dev.* **2023**, *5*, 15–20. [CrossRef]
49. Laidani, Y.; Henini, G.; Khatmi, B.; Dellal, A. Evaluation de la Pollution des Eaux du Sous Bassin Versant de l'oued Mina. In Proceedings of the 2nd Colloque International de Chimie, 1–3 December 2009; Volume 3. Available online: https://www.univ-chlef.dz/uhbc/seminaires_2009/commun_19_2009.pdf (accessed on 21 December 2023).
50. Rasekh, A.; Brumbelow, K. Drinking Water Distribution Systems Contamination Management to Reduce Public Health Impacts and System Service Interruptions. *Environ. Model. Softw.* **2014**, *51*, 12–25. [CrossRef]
51. Preoteasa, E.A.; Ionescu-Tirgoviste, C. A Case Study of Conductivity in Mineral Water. *Proc. Rom. Acad. Ser. B* **2016**, *18*, 15–23.
52. Naceradska, J.; Pivokonska, L.; Pivokonsky, M. On the Importance of pH Value in Coagulation. *J. Water Supply Res. Technol.* **2019**, *68*, 222–230. [CrossRef]
53. Drogui, P.; Blais, J.-F.; Mercier, G. Review of Electrochemical Technologies for Environmental Applications. *Recent Pat. Eng.* **2007**, *1*, 257–272. [CrossRef]
54. Chenini, I.; Khemiri, S. Evaluation of Ground Water Quality Using Multiple Linear Regression and Structural Equation Modeling. *Int. J. Environ. Sci. Technol.* **2009**, *6*, 509–519. [CrossRef]

55. Debbih, H.; Naili, B.E. Etude de Qualité des Eaux des Barrages de l'Est Algérien. Master's Thesis, Larbi Ben M'hidi University, Oum El Bouaghi, Algeria, 2015.
56. Ahuja, S.; Larsen, M.C.; Eimers, J.L.; Patterson, C.L.; Sengupta, S.; Schnoor, J.L. *Comprehensive Water Quality and Purification*; Elsevier: Amsterdam, The Netherlands, 2014; Volume 1.
57. Zogo, D.; Bawa, L.M.; Soclo, H.H.; Atchekpe, D. Elimination de La Couleur et de La Turbidité d'une Eau de Surface Par Coagulation-Floculation Au Sulfate d'aluminium: Cas de La Retenue d'eau de l'Okpara En République Du Bénin. *Int. J. Biol. Chem. Sci.* **2010**, *4*, 1667–1675. [CrossRef]
58. Maleiva, L.T.N.; Nugraheni, P.W.; Ramadhan, F.; Putra, L.S.A.; Kusumawardhani, E. Treatment of Dug Well Water in Kubu Raya, West Kalimantan, with Electrocoagulation. In *IOP Conference Series: Earth and Environmental Science—Proceedings of the International Conference on Sustainable Agriculture, Food, and Energy, Istanbul, Turkey, 17 October 2022*; IOP Publishing: Bristol, UK, 2023; Volume 1228, p. 012013.
59. Bendada, K.; Boulakradeche, M.W. Optimisation Des Conditions de Dosage Par Spectroscopie d'absorption Atomique (SAAF et SAAET) Application à La Détermination de la Pollution et de la Bioaccumulation des Métaux Lourds. Master's Thesis, Université des Sciences et de la Technologie Houari Boumediene d'Algérie, Bab Ezzouar, Algeria, 2011.
60. Harrat, N.; Achour, S. Pollution Physico-Chimique Des Eaux de Barrage de La Région d'El Tarf. Impact Sur La Chloration. *Larhyss J.* **2010**, *8*, 47–54.
61. Bora, J.; Saha, I.; Vaibhav, V.; Singh, M.; Malik, S. Combined Applications of Physico-Chemical Treatments in Treatment of Industrial Wastewater. In *Advanced and Innovative Approaches of Environmental Biotechnology in Industrial Wastewater Treatment*; Springer: Berlin/Heidelberg, Germany, 2023; pp. 399–418.
62. Achour, S.; Guesbaya, N. Coagulation-Floculation par le Sulfate d'aluminium de Composés Organiques Phénoliques et de Substances Humiques. *Larhyss J.* **2005**, *4*, 153–168.
63. Tahraoui, H.; Amrane, A.; Belhadj, A.-E.; Zhang, J. Modeling the Organic Matter of Water Using the Decision Tree Coupled with Bootstrap Aggregated and Least-Squares Boosting. *Environ. Technol. Innov.* **2022**, *27*, 102419. [CrossRef]
64. Wei, J.C.; Gao, B.Y.; Yue, Q.Y.; Wang, Y.; Lu, L. Performance and Mechanism of Polyferric-Quaternary Ammonium Salt Composite Flocculants in Treating High Organic Matter and High Alkalinity Surface Water. *J. Hazard. Mater.* **2009**, *165*, 789–795. [CrossRef]
65. Sharuddin, S.S.N. Removal of Ammonium through Adsorptive Coagulation Flocculatiom Process in Drinking Water Treatment. Master's Thesis, Universiti Teknologi Malaysia, Johor, Malaysia, 2015.
66. Lazo, D.E.; Dyer, L.G.; Alorro, R.D. Silicate, Phosphate and Carbonate Mineral Dissolution Behaviour in the Presence of Organic Acids: A Review. *Miner. Eng.* **2017**, *100*, 115–123. [CrossRef]
67. Attour, A.; Touati, M.; Tlili, M.; Amor, M.B.; Lapicque, F.; Leclerc, J.-P. Influence of Operating Parameters on Phosphate Removal from Water by Electrocoagulation Using Aluminum Electrodes. *Sep. Purif. Technol.* **2014**, *123*, 124–129. [CrossRef]
68. Owodunni, A.A.; Ismail, S.; Kurniawan, S.B.; Ahmad, A.; Imron, M.F.; Abdullah, S.R.S. A Review on Revolutionary Technique for Phosphate Removal in Wastewater Using Green Coagulant. *J. Water Process Eng.* **2023**, *52*, 103573. [CrossRef]
69. Bunting, S.W. *Principles of Sustainable Aquaculture: Promoting Social, Economic and Environmental Resilience*; Routledge: Abingdon, UK, 2013; ISBN 1-136-46251-1.
70. Pan, D.; Chen, P.; Yang, G.; Niu, R.; Bai, Y.; Cheng, K.; Huang, G.; Liu, T.; Li, X.; Li, F. Fe (II) Oxidation Shaped Functional Genes and Bacteria Involved in Denitrification and Dissimilatory Nitrate Reduction to Ammonium from Different Paddy Soils. *Environ. Sci. Technol.* **2023**, *57*, 21156–21167. [CrossRef] [PubMed]
71. Chabbi, M.; Nezli, N. Qualité Des Eaux de Consommation de La Vallée d'Oued Souf. 2018. Available online: http://dspace.univ-eloued.dz/handle/123456789/4277 (accessed on 21 December 2023).
72. Lacasa, E.; Cañizares, P.; Sáez, C.; Fernández, F.J.; Rodrigo, M.A. Removal of Nitrates from Groundwater by Electrocoagulation. *Chem. Eng. J.* **2011**, *171*, 1012–1017. [CrossRef]
73. Halkier, T. *Mechanisms in Blood Coagulation, Fibrinolysis and the Complement System*; Cambridge University Press: Cambridge, UK, 1991; ISBN 0-521-38187-8.
74. Bryan, N.S.; Loscalzo, J. *Nitrite and Nitrate in Human Health and Disease*; Springer: Berlin/Heidelberg, Germany, 2017; ISBN 3-319-46187-7.
75. Ding, J.; Wei, L.; Huang, H.; Zhao, Q.; Hou, W.; Kabutey, F.T.; Yuan, Y.; Dionysiou, D.D. Tertiary Treatment of Landfill Leachate by an Integrated Electro-Oxidation/Electro-Coagulation/Electro-Reduction Process: Performance and Mechanism. *J. Hazard. Mater.* **2018**, *351*, 90–97. [CrossRef] [PubMed]
76. Duan, J.; Wang, J.; Guo, T.; Gregory, J. Zeta Potentials and Sizes of Aluminum Salt Precipitates–Effect of Anions and Organics and Implications for Coagulation Mechanisms. *J. Water Process Eng.* **2014**, *4*, 224–232. [CrossRef]
77. Sarkar, S.K.; Saha, M.; Takada, H.; Bhattacharya, A.; Mishra, P.; Bhattacharya, B. Water Quality Management in the Lower Stretch of the River Ganges, East Coast of India: An Approach through Environmental Education. *J. Clean. Prod.* **2007**, *15*, 1559–1567. [CrossRef]
78. Wajih, N.; Alipour, E.; Rigal, F.; Zhu, J.; Perlegas, A.; Caudell, D.L.; Kim-Shapiro, D. Effects of Nitrite and Far-Red Light on Coagulation. *Nitric Oxide* **2021**, *107*, 11–18. [CrossRef]
79. Sharma, N.K.; Goswami, B.; Gajjar, B.; Jain, C.; Soni, D.; Patel, K. Utilization of Amul Dairy Effluent for Agriculture Practices. *Int. J. Environ. Sci.* **2011**, *2*, 22–32.

80. Watson, C.R. Risk-Based Decision Making during Public Health Emergencies Involving Environmental Contamination. Ph.D. Thesis, Johns Hopkins University, Baltimore, MD, USA, 2017.
81. Semerjian, L.; Ayoub, G.M. High-pH–Magnesium Coagulation–Flocculation in Wastewater Treatment. *Adv. Environ. Res.* **2003**, *7*, 389–403. [CrossRef]
82. Bratby, J. *Coagulation and Flocculation in Water and Wastewater Treatment*; IWA Publishing: London, UK, 2016; ISBN 1-78040-749-1.
83. Self, J.R. Domestic water quality criteria. Colorado State University Cooperative Extension. 1996. Available online: https://extension.colostate.edu/docs/pubs/crops/00513.pdf (accessed on 21 December 2023).
84. Tahraoui, H.; Belhadj, A.E.; Hamitouche, A.E. Prediction of the Bicarbonate Amount in Drinking Water in the Region of Médéa Using Artificial Neural Network Modelling. *Kem. Ind.* **2020**, *69*, 595–602. [CrossRef]
85. Chargaff, E. The Coagulation of Blood. *Adv. Enzymol. Relat. Areas Mol. Biol.* **1945**, *5*, 31–65.
86. Tahraoui, H.; Belhadj, A.-E.; Hamitouche, A.; Bouhedda, M.; Amrane, A. Predicting the Concentration of Sulfate (SO_4^{2-}) in Drinking Water Using Artificial Neural Networks: A Case Study: Médéa-Algeria. *Desalin. Water Treat.* **2021**, *217*, 181–194. [CrossRef]
87. Tahraoui, H.; Belhadj, A.-E.; Amrane, A.; Houssein, E.H. Predicting the Concentration of Sulfate Using Machine Learning Methods. *Earth Sci. Inform.* **2022**, *15*, 1023–1044. [CrossRef]
88. Edwards, M. Chemistry of Arsenic Removal during Coagulation and Fe–Mn Oxidation. *J. Am. Water Works Assoc.* **1994**, *86*, 64–78. [CrossRef]
89. Kohl, P.M.; Medlar, S.J. *Occurrence of Manganese in Drinking Water and Manganese Control*; American Water Works Association: Denver, CO, USA, 2006; ISBN 1-58321-504-2.
90. Srinivasan, P.T.; Viraraghavan, T.; Subramanian, K.S. Aluminium in Drinking Water: An Overview. *Water SA* **1999**, *25*, 47–55.
91. Srinivasan, P.T.; Viraraghavan, T. Characterisation and Concentration Profile of Aluminium during Drinking-Water Treatment. *Water SA* **2002**, *28*, 99–106. [CrossRef]
92. Black, A.P. Basic Mechanisms of Coagulation. *J. Am. Water Works Assoc.* **1960**, *52*, 492–504. [CrossRef]
93. Tahraoui, H.; Toumi, S.; Hassein-Bey, A.H.; Bousselma, A.; Sid, A.N.E.H.; Belhadj, A.-E.; Triki, Z.; Kebir, M.; Amrane, A.; Zhang, J.; et al. Advancing Water Quality Research: K-Nearest Neighbor Coupled with the Improved Grey Wolf Optimizer Algorithm Model Unveils New Possibilities for Dry Residue Prediction. *Water* **2023**, *15*, 2631. [CrossRef]

water

MDPI

Article

The Impact of Large-Scale Water Diversion Projects on the Water Supply Network: A Case Study in Southwest China

Kaiwen Song [1], Xiujuan Jiang [2], Tianye Wang [1,*], Dengming Yan [3], Hongshi Xu [1] and Zening Wu [1]

[1] School of Water Conservancy and Transportation, Zhengzhou University, Zhengzhou 450001, China; kevinsong1007@163.com (K.S.)
[2] Yunnan Water Conservancy and Hydroelectric Survey Design and Research Institute, Kunming 650021, China
[3] Yellow River Engineering and Consulting Co., Ltd., Zhengzhou 450000, China
* Correspondence: wangtianye@zzu.edu.cn; Tel.: +86-1669877861

Abstract: The uneven spatial and temporal distribution of water resources has consistently been one of the most significant limiting factors for social development in many regions. Furthermore, with the intensification of climate change, this inequality is progressively widening, posing a critical challenge to the sustainable development of human societies. The construction of large-scale water projects has become one of the crucial means to address the contradictions between water supply and demand. Thus, evaluating the functional aspects of water source network structures and systematically planning the layout of engineering measures in a scientifically reasonable manner are pressing issues that require urgent attention in current research efforts. Addressing this, our study takes the Erhai Lake basin and the surrounding areas in southwest China as the study area and combines landscape ecology and network analysis theory methods to propose a water supply network analysis method that takes into account both structure and node characteristics. Based on this methodology, we analyze the connectivity characteristics of water supply networks in the Erhai region under current (2020) and future (2035) planning scenarios. The results show that there were 215 nodes and 216 links in the water supply network of the Erhai Lake basin in 2020; with the implementation of a series of water conservancy projects, the planned 2035 water supply network will increase by 122 nodes and 163 links, and the connectivity of the regional water network will be significantly improved. Also, we identify some key nodes in the network, and the results show that the water supply network in 2035 will have obvious decentralization characteristics compared with that in 2020. And, based on the network degradation analysis, we find that with the implementation of engineering measures, the resilience of the water supply network will be significantly strengthened by 2035, with stronger risk tolerance. This study extends the quantitative representation of water source network characteristics, which can provide a useful reference for water network structure planning and optimization.

Keywords: water supply system; hydrological connectivity; landscape ecology; network analysis theory; failure risk analysis

1. Introduction

Water resources are among the most precious resources on Earth and represent a critical factor in sustaining human and ecosystem health. However, water resource management has been facing a big challenge in recent years. Population growth and the temporal–spatial mismatch of water resources exert immense pressure on the sustainable supply and quality of water. Particularly, extreme events like droughts and floods are becoming more frequent in the context of climate change, exacerbating the uneven distribution of water resources. In this situation, ensuring reliable water supply has become a key factor limiting economic and social development and ecological health [1,2] in many regions. Therefore, researching the coordination and complementarity between water sources and supply networks has become a crucial issue in water resource management and regulation.

The water resource system is a dual-cycle system that involves the deep coupling of natural hydrological processes and human-controlled management. Natural water network characteristics determine the endowment of water resources in a region. In humid areas, river networks are dense; water systems are well connected; and the spatial and temporal distribution of water resources is relatively balanced. Conversely, arid regions have poor endowment of water resources. Therefore, enhancing the connectivity among water sources and improving the coordination and complementarity among supply networks are essential strategies to address sustainable development issues in arid regions. To settle this, many regions have constructed reservoirs and water diversion projects to strengthen the interconnection and intercommunication between different water sources and supply systems, alleviating water scarcity problems, for example, the Central Valley project [3] in the United States and the construction of large-scale water diversion projects in California [4], the James Bay water diversion project [5] in Canada, the Churchill–Nelson water diversion project [6], and the Sarda Sarova project [7] in India. These projects have enhanced the connectivity of water resources among basins, changed the allocation and management of water resources, effectively improved water resource utilization, and increased the resilience of water supply systems [8]. In order to alleviate the water scarcity problem in northern China, the South-to-North Water Diversion Project has been implemented to transport water from the water-rich Yangtze River basin to the water-stressed North China Plain, connecting water sources across regions. The ecological and environmental benefits of the first phase of the Eastern Route alone have a total value of 6.233 billion RMB [9]. The ecological water transfer projects in the inland river basins of northwest China have alleviated water competition between human society and natural landscapes by coordinating water allocation between upstream and downstream areas, promoting basin ecological restoration [10]. However, it is noteworthy that such inter-basin water transfer practices are not without controversy globally. For instance, in Europe, similar water transfer projects are subject to stringent regulations. The EU Water Framework Directive (WFD), effective since 2000, places particular emphasis on cross-border cooperation in transregional water transfer projects, such as water competition, water rights trading, and cost–benefit analysis. These regulations provide an important reference point for the complexities and potential environmental impacts that must be considered when managing interregional water resources. By building massive water diversion projects, humans are creating "artificial rivers" on Earth [11], which have a profound impact on the global water supply network, alleviating the uneven distribution of water resources in time and space and increasing the availability of water resources [12]. In the future, we may face a world dominated by engineered water, which is the key measure to solve the contradiction between water supply and demand.

To address the exacerbation issue of the uneven temporal and spatial distribution of water resources against the backdrop of climate change and to enhance the resilience of supply systems, the central government of China has elevated the construction of a nationwide water network to a national strategic level [13]. While engineering measures are beneficial to enhancing the connectivity of a water network, the water resource conditions in a region are primarily determined by climate. The key challenge in water network planning and construction is how to match the scale of engineering projects with the level of social development. Therefore, analyzing the characteristics of water sources and supply networks to achieve interconnectivity, coordinated operation and supply, and collaborative prevention and control are scientific prerequisites for optimizing the structure of the water network and enhancing the resilience of the water network system.

The water supply network is a complex network system that combines natural water bodies with artificial engineering. Researchers from different disciplines have proposed various evaluation methods for studying the connectivity of water supply networks, such as graph theory [14,15], landscape analysis [16], and hydrodynamic modeling [17]. However, graph theory simplifies the water network and lacks temporal and spatial dimensions, making it unsuitable for complex terrains. Landscape analysis has strict data requirements and lacks analysis of hydrological processes, making it difficult to handle human interventions. Hydrodynamic modeling requires long-sequence data for parameter calibration and has high complexity and computational demands, and the determination of relevant parameters is not easy. In recent years, with the rapid development of society, artificial networks have become more complex and diverse [18–21], including transportation networks, electrical grids, social networks, etc. [22,23]. Research has shown that the topological structure of networks greatly influences their functionality, particularly certain critical network nodes that play a decisive role in network resilience [18,24,25]. A water supply network has the characteristics of a network, and network theory analysis, as a kind of graph theory, has already been used in research on water supply networks. Network analysis theory is applied to water supply networks, where nodes represent reservoirs, users, and connection points, while edges represent rivers and pipeline facilities [26]. Yazdani and Jeffrey [27] analyzed the vulnerability of water supply networks using the directed graph weighted by pipeline hydraulic capacity. Giudiciani [28] et al. studied the influence of topology on undirected and unweighted graphs in indicators based on network attributes such as connectivity and robustness. Thomas Anchita [29] used calculations of complex network indicators and associated hydraulic criteria and studied hydraulic performance and connectivity under various demand increase scenarios. Meng [30] used stress–strain testing to propose a general mapping framework of network elasticity and topological properties to analyze the key influencing factors of water supply network elasticity. Network analysis theory has also been used to analyze the connectivity among the physical components of the hydrological cycle [31]. In addition, node–node connectivity in network analysis theory has also been used in collaborative research on human–water systems [32], providing a new perspective for water resource management [33]. For large-scale water diversion projects, Erik Porse [8] conducted a connectivity and resilience analysis of California's waterway projects and a network degradation analysis. Xiang [34] proposed a system vulnerability assessment method for large-scale interregional construction projects based on complex network theory and proposed that the vulnerability of environmental and social factors was greater than that of economic factors in major transregional projects. Liu [35] used complex network theory to analyze the large water network project in Shanxi Province, China, including its transmission efficiency and important nodes, and put forward management suggestions. Wang [36] used network theory to analyze the changes in the importance of nodes before and after the construction of water conservancy projects in the Yongding River basin in China and found that water conservancy projects would not only cause the importance of some nodes to decrease but also lead to the importance of some nodes to increase. Network analysis provides a tool for analyzing water supply networks. When using network analysis methods to plan the structural elements of water supply networks, one can systematically understand the dynamics of complex networks, identify key nodes and connections to optimize resource allocation, and enhance the adaptability and resilience of the system. However, for complex networks, topological characteristics may not fully represent the impact of hydraulic performance and pipeline failure [37] and cannot reflect the characteristics of project scale, user demand, and economic benefits. Therefore, for complex water supply networks that combine natural features with artificial engineering, further research is needed to develop suitable analysis methods.

In response to the above-mentioned issues, this study focuses on the Erhai Lake basin in southwest China to explore methods for analyzing the characteristics of both natural and artificial supply networks. The Erhai Lake basin faces a series of problems, including water pollution, imbalanced water supply and demand, frequent droughts [38], and complex water resource management [39]. The Dianzhong Water Diversion Project and the Ludila Hydropower Station Water Resources Comprehensive Utilization Project (referred to as the Ludila Project) are two large-scale water engineering projects in Yunnan Province, China. They help improve water supply in the central part of Yunnan Province, enhance the reliability of water supply for drinking and agriculture, promote local economic and agricultural development, and strengthen water resource cooperation among different regions within Yunnan Province to facilitate regional development and resource sharing.

In order to quantitatively analyze the influence of large-scale water transfer projects on the water supply network of the Erhai Lake basin, it is difficult to select appropriate indicators to reasonably and quantitatively characterize the function of network system analysis; especially for water supply networks, the relevant theories and methods are still being explored. In this study, we combine landscape ecology and network analysis theory to construct a set of quantitative evaluation index systems of water supply networks and analyze the water supply network at the network layer and the node layer, respectively. Firstly, the water supply network of the Erhai Lake basin after the implementation of the project is compared with the existing water supply network of the Erhai Lake basin; the important nodes in the network are selected; and network degradation evaluation is carried out. The study hopes to provide new insights for water resource management and ecological conservation to address environmental challenges and ensure the sustainable development of water resources.

2. Research Area Background

The Erhai Lake basin is located in Dali Bai Autonomous Prefecture, western Yunnan Province, China, located in the three major water systems of the Jinsha River, the Lancang River, and the Red River. It encompasses Erhai Lake and the surrounding lakes, rivers, and mountains. The Erhai Lake basin belongs to a subtropical monsoon climate, with warm and humid summers and relatively low temperatures in winter. Precipitation is higher in the summer season, which is the main rainy season, while the winter season is relatively dry. Precipitation varies with elevation and geographic location, with mountainous area typically receiving more rainfall, while basins and low-lying areas receive relatively less. The main economic source in the Erhai Lake basin is agriculture, with agriculture accounting for a significant portion of the total economic output. Tourism and fisheries also play crucial roles in the local economy. Erhai Lake is the largest freshwater lake on the Yunnan Plateau and is one of the important freshwater resources in Southwest China. It not only provides essential freshwater resources but also holds a special position in local culture and ecosystems. The study area covers 20,061 km^2, with a total population of 2.825 million. The annual average rainfall is 995 mm; the annual average water resources are 5.858 billion m^3; and the per capita water resources are 2074 m^3, which is far lower than the provincial average. The actual total water consumption in 2019 was 1.162 billion cubic meters, of which 97 million cubic meters were for urban domestic use; 69 million cubic meters, for rural domestic use; 111 million cubic meters, for industrial use; and 885 million cubic meters, for agricultural irrigation. These accounted for 10%, 7%, 9%, and 74% of the total water consumption, respectively. Because the study area is located in a mountainous area, the distribution of water resources is uneven in time and space, which is very inconsistent with the distribution of urban population, cultivated land, etc., and the utilization of groundwater is lower. Therefore, compared with the surrounding areas, many water storage facilities have been built in the study area for water storage, including many small, independent water storage facilities, such as ponds, etc., and there are 1573 water storage facilities in the study area. In view of the fact that many small reservoirs and small dams only exist due to water storage projects and are not connected with the water system, this paper only considers large- and medium-sized water storage facilities and some small reservoirs. The specific distribution of reservoirs and rivers is shown in Figure 1.

When the highest operating water level of Erhai Lake is 1966 m, the lake area is 252 km^2, and the lake capacity is 2.92 billion m^3. The lowest operating water level is 1964.3 m, and the variation between the legal maximum and minimum water levels is 1.7 m; the lake capacity is adjusted accordingly to 427 million m^3. The average annual natural runoff in the Erhai Lake basin is 1.145 billion m^3, and the utilization rate of water resources reaches 54.3% after deducting evaporation loss from the lake surface; the problem of water shortage in the basin is becoming increasingly prominent. Due to the constraints of water resources, it is important to maintain the basic balance of water volume in Erhai Lake as a last resort. Under the background conditions of developed planting industry, abundant tourism resources, and distinct dry and wet seasons in the Erhai Lake basin, agricultural non-point source pollutants are concentrated into the lake with rainfall runoff in June–October, which is the key environmental driving factor for the water quality exceeding the standard in the rainy season in the Erhai Lake in recent years, resulting in the phenomenon whereby the rise and fall in the Erhai water level and the change in water quality concentration in the lake area are basically synchronized.

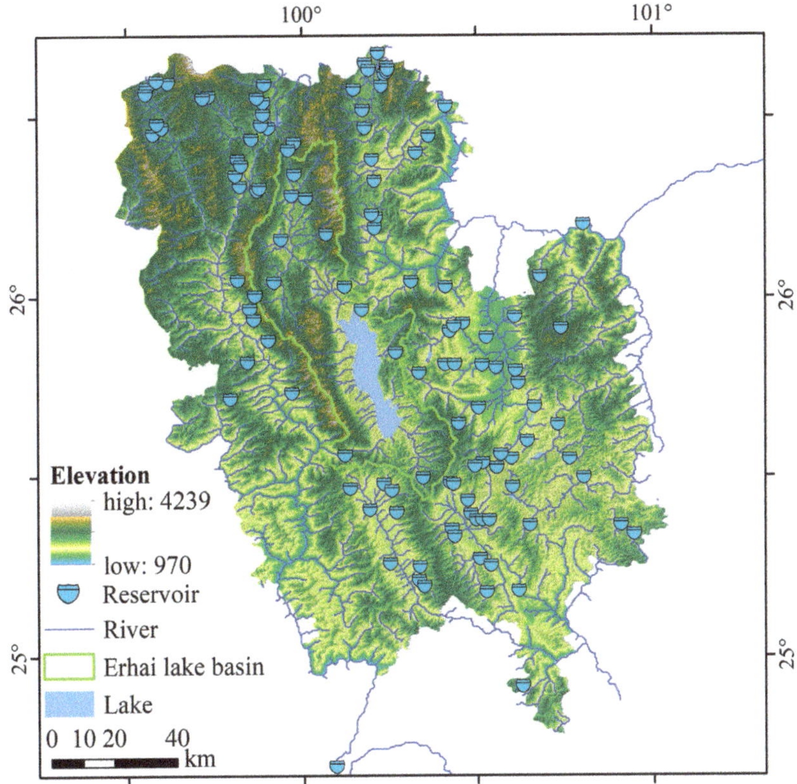

Figure 1. The Erhai Lake basin.

The Dianzhong Water Diversion Project is a significant hydraulic engineering project in Yunnan Province, China, with substantial implications. It diverts the abundant water resources from Dianchi Lake to Kunming City and its surrounding areas, meeting the urgent need for water supply and irrigation. Its total water supply is 342 million m^3, of which 74 million m^3 is for urban life; 42 million m^3, for the industry; and 225 million m^3, for agricultural irrigation. The current population of the receiving area is 2.02 million, and the irrigated area is 80,666 km^3. This project has not only essential implications in terms of technology and infrastructure but also profound impacts on ecology, society, and the economy. Particularly, in the Erhai Lake basin, the implementation of this project will replace Erhai Lake as the water source for the Binchuan and Xiangyun irrigation areas, alleviating the water stress in the Erhai Lake basin while preserving its ecological environment.

The Ludila Hydropower Station Water Resources Comprehensive Utilization Supporting Project is another crucial hydraulic engineering project in Yunnan Province, China. This project not only provides electricity resources to the region but also achieves the rational distribution of water resources through water regulation. The Ludila Project facility can divert water from the Jinsha River to Binchuan County and Xiangyun County, increasing the water supply for these areas. The Ludila Project facility is designed to supply 189.71 million m^3 of water to the Binchuan irrigation district and 138.61 million m^3 to the Xiangyun irrigation district. Additionally, the Ludila Project and Dianzhong Water Diversion Project facilities can be interconnected, serving as backup water sources for each other. Before the completion of the Dianzhong Water Diversion Project, the Ludila Project can solve the water demand of urban life, industry, and efficient agricultural development in Binchuan County, which has recently built the industrial and economic center of Dali Prefecture in Xiangyun County. After the completion of the Dianzhong Water Diversion Project, the Ludila Project can further guarantee the urban living and industrial water demand of Xiangyun County to sustain the industrial and economic center of Dali Prefecture in the long term and, at the same time, create conditions for the Dianzhong Water Diversion Project to supply water to the central Yunnan urban agglomeration, with Kunming as the core.

3. Materials and Methods

In this study, we use the combination of two methods to analyze the hydrological connectivity of the Erhai Lake basin. Landscape ecology pays attention to the analysis of water networks at the network level, and network analysis theory can analyze both the network characteristics and the node characteristics of water networks. First, we use the hydrological loop degree, node connectivity rate, and network connectivity in landscape ecology, and the average path length and clustering coefficient in network analysis theory to analyze the connectivity characteristics of the entire water supply network. Then, we apply degree centrality, betweenness centrality, closeness centrality, and improved K-shell decomposition method in network analysis theory to analyze the importance of nodes in the water supply network. The combination of these two methods not only helps us to fully understand the connectivity characteristics of the whole network but also analyzes the connectivity function of each node in the network and simulates the fault changes after node deletion.

This study first evaluates the connectivity of the water system in the Erhai Lake basin in 2020 and 2035 to analyze the impact of the Dianzhong Diversion Project, the Ludila Project, and other water projects. Then, the node importance of the water supply network is analyzed to select important nodes in the system for 2020 and 2035. The study also simulates the scenario where the water supply network is damaged by removing certain important nodes.

A water resource system is a network that includes water sources, links, and users (both natural and artificial), which can be mathematically represented as graphs with vertices (nodes) connected by edges (links). Based on engineering information on the Erhai Lake basin and the generated dam–river information provided by the Yunnan Water Conservancy and Hydroelectric Survey Design and Research Institute, Kunming, China, we first established the topological relationship network model of the water supply network of the Erhai Lake basin, as shown in Figures A1 and A2, and node information is shown in Tables A1 and A2. Then, the constructed network model was imported into Cytoscape software v.3.8.1 [40], and the Network Analyzer module was utilized to calculate some indicators. Cytoscape is open-source software for biological network analysis and visualization. It provides powerful tools and functionalities that help biologists, bioinformaticians, and systems biologists study and understand the structure and function of biological networks. Cytoscape has a wealth of network analysis tools for identifying topological features, centrality metrics, network modules, etc. [41], which can be employed for calculating network metrics in this research. However, there are some limitations in the process of model construction. We only consider surface water sources and engineering nodes, ignoring groundwater. Additionally, this is a purely topological study, lacking hydraulic indicators.

3.1. Hydrological Connectivity

This study evaluates the structural connectivity of the water system in the region's river network based on the "node-edge" relationship in landscape ecology [42] and graph theory [43]. Three hydrological connectivity evaluation indicators are employed: hydrological loop degree (α), node connectivity rate (β), and network connectivity (χ) [44–46]. Additionally, the average path length and clustering coefficient from network analysis theory are also introduced [47] to quantify the hydrological connectivity of the water system. The water system connectivity evaluation indicators are established as follows:

- Hydrological loop degree: This indicator is used to quantify the degree to which nodes in a river network form loops, reflecting the capacity of each node in the river network to exchange material and energy [45].

$$\alpha = (n - v + 1)/(2v - 5) \tag{1}$$

- Node connectivity rate: Used to quantify the ease or difficulty with which nodes in a river network connect with other nodes, reflecting the ability of each node in the river network to establish and maintain connections within the water system [46].

$$\beta = v/n \tag{2}$$

- Network connectivity: It represents the ratio of the existing number of connections among corridors within the river network to the maximum possible number of connections, reflecting the strength of connectivity between river network systems and their capacity for water transport [44].

$$\chi = n/3(v - 2) \tag{3}$$

where "n" represents the number of connecting links in the river network hydrological model and "v" represents the number of nodes.

- Average path length: It refers to the average shortest path length between nodes "i" and "j" in the graph. This metric measures the "closeness" of the graph and can be used to understand the speed of flow of certain elements in this network. A smaller average path length indicates higher efficiency of water transfer and complementarity in the water supply network [48].

$$\mathrm{L} = \frac{1}{\mathrm{n}(\mathrm{n}-1)} \sum\nolimits_{\mathrm{i} \neq \mathrm{j}} \mathrm{d_{ij}} \tag{4}$$

where "$\mathrm{d_{ij}}$" represents the distance between node "i" and node "j" and "n" stands for the number of nodes in the network.

- Clustering coefficient: It reflects the density and clustering of connections among nodes in a network [49].

$$CC(n) = \frac{2R_n}{k_n(k_n - 1)} \tag{5}$$

$$CC = \frac{\sum_{n=1}^{n} CC(n)}{n} \tag{6}$$

n: node.
CC(n): clustering coefficient of node *n*.
CC: clustering coefficient of the entire network.
R_n: number of relationships (triangles counted) among n's neighboring nodes.
K_n: number of first-order neighboring nodes of *n*.

3.2. Node Importance

A water network is the physical carrier of water circulation and water resource allocation. As the hub of a river, the node of a water network plays an important role in the connection of a water system. Node importance refers to the importance of nodes in the entire water supply network. It is classified according to different network attributes of nodes. In this study, node importance is weighted with four indicators: degree centrality, betweenness centrality, closeness centrality, and improved K decomposition in network analysis theory. The higher the value, the higher the importance of nodes in the entire water supply network. Identifying key nodes in complex water networks is of great significance to the comprehensive planning and management of water networks.

3.2.1. Node Importance Evaluation Index

Network analysis is a commonly used method in the study of transportation, electricity, and other networks. When applied to analyze water supply networks, a water supply network can be generalized as a network consisting of water sources, links, and users, which can be mathematically expressed as a set of vertices (nodes) connected by edges (links). This approach not only considers the relationships among various components of the water resource system but also reveals key nodes and critical paths in the network. In terms of studying key nodes in the network, Liu [35] defined node importance by four dimensions: local properties, global properties, propagation properties, and network position. Wang [50] pointed out the characteristics and application scope of these four indicators. When [51] conducted node importance analysis, Schick not only considered node connectivity and centrality but also added social functions characterized by indicators such as node population size and reproductive rate to quantify the node's conservation value from a social perspective. Segurado [52] used graph theory methods to identify obstacles that affect structural connectivity in a watershed, thereby determining important nodes that can improve node connectivity and their connection modes. Bodin [53] established a joint connectivity index based on fish population protection value, habitat area, and connectivity relationships to prioritize nodes that need protection. For specific problems, degree centrality, betweenness centrality, closeness centrality, Laplacian operator, and other indicators are often used to characterize node importance from different perspectives [54,55]. When evaluating the importance of nodes in the water network, it is necessary to first select the existing evaluation indicators. In the selection process, the principles of reasonability, comprehensiveness, and ease of operation should be followed. Due to the limitation of evaluating node importance using a single indicator, this study selected four indicators, namely, degree centrality, betweenness centrality, closeness centrality, and improved K-core decomposition, in the dimensions of local properties, propagation properties, global properties, and network position, to evaluate the importance of nodes in the Erhai Lake basin water network.

- Degree centrality: Degree centrality is the most direct measure of node centrality in network analysis. The degree centrality value represents the ability of a network node to connect with its neighboring nodes. Higher degree centrality indicates a higher importance of the node in the network, meaning that the water source is more important in the water supply network. The formula for calculating degree centrality for a node is as follows:

$$DC_i = \frac{k_i}{N-1} \tag{7}$$

  where k_i represents the number of existing edges connected to node i and $(N - 1)$ represents the number of edges through which node i is connected to all other nodes.

- Betweenness centrality: Node betweenness refers to the number of shortest paths passing through a node in a network. The higher the betweenness centrality, the more paths pass through that node in the water supply system network, indicating that the node has stronger hydraulic connectivity and is more important in the network. The formula for calculating betweenness centrality for a node is as follows [56]:

$$BC_i = \frac{1}{(n-1)(n-2)/2}\sum_{s\neq i\neq t}\frac{n_{st}^i}{g_{st}} \tag{8}$$

where n_{st}^i represents the number of paths, being the shortest paths, passing through node I; g_{st} represents the number of shortest paths connecting s and t; and n is the number of nodes in the network.

- Closeness centrality: Closeness centrality is used to measure the ability of a node to influence other nodes through network connections, i.e., the impact of one water source on other water sources in a water supply system network. The closeness centrality (CC_i) of a node is calculated as follows [57]:

$$CC_i = \frac{1}{d_i} \tag{9}$$

$$d_i = \frac{1}{N-1}\sum_{j=1}^{N} d_{ij} \tag{10}$$

where "d_i" represents the average distance from node i to all other points and the reciprocal of the average distance is the closeness centrality.

- Network position—improved K-shell decomposition: The importance of a node in the overall network is determined by its position in the network. The K-shell method [58] can be used to measure the positional attributes of nodes, as indicated by the Ks metric. According to the K-shell method, nodes with degrees lower than or equal to k are sequentially removed from the network, resulting in the Ks value for each node. The procedure of the K-shell method is as follows: decompose the network graph into K-shells, where the maximum subnetwork in S = (G,E | G) with degrees greater than or equal to k is the K-core; nodes with Ks = k are the k-shell. However, when applying the K-shell method to evaluate node importance in the water network of the Erhai Lake basin, the evaluation results were not satisfactory, as all nodes had a Ks value of 1. Therefore, this study adopts an improved K-shell method (IKs).

The procedure of the improved K-shell method is as follows: Let the original global network be T_0. For the node with the smallest degree in T_0, set $Ks_1 = 1$. Remove the node with the smallest degree in T_0, resulting in subgraph T_1. For the node in T_1 with the smallest degree, set $Ks = Ks_1 + 1$. Repeat this process until all nodes are removed. After the k-th node is removed, for nodes in the subgraph with a degree of 0, set their Ks value of k + 1. The improved K-shell method can reflect both the global network position attributes of nodes and the local differences among nodes.

3.2.2. Standardization of Metrics

Due to the measurement units and scales of the various indicators being different, it is difficult to conduct a comprehensive analysis. To facilitate comparisons of the importance of nodes under different evaluation systems, this study standardizes each metric as shown in the formula below:

$$P_i = \frac{Q_i}{max\{Q_i, i\in\{1,2,\ldots N\}\}}, Q\in\{DC, BC, CC, IK_s\} \tag{11}$$

where Q_i represents the original evaluation metric, and P_i represents the standardized evaluation metric.

3.2.3. Comprehensive Node Importance Assessment

This study conducts a comprehensive evaluation of node importance using the analytic hierarchy process (AHP). The weights of the DC, BC, CC, and IKs indicators in the evaluation system are calculated using the AHP method.

The three-scale method (0, 1, 2) is used to compare each indicator pairwise and establish a comparison matrix. Since DC reflects the least abundant global topological structure information, it is considered relatively less important compared with other indicators. BC reflects the connectivity of nodes in the network, while CC reflects the degree of proximity between nodes and the network center. Both indicators characterize the global attributes of nodes and are considered equally important. IKs not only characterizes the global properties of nodes but also reflects local characteristics, making it more important than other indicators. Based on the assigned formulas below, the importance evaluation indicators for each node are obtained as shown in Table 1.

$$A = (a_{ij}) = \begin{cases} 2, \text{Indicator i is more important than indicator j.} \\ 1, \text{Indicator i is equally important as indicator j}; \mathrm{t_i} = \sum_{j=4}^{4} \mathrm{a_{ij}} \\ 0, \text{Indicator j is more important than indicator i.} \end{cases} \quad (12)$$

Table 1. Comparison of node importance evaluation indicators.

A	DC	CC	BC	IKs	t_i
DC	1	0	0	0	1
CC	2	1	1	0	4
BC	2	1	1	0	4
IKs	2	2	2	1	7

We construct a judgment matrix using the range method:

$$E = (E_{ij}) = \begin{bmatrix} e & DC & CC & BC & IK_s & M_i & W_i & W \\ DC & 1 & 1/3 & 1/9 & 1/9 & 1/9 & 1/3 & 0.0625 \\ CC & 3 & 1 & 1/3 & 1/3 & 1/3 & 1 & 0.1875 \\ BC & 3 & 1 & 1/3 & 1/3 & 1 & 1 & 0.1875 \\ IK_s & 9 & 3 & 1 & 1 & 81 & 3 & 0.2625 \end{bmatrix} \quad (13)$$

where $\mathrm{T} = \max(\mathrm{t_1} \ldots, \mathrm{t_4}) - \min(\mathrm{t_1} \ldots, \mathrm{t_4})$, $\mathrm{E_{ij}} = \mathrm{E_t^{t_i - t_j/T}}$, $\mathrm{E_t} = 9$, $\mathrm{M_i} = \prod_{j=1}^{4} \mathrm{e_{ij}}$, $\mathrm{W_i} = \sqrt[4]{\mathrm{M_i}}$, and $\mathrm{W} = \mathrm{W_i} / \left(\sum_{i=1}^{4} \mathrm{W_i}\right)$.

Upon verification, consistency ratio $CR < 1$, indicating that the judgment matrix passes the consistency test. The weight vector for each indicator is as follows:

$$W = [0.0625, 0.1875, 0.1875, 0.5625] \quad (14)$$

Node importance is

$$NI = 0.0625DC + 0.1875CC + 0.1875BC + 0.5625IK_s \quad (15)$$

4. Results

4.1. Analysis of Changes in Water Network Structural Connectivity

Based on Table 2 above, the following changes in the water network structure in the Erhai Lake basin are observed after the construction of the Dianzhong Water Diversion Project, the Ludila Project, and a series of reservoirs for the period from 2020 to 2035.

Table 2. Water network structural connectivity.

Indicator	n	v	α	β	χ	L	CC
2020	215	216	0.047	1.005	0.338	18.423	0.006
2035	337	379	0.643	1.125	0.377	11.680	0.024

Change in water network characteristics: From 2020 to 2035, significant changes occur in the water network characteristics of the Erhai Lake basin. The number of nodes increases from 215 to 337, and the number of links increases from 216 to 379, indicating a notable increase in the complexity and connectivity of the water network.

Increase in hydrological loopiness: Hydrological loopiness increases from 0.047 to 0.643, indicating increased complexity and diversity in the hydrological loops. This could be attributed to the construction of new water diversion projects and reservoirs, which results in more intersections of water sources and pathways.

Improvement in node connectivity: Node connectivity increases from 1.005 to 1.125, indicating a closer connection among different nodes within the basin and more efficient flow and distribution of water resources.

Changes in network connectivity: Hydrological connectivity slightly increases from 0.338 to 0.377, indicating stronger connections among different hydrological systems within the basin, potentially contributing to a more balanced distribution of water resources.

Reduction in characteristic path length: The characteristic path length decreases from 18.423 to 11.680, indicating that water resources' transmission paths within the basin become, making it easier for water to flow from one location to another.

Increase in clustering coefficient: The clustering coefficient increases from 0.006 to 0.024, indicating that by 2035, water resource distribution within the basin will tend to aggregate in certain specific areas rather than being dispersed throughout the entire basin.

In summary, significant changes will occur in the water network characteristics of the Erhai Lake basin by 2035, with profound impacts on the distribution and flow of water resources due to the construction of new water diversion projects and reservoirs. These findings have significant reference value for watershed water resource management and planning for future water resource utilization. To better understand the impacts of these changes and the challenges that need to be faced in the future, further in-depth research is still needed.

4.2. Node Importance Analysis

4.2.1. The Importance of Nodes in the Year 2020

Following calculations, node importance in the Erhai Lake basin in the 2020 scenario is shown in the following Figures 2 and 3. The names corresponding to the Cytoscape graph and node numbers can be seen in Figure A1 and Table A1.

The top ten nodes in terms of importance in 2020 in the Erhai Lake basin are as follows: Xiaoguan Village reservoir, Erhai Lake, Pindianhai reservoir, Qinghai Lake reservoir, Zhonghe 0, Mici River 1, Zhonghe 1, Er Dian Qing, Hunsu Lake reservoir, and Zhonghe-Yupao River. These include one engineering node, one water body, four reservoirs, and four river nodes.

In the 2020 scenario, the Xiaoguan Village reservoir is considered one of the most important nodes in the Erhai Lake basin. The Xiaoguan Village reservoir, Pindianhai reservoir, Hunsu Lake reservoir, Qinghai Lake reservoir, and other reservoirs collectively form a connected water diversion system. The Xiaoguan Village reservoir not only supplies water but also stores water to meet downstream the Pindianhai reservoir's and the Hunsu Lake reservoir's water storage needs. The Pindianhai reservoir and Hunsu Lake reservoir can also supply water to the Qinghai Lake reservoir, forming a jointly operated Xiaoguan Village water source system, with the Xiaoguan Village reservoir at its core. The Xiaoguan Village reservoir plays a crucial role in water resource storage and distribution, storing water during the dry season to alleviate water pressure for usage and releasing water during the rainy season to prevent flooding disasters. It holds significant importance in watershed water resource management.

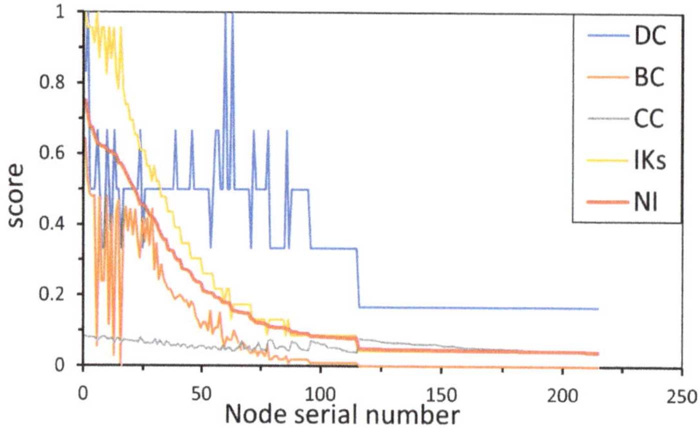

Figure 2. Node importance in 2020 scenario.

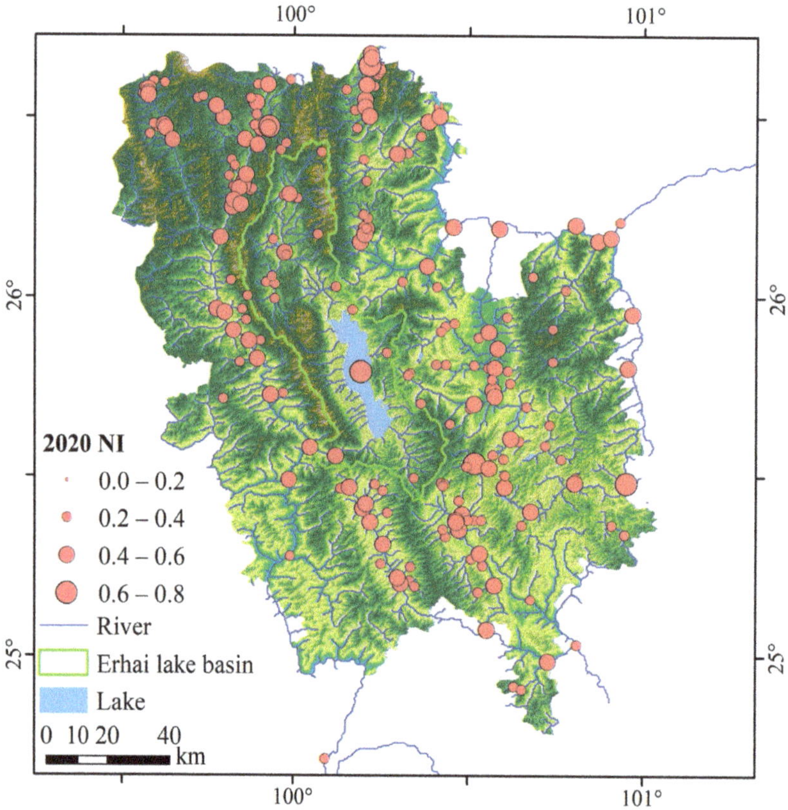

Figure 3. Spatial distribution of node importance in 2020 scenario.

Erhai Lake, as an important water body node in the basin, has significant water storage capacity and plays a critical role in meeting water resource demand and supply.

Its water quality and quantity directly impact the sustainability of the ecosystem and society. Erhai Lake provides conditions for the existence of multiple ecosystems and species and plays an important role in maintaining the ecological balance of the watershed. Its importance is closely related to ecological protection, biodiversity maintenance, and wetland function restoration.

River nodes such as Zhonghe 0, Mici River 1, Zhonghe 1, and Zhonghe-Yupao River have high importance. These nodes are convergence points for different water bodies and water flows and typically have complex water resource flow networks. They enhance the connectivity of water resources among different regions within the watershed, and these nodes play a crucial role in the transmission and distribution of water resources, thus holding high importance in the water resource network of the watershed.

Er Dian Qing, as an important node, is the entry point for the Yin'er Diversion Project. This project aims to provide water from Erhai Lake to fulfill the irrigation needs of the Binchuan irrigation area. It is also the intersection point for the future Ludila Project and Qinghai Lake Phase II Project, playing an important role in water supply within the watershed.

Overall, the water network characteristics of the Erhai Lake basin will undergo significant changes by 2035 due to the construction of new water diversion projects and reservoirs. These findings have important implications for watershed water resource management and future water resource utilization planning, and further research may be needed to better understand the impacts of these changes and future challenges.

4.2.2. The Importance of Nodes in the Year 2035

Following calculations, node importance in the Erhai Lake basin in the 2035 scenario is shown in the following Figures 4 and 5. The names corresponding to the Cytoscape graph and node numbers can be seen in Figure A2 and Table A2.

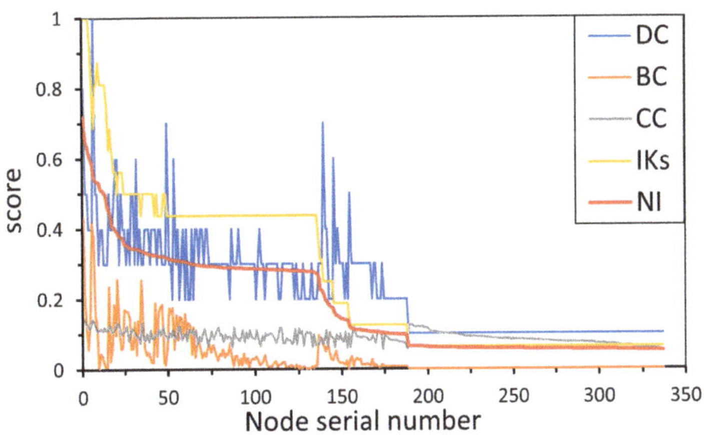

Figure 4. Node importance in 2035 scenario.

The top ten nodes in terms of importance are Er Dian Qing, Da Yin Dian Node, Dian Lu Er, Lu Di La Phase II Node, Lu Di La Bin Chuan Node, Dian Zhong 1 Node, Erhai Lake, Lu Di La East Line Project, Sang Yuan River–Jinsha River, and Xian E reservoir. There are a total of seven engineering nodes, one reservoir, one water body, and one river intersection in this list.

Erhai Lake, as a crucial water body in the basin, will continue to hold high importance even after the construction of the projects, maintaining its significant impact on the basin's water resources and ecosystem.

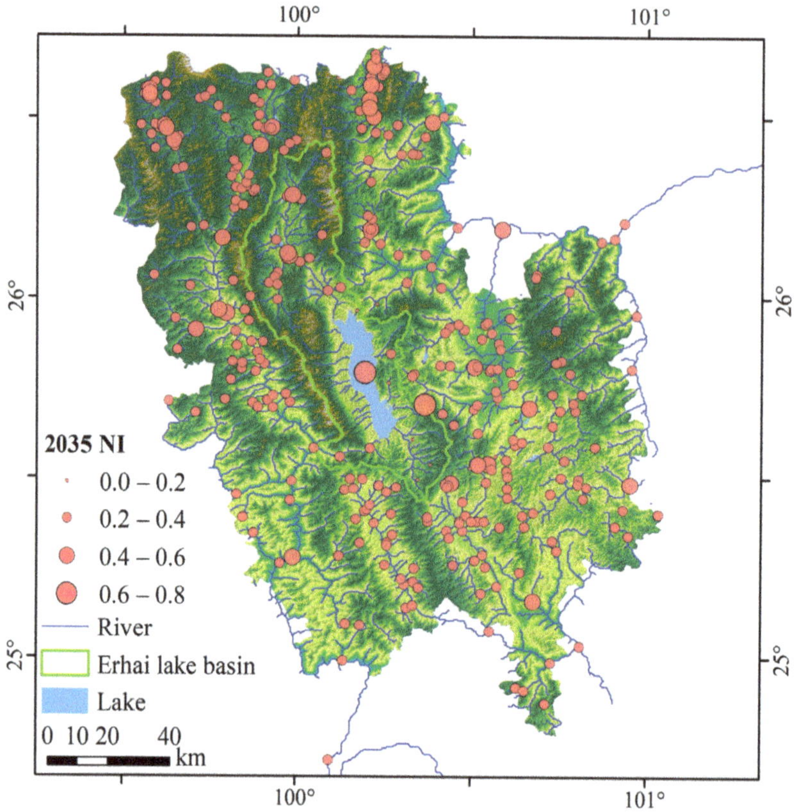

Figure 5. Spatial distribution of node importance in 2035 scenario.

With the construction of these projects, seven of the ten important nodes are engineering nodes related to the Dian Zhong Water Diversion Project or the Lu Di La Project, which proves that these projects will largely improve the connectivity of the water system in the Erhai basin. The Dian Zhong Water Diversion Project and Lu Di La Project will not only introduce external water sources but also establish interconnectivity among various projects, such as the Dian Zhong Water Diversion Project, Lu Di La Project, Qinghai Lake Project, and Yinyu Ruins Project. This will significantly enhance the connectivity of the water supply network and facilitate balanced water supply throughout the entire watershed.

The Xian E reservoir is a medium-sized reservoir in Binchuan County which not only connects the Yimin Lake reservoir and the Sangyuan River but also the Binchuan node of the Lu Di La Project, playing a crucial role in supplying water to the Binchuan irrigation area.

The Sang Yuan River–Jinsha River node serves as an important river confluence point and a water intake node for the Lu Di La East Line Project. Water taken from this node can be transported through the Lu Di La East Line Project to meet water demands within the basin. Simultaneously, it can flow naturally back to this node through the Sang Yuan River, facilitating water recycling. Therefore, it holds high importance.

Compared with 2020, the number of engineering nodes ranked in the top ten in terms of importance significantly will increase in 2035. The number of reservoir nodes and river nodes will decrease. This indicates that newly constructed water diversion projects have remarkable significance in the water supply network, making the basin water network larger and more complex, increasing water network connectivity, and improving resilience against risks. Nodes of different types are considered highly important because they play a critical role in water resource management, ecological conservation, flood control, and water resource distribution. Their high importance reflects their contributions and impacts on the overall water resource system within the basin. Therefore, these nodes require special attention and management to ensure the sustainable utilization of water resources and the sustainable development of the basin.

4.3. Network Degradation Analysis

When the inflow points of the Erhai Lake basin water system, control hubs, and hydraulic engineering nodes are damaged, the connectivity of the entire water network decreases. The impact of the destruction of nodes with different levels of importance on the water supply network's efficiency varies. This article selectively removes some nodes based on their importance in the 2020 and 2035 water supply networks to analyze the changes in connectivity when these nodes are damaged.

4.3.1. Changes in the 2020 Water Network after Node Losses

Reduction in the number of nodes and links: After removing the selected nodes, the total number of nodes in the water network decreased by two–three units, and the number of links decreased by two–six units. This is likely because these nodes have more connections in the network, and their removal results in the breakage of some links. As a result, some secondary nodes lose those connections, leading to a decrease in the number of nodes. The specific changes of network characteristics are shown in Table 3.

Table 3. Changes in the water supply network after the removal of some nodes in 2020 scenario.

Removed Nodes	n	v	α	β	χ	L	CC
No nodes	215	216	0.047	1.005	0.338	18.423	0.006
Xiaoguan Village reservoir	213	211	0.007	0.991	0.338	9.524	0.000
Erhai Lake	211	210	0.005	0.995	0.336	14.558	0.000
Pindianhai reservoir	214	214	0.002	1.000	0.334	18.481	0.006
Qinghai Lake reservoir	214	213	0.005	0.995	0.336	13.995	0.010
Zhonghe 0	213	213	0.002	1.000	0.334	13.966	0.010
Micai River 1	213	212	0.005	0.995	0.336	18.604	0.000
Zhonghe 1	213	213	0.002	1.000	0.334	13.950	0.010
Er-Dian-Qing	213	213	0.002	1.000	0.334	14.591	0.000
Hunshuihai reservoir	214	214	0.002	1.000	0.334	18.481	0.006
Zhonghe-Yupaojiang	213	212	0.005	0.995	0.336	13.959	0.010

Changes in hydrological loopiness: The hydrological system experienced a significant decline in overall water network cyclicality, indicating a reduction in the circular structure. This indicates that these nodes play an important role in forming circular connections within the water network, and the disruption of these nodes results in a substantial decrease in the number of circular patterns.

Decreased node connectivity: The connectivity rate of the nodes experienced a slight decline, suggesting a minor weakening of the interconnectivity among the nodes upon the removal of each of these nodes. This could be attributed to the critical role these nodes play in maintaining the connections between them.

Slight decrease in network connectivity: There was a slight decrease in the connectivity of the water system, with minimal changes being observed. This suggests that the overall connectivity of the entire water system remains relatively stable, although there is still a slight impact when certain nodes are removed.

Changes in characteristic path length: After removing the Xiaoguan Village reservoir, the characteristic path length decreased from 18.423 to 9.524, indicating a significant reduction in the average path length between nodes. This is because the Xiaoguan Village node is an important intermediate node, and its removal makes the transmission efficiency among nodes more direct and rapid. However, after removing the Mi Ci River 1 node, the characteristic path length increased from 18.423 to 18.604, resulting in a slight increase in network transmission efficiency. This node is located at the intersection of multiple paths, and its removal leads to longer paths that previously passed through this node, thus causing an increase in the average path length.

Significant change in clustering coefficient: The significant decrease in the clustering coefficient from 0.006 to 0 indicates a major change in the network structure. The clustering coefficient measures the density of connections among a node's neighbors, so the decrease from 0.006 to 0 means a significant reduction in the density of connections among a node's neighbors. These nodes may have multiple neighbor nodes in the network, and their removal results in the loss of connections for other nodes associated with them, leading to a sparser network of connections and affecting the clustering coefficient of the entire network.

4.3.2. Changes in the 2035 Water Network after Node Losses

In the 2035 scenario, the water supply network in the Erhai Lake basin exhibited different response patterns, with a range of changes in the number of nodes and links, mainly due to the evolution and adjustments in the network structure, making certain nodes more critical components with multiple links. The changes in the water system's degree of connectedness were more significant compared with the 2020 scenario, indicating an increased reliance on key nodes within the network. Nonetheless, overall water system connectivity and node connectivity saw relatively modest changes, reflecting the resilience and stability of a mature network. However, changes in the clustering coefficient indicate significant adjustments in the local structure of the network, with variations in the density of connections among neighboring nodes, reflecting the network's dynamic evolution in adapting to new pressures and challenges. The changes in characteristic path length were relatively small compared with the 2020 scenario, with the greatest increase being observed after the removal of the Erhai Lake node, signifying a significant reduction in network transmission efficiency. This is because the Erhai Lake node serves as an intersection point for multiple pathways, connected to various water diversion projects such as the Erhai-to-Binchuan Water Diversion Project, Dianzhong Water Diversion Project, and Ludila Project. Its removal leads to longer pathways that were originally routed through this node, resulting in an increase in characteristic path length. The specific changes of network characteristics are shown in Table 4.

Table 4. Changes in the water supply network after the removal of some nodes in 2035 scenario.

Remove Nodes	n	v	α	β	χ	L	CC
No nodes	337	379	0.643	1.125	0.377	11.680	0.024
Er-Dian-Qing	333	371	−0.050	1.114	0.300	13.850	0.024
Dayingdian Node	336	374	−0.050	1.113	0.300	12.170	0.024
Dianlu Er	336	374	−0.050	1.113	0.300	11.862	0.024
Ludi La Phase Two Node	336	375	−0.051	1.116	0.299	11.833	0.024
Ludi La Binchuan Node	336	375	−0.051	1.116	0.299	11.878	0.021
Dianzhong 1 Node	336	375	−0.051	1.116	0.299	12.864	0.024
Erhai Lake	330	369	−0.052	1.118	0.299	15.040	0.024
Lu Diladong Line Project	335	374	−0.051	1.116	0.299	11.990	0.025
Sangyuan River–Jinsha River	336	374	−0.050	1.113	0.300	12.279	0.022
Xiane reservoir	336	376	−0.052	1.119	0.298	11.697	0.021

In summary, according to the comparison of the 2020 and 2035 scenarios, the water supply network in the Erhai Lake basin underwent significant changes in both structure and function, increasing in complexity and dependence on key nodes. These changes indicate that over time and with engineering developments, the supply network will be adjusted to accommodate new challenges and demands. These changes are evident not only in alterations in network size and connectivity but also in its response to the removal of key nodes. Understanding these changes is crucial to assessing long-term network stability and resilience and provides important insights for future water supply network design and management. By analyzing these changes, we can better plan for and address potential risks and challenges in the future, ensuring the sustainability and efficiency of the water supply network.

5. Discussion

This study combines landscape ecology theory and the network analysis methodology to analyze the structural characteristics of the water supply network in the Erhai Lake basin. This study compares the characteristics of the water supply network under future planning scenarios with those of the current scenario, revealing the essential structural characteristics of the water supply network in this region and conducting failure simulations. According to the government's regional development plan "Yunnan Water Network Construction Plan", in order to alleviate the contradiction between the supply and demand of water resources in the region, it is expected that large-scale water conservancy projects, such as the Dianzhong Water Diversion Project and the Ludila hydropower station, as well as many other reservoirs and water transfer projects, will be completed by 2035. Therefore, we use the planned water network for 2035 as the future analysis scenario. However, as this is a long-term planning project, some adjustments may be made during the construction process due to the impact of real climate change and human activities. For example, the construction of some reservoirs may be canceled, and some water transfer pipeline routes may be changed, so the real water network in 2035 may be somewhat different from the expected water network.

5.1. Impact of Large-Scale Water Transfer Projects on the Water Supply Network

Large-scale water diversion and transfer projects will increase the number of nodes and connecting lines in the water supply network, and new engineering projects may create more water flow paths, not only providing additional water to the receiving area but also making the overall water network structure more complex, improving the flexibility and regulation of the water supply system. The addition of paths and nodes increases the system's resilience to disturbances, such as drought and pollution, and ensures water supply through alternative paths. The California State Water Project [4] has significantly changed the topology of California's water supply. It has not only increased the water transfer paths and nodes in the water supply network but also improved the stability of the system [8] during drought. However, it has a negative impact on the ecological environment of the delta region and has triggered the political controversy over water resources. After the construction of water conservancy projects in the Yongding River basin in China, the importance of nodes has changed, and it has been found that water conservancy projects will not only lead to the decrease in the importance of some nodes [36] but also lead to the increase in the importance of some nodes. The results of this study show that by 2035, the water supply network in the Erhai Lake basin will be more complex, with higher connectivity and higher transmission efficiency among nodes compared with 2020. The newly constructed engineering nodes have high importance in the entire water network, indicating that the construction of these projects is necessary. The addition of these new engineering nodes will allow the 2035 network to not only improve the utilization of local water sources but also bring about external water diversion. It reduces the excessive dependence of local users on local water sources, improves the ability of the water supply system to cope with risks, and guarantees the steady development of regional economy and society. But it also requires careful planning and management to ensure ecological and socio-economic balance.

5.2. Impact of Node Failure on Water Supply Network

Through network degradation analysis, it can be observed that damage to key nodes significantly affects the connectivity of the entire water network, and the degree and category of impact may vary among different nodes [53]. This may lead to the unuse of planned water sources, which would not only affect the normal use of industrial, agricultural, and domestic water but also lead to huge losses in some industries that are highly dependent on water. At the same time, there are ecological protection requirements of the Erhai Lake basin regarding the flow. The interruption of some nodes may lead to the decrease in the inflow to Erhai Lake and the decrease in the purification capacity of Erhai Lake, thus suffering ecological damage. The planned hydraulic engineering projects have increased new connectivity links, providing more alternative transmission pathways among water sources and even connecting previously disconnected networks. This can also lead to the emergence of new, shorter paths, improving transmission efficiency and avoiding multiple water sources sharing a single link. On the other hand, the complexity of the network increases, and even if some links are damaged, transmission can still be achieved through other paths. However, despite the construction of new facilities, the network still exhibits a high level of centralization, with most of the newly constructed facilities being located at key nodes and key links. Although this improves the ability to withstand risks to some extent, the impact of damage to key links and nodes should not be underestimated [8,59].

5.3. Deficiencies and Prospects

In an ideal water supply network, there should be multiple alternative transmission paths between nodes, and the importance of each node should not differ significantly, thereby preventing significant disruptions to the entire network due to damage to certain links. The removal of important nodes from the water supply network of the Erhai Lake basin has a significant impact on the network. Therefore, it is necessary to distribute important resources throughout the network rather than relying excessively on certain nodes. By decentralizing nodes and reducing their importance, the impact of a single node on the entire network can be reduced, potentially enhancing the network's resilience and ability to handle failures. Although long-distance water transfer cannot avoid losses during the transportation process, excessive reliance on local water supplies by users makes them vulnerable to local hydrological changes. For instance, the drought that has occurred in the Erhai Lake basin in recent years has caused significant socio-economic losses. Inter-basin water transfer projects can divert water from abundant areas to water-deficient areas, reducing the degree to which users are affected by local hydrological changes and improving their ability to cope with natural risks. Due to the analysis of the water supply network being a purely topological study, it also has some limitations based on topological theory [27,60]. Firstly, the analysis conducted in this study primarily focuses only on surface water infrastructure and does not consider crucial resources such as groundwater, which also plays a significant role in human economic and social activities, especially during drought periods [61]. Secondly, the analysis of the entire water supply network in the Erhai Lake basin does not consider the weight factors of specific nodes or links, such as considering flow rates or construction costs. Thirdly, treating the network as a set of equidistant points and assuming equal distances between adjacent points simplifies the actual water supply network and does not consider factors such as losses during transportation over distance. In future research, the influence of water transport losses should be considered, such as the distance between nodes, losses during transport, and weighting nodes based on factors like flow rates or economic costs. Additionally, when setting up fault simulations, considering scenarios involving the simultaneous disruption of multiple nodes could provide a more comprehensive analysis [8,37]. Furthermore, integration with other system analysis techniques in future research can ensure that network analysis aligns more closely with real-world conditions. Water diversion projects play an important role in strengthening water supply networks, helping to expand and complicate basin water networks, increase connectivity within water networks, and enhance resilience to various risks [62]. However, the construction of inter-basin water transfer project facilities also needs scientific and reasonable technical support and policy guarantee, involving multiple factors, such as project investment, environmental protection, and social stability [63], and has a huge impact on human beings. In order to improve construction rationality and maximize the function of a water transfer project, the connectivity attribute of the water supply network in the graph theory layer may also be considered as one of the factors in project planning in the future.

6. Conclusions

In the southwestern region of China, the total water resources are abundant, but their spatial and temporal distribution is uneven. Drought has become a significant factor limiting the economic and social development of this area. Addressing this issue, this study focuses on the Erhai Lake basin and combines landscape ecology theory and network analysis theory to analyze the structural characteristics of the water supply network in both the current and future planning scenarios. The main conclusions of the study are as follows.

In 2020, the water supply network of the Erhai Lake basin reached 215 nodes and 216 links, forming a complex network system. The average path length was 18.423, and the hydrological loop degree was 0.047, indicating poor connectivity of the water supply network and low water transfer efficiency. In 2035, it is expected that after the completion of the Ludila Project and the Dianzhong Water Diversion Project, the water supply network of the Erhai Lake basin will have 337 nodes and 379 links, the average path length will decrease to 11.680, the hydrological loop degree will increase to 0.643, and the node connectivity rate will also increase slightly in terms of network connectivity. This shows that these large-scale water diversion projects have greatly increased the hydrological connectivity of the water supply network, and the water flow transfer efficiency is higher. At the same time, among the top 10 nodes in importance, the number of engineering nodes increases from 1 in 2020 to 7 in 2035, which also validates the rationality of planned works having a very positive impact on the hydrological connectivity of the water supply network. Node failure scenario settings for 2020 and 2035 show that node failure leads to a significant decline in the water system circulation index and other indicators, reducing the hydraulic connectivity of the water supply network.

This research provides new insights for water resource management and planning to better address environmental challenges and promoting the sustainable development of water resources.

Author Contributions: Conceptualization and methodology, K.S. and T.W.; validation, X.J. and H.X.; resources, X.J. and D.Y.; writing—original draft preparation, K.S.; supervision, T.W. and Z.W. All authors have read and agreed to the published version of the manuscript.

Funding: This work was supported by grants from National Key Research and Development Program of China (No. 2021YFC3000204), Natural Science Foundation of Henan (No. 222300420327), Natural Science Foundation of China (NSFC) (No. 52209038).

Data Availability Statement: The data presented in this study are available in Appendix A.

Acknowledgments: Thanks to Yunnan Institute of Water Conservancy and Hydropower Survey, Design, and Research for their material support for this study.

Conflicts of Interest: Author Dengming Yan was employed by the company Yellow River Engineering and Consulting Co. The remaining authors declare that the research was conducted in the absence of any commercial or financial relationships that could be construed as potential conflicts of interest.

Appendix A

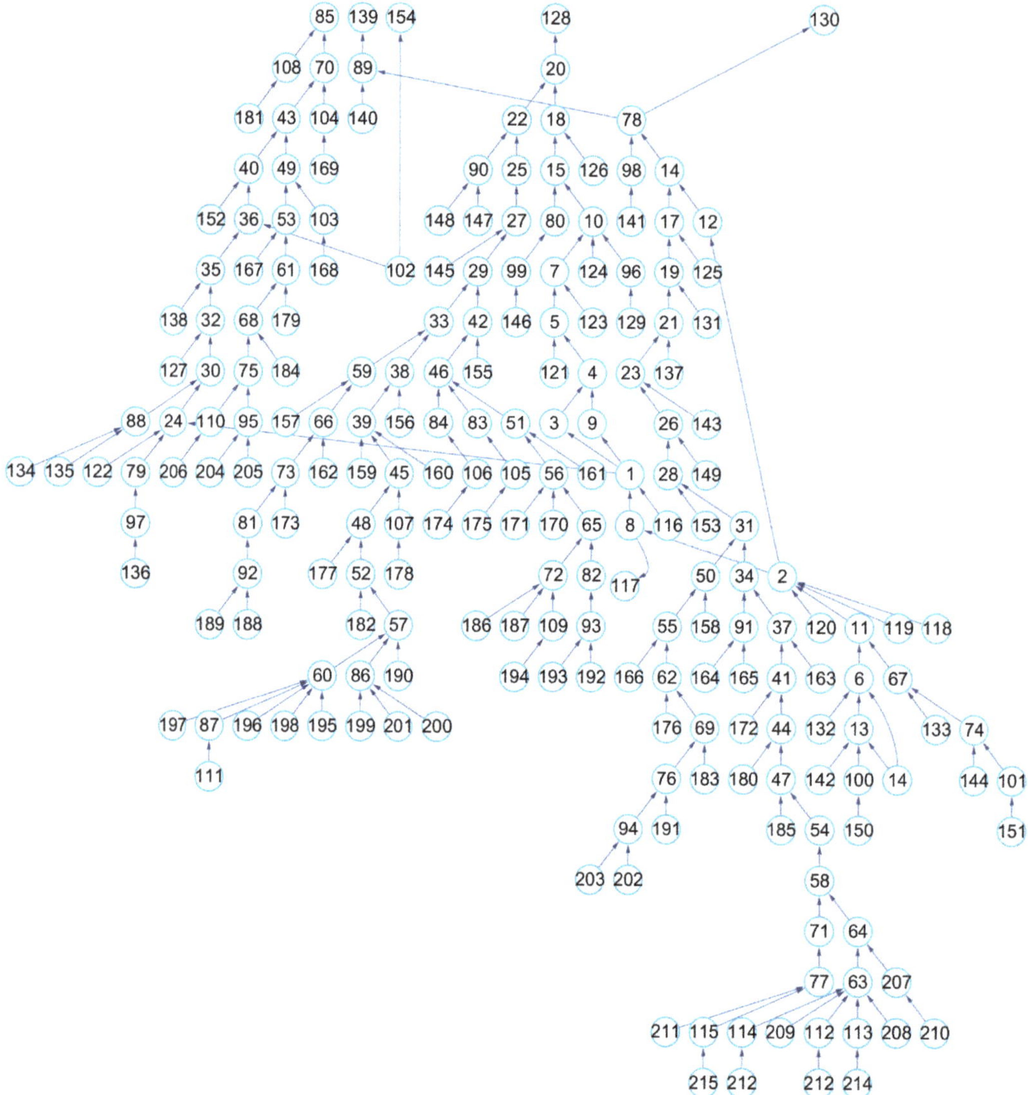

Figure A1. Water system modeling for 2020 scenario produced with Cytoscape. The numbers represent the node numbers, as specified in Schedules 1 and 2, and the arrows represent the direction of the water flow.

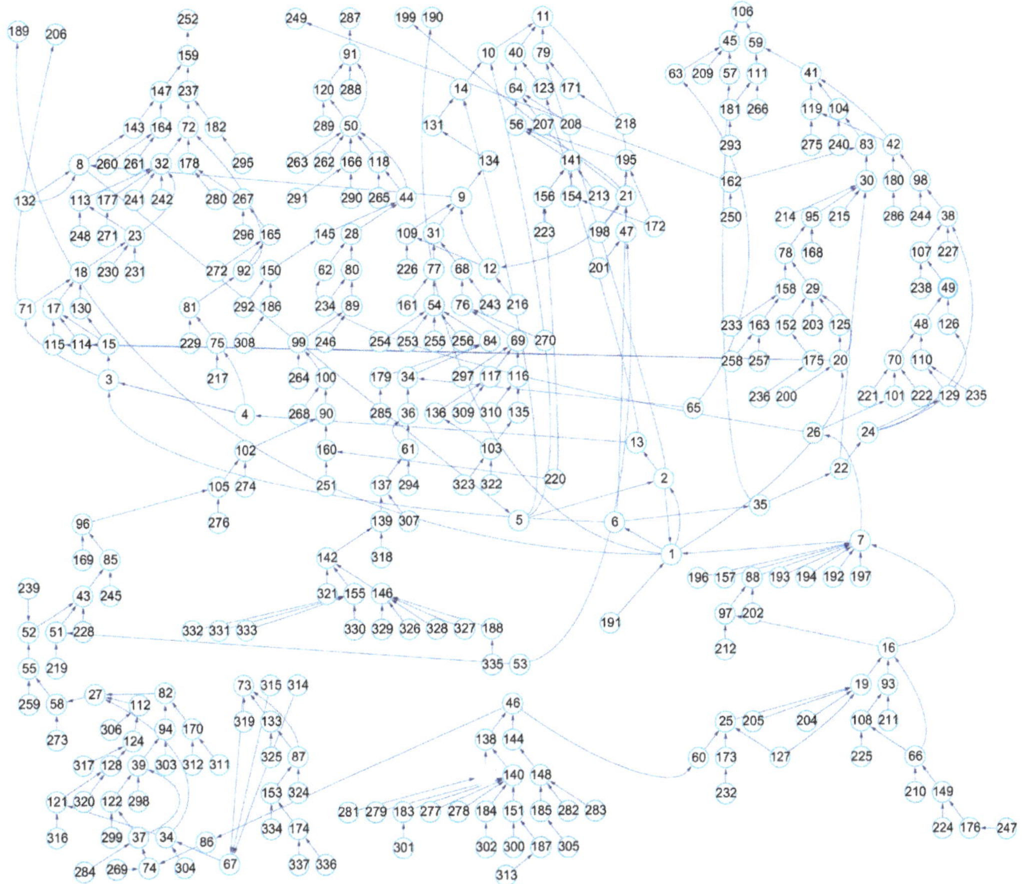

Figure A2. Water system modeling for 2035 scenario produced with Cytoscape. The numbers represent the node numbers, as specified in Schedules 1 and 2, and the arrows represent the direction of the water flow.

Table A1. Node ordinal numbers and node names for 2020 scenario.

Node	Name	Node	Name
1	Xiaoguan Village reservoir	109	Bingju River 1
2	Erhai Lake	110	Yuanjiang 0
3	Pindian Sea reservoir	111	Dalongtan reservoir
4	Qinghai Lake reservoir	112	Dalitang reservoir
5	Zhonghe 0	113	Manxianlin reservoir
6	Mici River 1	114	Golden Phoenix River
7	Zhonghe 1	115	Taoyuan River 0
8	Er Dian Qing	116	Youfengba reservoir
9	Hunsu Lake reservoir	117	Binchuan irrigation district
10	Zhonghe–Yupao River	118	Three Sentinel reservoir
11	Mi Tho River 0	119	Ganhaizi reservoir
12	Xi'er River reservoir (hydropower station)	120	Madian reservoir
13	Mitz River 0	121	Sanjia reservoir
14	Heihuijiang–Xi'er River	122	Mulberry Basket reservoir
15	Chuchang River–Yubao River	123	Luojia Village

Table A1. *Cont.*

Node	Name	Node	Name
16	Sancha River reservoir	124	Upper reaches of the Middle River
17	Heihuijiang 11	125	Koharuji reservoir
18	Qingshui River–Yubao River	126	Upper reaches of the Qingshui River
19	Heihuijiang 10	127	Dayokogi reservoir
20	Yubaojiang–Jinsha River	128	Jinsha River exit
21	Heihuijiang 9	129	Shaojia reservoir
22	Pingcheon River–Jinsha River	130	Lancang River
23	Heihuijiang 8	131	Shuangjian reservoir
24	Shipai Village reservoir	132	Lake Jubi reservoir
25	Rudila hydroelectric power station	133	Wash Matang reservoir
26	Heihuijiang 7	134	Dam reservoir
27	Ludila Eastern Line Project	135	Ba Chong Ji reservoir
28	Heihuijiang 6	136	Yamataka Village reservoir
29	Sangyuan River–Jinsha River	137	Luoguping reservoir
30	Boli River 1	138	Menghuaqi reservoir
31	Misa River–Heihui River	139	Xiaowan hydropower station
32	Buli River 2	140	Fuqing reservoir
33	Luoyu River–Jinsha River	141	Caryophyllum reservoir
34	Heihuijiang 5	142	East Great River
35	Boli River 3	143	Moon Ping reservoir
36	Boli River 4	144	Dabuntang reservoir
37	Heihuijiang 4	145	Longdushan reservoir
38	Yanggongjiang–Jinsha River	146	Xinxing Tho reservoir
39	Liuhe	147	Songping Whistle reservoir
40	Boli River 5	148	Celery Pond reservoir
41	Heihuijiang 3	149	Dayeping reservoir
42	Mulberry River 6	150	Forty Mile Pu reservoir
43	Wonjiang–Buli River	151	Fengyu River reservoir
44	Heihuijiang 2	152	Osmanthus reservoir
45	Songgui River–Yanggong River	153	Pine Garden reservoir
46	Mulberry River 5	154	Rear sea reservoir
47	Heihuijiang 1	155	Small river-bottom reservoir
48	Yanggongjiang 4	156	Longkaikou hydropower station
49	Yuanjiang 5	157	Longzikou reservoir
50	Misa River 3	158	Yulong reservoir
51	Mulberry River 4	159	Five Star reservoir
52	Yanggongjiang 3	160	Nanpo Qiao reservoir
53	Yuanjiang 4	161	Yanggongqi reservoir
54	Heihuijiang 0	162	Three-pot pile reservoir
55	Misa River 2	163	Shuangfeng reservoir
56	Mulberry River 3	164	Tianshengtang reservoir
57	Yanggongjiang 2	165	Peak reservoir
58	Taoyuan River–Heihui River	166	Huanghua reservoir
59	Leaky River 2	167	East River reservoir
60	Yanggongjiang 1	168	Weibaoshan reservoir
61	Yuanjiang 3	169	Baoyi reservoir
62	Misa River 1	170	Liliu reservoir
63	Sword Lake	171	Unity reservoir
64	Sword peach	172	Changle reservoir
65	Sangyuan River–Bingju River	173	New Ma reservoir
66	Leaky River 1	174	Shimen River reservoir
67	Phoenix Feather River 2	175	Flower Palanquin reservoir
68	Yuanjiang 2	176	Taiping reservoir
69	Misa River 0	177	Yanglongtan reservoir
70	Cow Street River	178	Nanzhuang reservoir
71	Taoyuan reservoir	179	Large slate reservoir
72	Bingju River 2	180	Zhongshan reservoir
73	Leaky River 001	181	Unity reservoir (Midu Prefecture)

Table A1. *Cont.*

Node	Name	Node	Name
74	Phoenix River 1	182	Horse wash pond reservoir
75	Yuanjiang 1	183	Laojunshan reservoir
76	Shin Misugi	184	Dianzhong River reservoir
77	Taoyuan River 1	185	Shilong reservoir
78	Shunxi River–Heihui River	186	Wulongba reservoir
79	Black Mud reservoir	187	Renchi Lake reservoir
80	Qingjianmei reservoir	188	Qiping reservoir
81	Drop Leak River 0	189	Shamo River reservoir
82	Haiyo reservoir	190	Wulongtan reservoir
83	Xiange reservoir	191	Meishui reservoir
84	Great Yindian reservoir	192	Tuguan Village reservoir
85	Luwo River Yuanjiang	193	Cuijiaqi reservoir
86	Longtan	194	Thatched lawn reservoir
87	Yanggongjiang 0	195	Hanlongtan reservoir
88	Buli River 0	196	General Temple reservoir
89	Heihuijiang 12	197	Haiyan Pond reservoir
90	Hirakawa Daigawa 1	198	Meilongtan reservoir
91	Sky high	199	Shizhuang Longtan reservoir
92	Shechasi reservoir	200	Shizhaizi Longtan reservoir
93	Mulberry River 1	201	Xilongtan reservoir
94	New Cedar	202	New reservoir
95	Five locks	203	Fir Tree reservoir
96	Pu Peng reservoir	204	Wumaolin reservoir
97	Changpoling reservoir	205	Suoshuige reservoir
98	Shunxi River	206	Mill Hoop reservoir
99	Chuchang River 1	207	Yuhua reservoir
100	Haixi Sea reservoir	208	Upper reaches of the Mantis River
101	Phoenix Feather River 0	209	Longmen reservoir
102	Chestnut Camp reservoir	210	Gaopingba reservoir
103	Yellow gravel mouth reservoir	211	Baiyiqi reservoir
104	Cow Street River 1	212	Twin Rivers reservoir
105	Yimin Sea reservoir	213	Houqi reservoir
106	Li Dazhuang reservoir	214	Yongfeng reservoir
107	Matsugui reservoir	215	Daganchang reservoir
108	Luwo River 1		

Table A2. Node ordinal numbers and node names for 2035 scenario.

Node	Name	Node	Name
1	Er Dian Qing	170	Sky high
2	Da Yin Dian Node	171	Li Dazhuang reservoir
3	Dian Lu Er	172	Xiangshui reservoir
4	Lu Di La Phase II Node	173	Haixi Sea reservoir
5	Lu Di La Bin Chuan Node	174	New Cedar
6	Dian Zhong 1 Node	175	Changpoling reservoir
7	Erhai Lake	176	Phoenix Feather River 0
8	Lu Di La East Line Project	177	Pu Peng reservoir
9	Sang Yuan River–Jinsha River	178	Chuchang River 4
10	Xian E reservoir	179	Matsugui reservoir
11	Mulberry River 5	180	Yellow gravel mouth reservoir
12	Mulberry River 6	181	Cow Street River 0
13	Yang Gongqi node	182	Upper reaches of the Qingshui River
14	Yimin Sea reservoir	183	Dalitang reservoir
15	Qinghai Lake node	184	Manxianlin reservoir
16	Mitho River 0	185	Taoyuan River 0
17	Qinghai Lake reservoir	186	Shunxi River 0

Table A2. *Cont.*

Node	Name	Node	Name
18	Middle River 0	187	Golden Phoenix River 0
19	Mitz River 1	188	Yanggongjiang 0
20	Xiaoguancun reservoir	189	Fengyi Town irrigation district
21	Haiyu reservoir (expansion)	190	Binchuan irrigation district
22	Central Yunnan 1, Midu County 2	191	Imported from Yunnan
23	Nakawa 1	192	Three Sentinel reservoir
24	Central Yunnan 1 Weishan County	193	Fengweiqi reservoir
25	Mitz River 0	194	Ganhaizi reservoir
26	Xi'er River Comprehensive Utilization Project	195	Mulberry River 2
27	Misa River–Heihui River	196	Hongqi reservoir
28	Heihuijiang–Xi'er River	197	Elephant fly reservoir
29	Shipai Village reservoir	198	Kumura reservoir
30	Boli River 3	199	Upper reaches of the Mulberry River
31	Luoyu River–Jinsha River	200	Youfengba reservoir
32	Middle River–Yubaojiang	201	Cuijiaqi reservoir
33	Yanggongjiang 5	202	Scorched stone reservoir
34	Misa River 3	203	Mulberry Basket reservoir
35	Central Yunnan 1, Midu County 1	204	Nantangzi reservoir
36	Maple River reservoir	205	Lake Jubi reservoir
37	Heihuijiang 2	206	Central Yunnan export
38	Yuanjiang 3	207	Liliu reservoir
39	Heihuijiang 3	208	Unity reservoir
40	Mulberry River 4	209	Wupanma reservoir
41	Wonjiang–Buli River	210	Wash Matang reservoir
42	Yuanjiang 5	211	Kamimura reservoir
43	Heihuijiang 7	212	Caohaizi reservoir
44	Shunxi River–Heihui River	213	Wulongba reservoir
45	Luwo River 1	214	Daisan Village reservoir
46	Taoyuan River–Heihui River	215	Menghuaqi reservoir
47	Mulberry River 1	216	Small river-bottom reservoir
48	Yuanjiang 1	217	Mozi buy reservoirs
49	Yuanjiang 2	218	Shimen River reservoir
50	Heihuijiang 12	219	Liping reservoir
51	Soil power	220	Thatched lawn reservoir
52	Heihuijiang 6	221	Mill Hoop reservoir
53	Tuguan Village reservoir	222	West River reservoir
54	Liuhe	223	Maanshan reservoir
55	Heihuijiang 57	224	Dabuntang reservoir
56	Sangyuan River–Bingju River	225	Green Pond reservoir
57	Unity reservoir (Midu Prefecture)	226	Yanjianqiao reservoir
58	Heihuijiang 56	227	Large slate reservoir
59	Cow Street River	228	Dayeping reservoir
60	East Great River	229	Xuchang reservoir
61	Yanggongjiang 4	230	Large seawater reservoir
62	Xi'er River reservoir (hydropower station)	231	Luojia Village
63	Luwo River 0	232	Forty Mile Pu reservoir
64	Mulberry River 3	233	Nangouqi reservoir
65	Sharpening Basket reservoir	234	Snow Mountain River reservoir
66	Phoenix Feather River 2	235	Wumaolin reservoir
67	Misa River 2	236	Yamataka Village reservoir
68	Leaky River 2	237	Qingshui River–Yubao River
69	Leaky River 1	238	Temple Street River reservoir
70	Yuanjiang 0	239	Pine Garden reservoir
71	2 nodes in Yunnan	240	Osmanthus reservoir
72	Chuchang River–Yubao River	241	Upper reaches of the Middle River
73	Misa River 1	242	Bijiaqi reservoir
74	Heihuijiang 1	243	Longzikou reservoir
75	Xinxing Tho reservoir	244	East River reservoir
76	Leaky River 12	245	Moon Ping reservoir

Table A2. *Cont.*

Node	Name	Node	Name
77	Yanggongjiang–Jinsha River	246	Pulling reservoir
78	Boli River 1	247	Fengyu River reservoir
79	Great Yindian reservoir	248	Jindan reservoir
80	Heihuijiang 1113	249	Rear sea reservoir
81	Chuchang River 1	250	Pengjiazhuang reservoir
82	Heihuijiang 5	251	Hubanchang reservoir
83	Boli River 4	252	Jinsha River exit
84	Songgui River–Yanggong River	253	Five Star reservoir
85	Heihuijiang 8	254	Nanpo Qiao reservoir
86	Heihuijiang 0	255	Backyard Hoop reservoir (expansion)
87	Misa River 0	256	Baiyiqi reservoir
88	Mitho River 2	257	Dam reservoir
89	Heihuijiang 1112	258	Ba Chong Ji reservoir
90	Heihuijiang 110	259	Sydney Tree reservoir
91	Heihuijiang 13	260	Songping Whistle reservoir
92	Qingjianmei reservoir	261	Celery Pond reservoir
93	Phoenix River 4	262	Shifang River reservoir
94	Heihuijiang 4	263	Fuqing reservoir
95	Buli River 2	264	Koharuji reservoir
96	Heihuijiang 9	265	Dahuofang reservoir
97	Mitho River 1	266	Baoyi reservoir
98	Yuanjiang 4	267	Chuchang River 3
99	Heihuijiang 11	268	Little Nishigo reservoir
100	Heihuijiang 111	269	Shilong reservoir
101	Houzhuang River reservoir	270	Thunder Temple reservoir
102	Heihuijiang 10	271	Shaojia reservoir
103	Shechasi reservoir	272	Longwangmiao reservoir
104	Boli River 5	273	Xinping reservoir
105	Heihuijiang 910	274	Shuangjian reservoir
106	Luwo River Yuanjiang	275	Dressing River
107	Yuanjiang 23	276	Gooden River reservoir
108	Phoenix River 3	277	White Mountain Mother reservoir
109	Mulberry River 7	278	Upper reaches of the Mantis River
110	Five locks	279	Longmen reservoir
111	Cow Street River 1	280	Wanhuaxi reservoir
112	Misa River 7	281	Gaopingba reservoir
113	Kokura reservoir	282	Baiyiqi reservoir
114	Pindian Sea reservoir	283	Baiyiqi reservoir (expansion)
115	Muddy water reservoir	284	Zhongshan reservoir
116	Leaky River 001	285	Nanzhuang reservoir
117	Leaky River 01	286	Weibaoshan reservoir
118	Heihuijiang 11–12	287	Xiaowan hydropower station
119	Yuanjiang 6	288	White Stone River reservoir
120	Heihuijiang 12–13	289	Lu River reservoir
121	Misa River 4	290	Wuligang reservoir
122	Heihuijiang 23	291	Pear orchards and reservoirs
123	Yanggongqi reservoir	292	Caryophyllum reservoir
124	Misa River 6	293	Wuben reservoir
125	Xigou River reservoir	294	Yanglongtan reservoir
126	Dianzhong River reservoir	295	Qingshui River reservoir
127	Sancha River reservoir	296	Bowl Bowl Hoop reservoir
128	Misa River 5	297	Three-pot pile reservoir
129	Suoshuige reservoir	298	Changle reservoir
130	Sanjia reservoir	299	Guiziqi reservoir
131	Flower Car reservoir (expansion)	300	Twin Rivers reservoir
132	Longdushan reservoir	301	Houqi reservoir
133	Misa River 01	302	Yongfeng reservoir
134	Rudila Phase I	303	Shuangfeng reservoir
135	Drop Leak River 0	304	Yulong reservoir

Table A2. *Cont.*

Node	Name	Node	Name
136	Pot Factory River reservoir	305	Daganchang reservoir
137	Yanggongjiang 34	306	Wenkai reservoir
138	Sword peach	307	Backyard Hoop reservoir
139	Yanggongjiang 3	308	Dutian reservoir
140	Sword Lake	309	Xinfa reservoir
141	Bingju River 2	310	New Ma reservoir
142	Yanggongjiang 2	311	Tianshengtang reservoir
143	Rudila hydroelectric power station	312	Peak reservoir
144	Taoyuan reservoir	313	Mantis River reservoir
145	Qinghe reservoir	314	Tailaping reservoir
146	Yanggongjiang 1	315	Huanghua reservoir
147	Pingcheon River–Jinsha River	316	Iron Gate reservoir
148	Taoyuan River 1	317	Snow Field reservoir
149	Phoenix River 1	318	Horse wash pond reservoir
150	Shunxi River 1	319	Taiping reservoir
151	Golden Phoenix River	320	Jiangchangqi reservoir
152	Black Mud reservoir	321	Wulongtan reservoir
153	Shin Misugi	322	Qiping reservoir
154	Renchi Lake reservoir	323	Shamo River reservoir
155	Longtan	324	Laojunshan reservoir
156	Bingju River 1	325	Fumin reservoir
157	Madian reservoir	326	Hanlongtan reservoir
158	Tholy River 01	327	General Temple reservoir
159	Yubaojiang–Jinsha River	328	Haiyan Pond reservoir
160	Hu Mao	329	Meilongtan reservoir
161	Longkaikou hydropower station	330	Shizhuang Longtan reservoir
162	Chestnut Camp reservoir	331	Shizhaizi Longtan reservoir
163	Buli River 0	332	Caohai Dalongtan reservoir
164	Hirakawa Daigawa 1	333	Xilongtan reservoir
165	Chuchang River 2	334	Meishui reservoir
166	Pear five	335	Dalongtan reservoir
167	Yuhua reservoir	336	New reservoir
168	Dayokogi reservoir	337	Fir Tree reservoir
169	Luoguping reservoir		

References

1. Wang, T.; Wu, Z.; Wang, P.; Wu, T.; Zhang, Y.; Yin, J.; Yu, J.; Wang, H.; Guan, X.; Xu, H.; et al. Plant-groundwater interactions in drylands: A review of current research and future perspectives. *Agric. For. Meteorol.* **2023**, *341*, 109636. [CrossRef]
2. Wang, T.; Wang, P.; Wu, Z.; Yu, J.; Pozdniakov, S.P.; Guan, X.; Wang, H.; Xu, H.; Yan, D. Modeling revealed the effect of root dynamics on the water adaptability of phreatophytes. *Agric. For. Meteorol.* **2022**, *320*, 108959. [CrossRef]
3. Sugg, Z. An Equity Autopsy: Exploring the Role of Water Rights in Water Allocations and Impacts for the Central Valley Project during the 2012–2016 California Drought. *Resources* **2018**, *7*, 12. [CrossRef]
4. Grigg, N.S. Large-scale water development in the United States: TVA and the California State Water Project. *Int. J. Water Resour. Dev.* **2023**, *39*, 70–88. [CrossRef]
5. Rouse, W.R.; Woo, M.-K.; Price, J.S. Damming james bay: I. potential impacts on coastal climate and the water balance. *Can. Geogr.* **1992**, *36*, 2–7. [CrossRef]
6. Shay, C.T.; Shay, J.M.; Johnston, B. Evaluating the impacts of a hydro-electric development in northern Manitoba, Canada. *Environmetrics* **1991**, *2*, 217–226. [CrossRef]
7. Fisher, W.F. Diverting Water: Revisiting the Sardar Sarovar Project. *Int. J. Water Resour. Dev.* **2001**, *17*, 303–314. [CrossRef]
8. Porse, E.; Lund, J. Network Analysis and Visualizations of Water Resources Infrastructure in California: Linking Connectivity and Resilience. *J. Water Resour. Plan. Manag.* **2016**, *142*, 04015041. [CrossRef]
9. Yang, A.-M.; Zhang, L.; Gan, H.; Wang, H. Evaluation on eco-environmental benefits in water reception areas of the east-route phase 1 project of the south-to-north water diversion. *Shuili Xuebao/J. Hydraul. Eng.* **2011**, *42*, 563–571.
10. Wang, S.; Wang, J.; Zhou, K.; Wang, W.; Wan, Y. Response of land-use/land cover change to ecological water conveyance in the lower reach of Tarim River. *Water Resour. Prot.* **2021**, *37*, 69–74+80. [CrossRef]
11. Shumilova, O.; Tockner, K.; Thieme, M.; Koska, A.; Zarfl, C. Global Water Transfer Megaprojects: A Potential Solution for the Water-Food-Energy Nexus? *Front. Environ. Sci.* **2018**, *6*, 150. [CrossRef]

12. Wang, Z.L.; Nixon, R.; Erwin, A.; Ma, Z. Assessing the Impacts of Large-Scale Water Transfer Projects on Communities: Lessons Learned from a Systematic Literature Review. *Soc. Nat. Resour.* **2021**, *34*, 820–841. [CrossRef]
13. Outline of National Water Network Construction Plan. *China Water Conservancy* **2023**, *11*, 1–7. Available online: https://link.cnki.net/urlid/11.1374.TV.20230614.1915.002 (accessed on 1 January 2024).
14. Kesavan, H.K.; Chandrashekar, M. Graph-theoretic models for pipe network analysis. *J. Hydraul. Div.* **1972**, *98*, 345–364. [CrossRef]
15. Pien, K.-C.; Han, K.; Shang, W.; Majumdar, A.; Ochieng, W. Robustness analysis of the European air traffic network. *Transp. A-Transp. Sci.* **2015**, *11*, 772–792. [CrossRef]
16. Pascual-Hortal, L.; Saura, S. Comparison and development of new graph-based landscape connectivity indices: Towards the priorization of habitat patches and corridors for conservation. *Landsc. Ecol.* **2006**, *21*, 959–967. [CrossRef]
17. Jackson, C.R.; Pringle, C.M. Ecological Benefits of Reduced Hydrologic Connectivity in Intensively Developed Landscapes. *Bioscience* **2010**, *60*, 37–46. [CrossRef]
18. Albert, R.; Jeong, H.; Barabási, A.L. Error and attack tolerance of complex networks. *Nature* **2000**, *406*, 378–382. [CrossRef]
19. Bompard, E.; Napoli, R.; Xue, F. Analysis of structural vulnerabilities in power transmission grids. *Int. J. Crit. Infrastruct. Prot.* **2009**, *2*, 5–12. [CrossRef]
20. He, S.; Li, S.; Ma, H.R. Integrating fluctuations into distribution of resources in transportation networks. *Eur. Phys. J. B* **2010**, *76*, 31–36. [CrossRef]
21. Yazdani, A.; Jeffrey, P. A complex network approach to robustness and vulnerability of spatially organized water distribution networks. *arXiv* **2010**, arXiv:1008.1770. [CrossRef]
22. Kurant, M.; Thiran, P.; Hagmann, P. Error and attack tolerance of layered complex networks. *Phys. Rev. E* **2007**, *76*, 026103. [CrossRef] [PubMed]
23. Strogatz, S.H. Exploring complex networks. *Nature* **2001**, *410*, 268–276. [CrossRef] [PubMed]
24. Lai, Y.-C.; Motter, A.E.; Nishikawa, T. Attacks and Cascades in Complex Networks. *Complex Netw.* **2004**, *650*, 299–310.
25. Ren, X.; Lyu, L. Review of ranking nodes in complex networks. *Chin. Sci. Bull.* **2014**, *59*, 1175–1197. [CrossRef]
26. Castro-Gama, M.E.; Pan, Q.; Jonoski, A.; Solomatine, D. A Graph Theoretical Sectorization Approach for Energy Reduction in Water Distribution Networks. *Procedia Eng.* **2016**, *154*, 19–26. [CrossRef]
27. Yazdani, A.; Jeffrey, P. Water distribution system vulnerability analysis using weighted and directed network models. *Water Resour. Res.* **2012**, *48*, 10. [CrossRef]
28. Giudicianni, C.; Di Nardo, A.; Di Natale, M.; Greco, R.; Santonastaso, G.; Scala, A. Topological Taxonomy of Water Distribution Networks. *Water* **2018**, *10*, 444. [CrossRef]
29. Anchieta, T.; Meirelles, G.; Carpitella, S.; Brentan, B.; Izquierdo, J. Water distribution network expansion: An evaluation from the perspective of complex networks and hydraulic criteria. *J. Hydroinform.* **2023**, *25*, 628–644. [CrossRef]
30. Meng, F.; Fu, G.; Farmani, R.; Sweetapple, C.; Butler, D. Topological attributes of network resilience: A study in water distribution systems. *Water Res.* **2018**, *143*, 376–386. [CrossRef]
31. Sivakumar, B. Networks: A generic theory for hydrology? *Stoch. Environ. Res. Risk Assess.* **2015**, *29*, 761–771. [CrossRef]
32. Frota, R.L.; Souza Filho, F.d.A.; Barros, L.S.; Silva, S.M.O.; Porto, V.C.; Rocha, R.V. "Network" socio-hydrology: A case study of causal factors that shape the Jaguaribe River Basin, Ceará-Brazil. *Hydrol. Sci. J.* **2021**, *66*, 935–950. [CrossRef]
33. Castro, C.V.; Carney, C.; de Brito, M.M. The role of network structure in integrated water management: A case study of collaboration and influence for adopting nature-based solutions. *Front. Water* **2023**, *5*, 1011952. [CrossRef]
34. Xiang, P.; Li, J. Research on system vulnerability of interregional large-scale construction projects. *Syst. Eng.-Theory Pract.* **2016**, *36*, 2383–2390.
35. Liu, Z.H.; Yu, H.; Yang, F.T. Evaluate the node importance for water network based on complex network theory. *Sci. Sin. Technol.* **2014**, *44*, 1280–1294. [CrossRef]
36. He, W. Research on the Impact of Water Conservancy Projects on the Accessibility, Importance and Connectivity of Water Networks in River Basins. Master Thesis, Tianjin University, Tianjin, China, 2022.
37. Pagano, A.; Sweetapple, C.; Farmani, R.; Giordano, R.; Butler, D. Water Distribution Networks Resilience Analysis: A Comparison between Graph Theory-Based Approaches and Global Resilience Analysis. *Water Resour. Manag.* **2019**, *33*, 2925–2940. [CrossRef]
38. Yang, X.; Yuan, L.; Jiang, Z.; Feng, X.; Tang, M.; Li, C. Quantitative analysis of abnormal drought in Yunnan Province from 2011 to 2020 using GPS vertical displacement observations. *Acta Geophys. Sin.* **2022**, *65*, 2828–2843. [CrossRef]
39. Ma, W.; Jiang, R.C.; Zhou, Y.; Su, J.G.; Xin, L.Y. Study on diagnosis of water environmental problems and water qualityprotection countermeasures in Erhai Lake Basin. *Yangtze River* **2021**, *52*, 45–53. [CrossRef]
40. Shannon, P.; Markiel, A.; Ozier, O.; Baliga, N.S.; Wang, J.T.; Ramage, D.; Amin, N.; Schwikowski, B.; Ideker, T. Cytoscape: A software environment for integrated models of biomolecular interaction networks. *Genome Res.* **2003**, *13*, 2498–2504. [CrossRef]
41. Assenov, Y.; Ramírez, F.; Schelhorn, S.E.; Lengauer, T.; Albrecht, M. Computing topological parameters of biological networks. *Bioinformatics* **2008**, *24*, 282–284. [CrossRef]
42. Xiao, D.; Li, X.; Gao, J. *Landscape Ecology*; Science Press: Beijing, China, 2010; p. 351.
43. Zhao, J.Y.; Dong, Z.R.; Yang, X.M.; Zhang, J.; Ma, D.; Xu, Z.H. Connectivity Evaluation Technology for Plain River Network Regions basedon Edge Connectivity from Graph Theory. *J. Hydroecology* **2017**, *38*, 1–6. [CrossRef]

44. Min, X.; Zhen, Z.; Haixia, Z. Evaluation of Water System Connectivity of the District around Chaohu LakeBased on Comprehensive Indexes. *Geogr. Geo-Inf. Sci.* **2017**, *33*, 73–74. [CrossRef]
45. Huang, C.; Chen, Y.; Li, Z.; Shen, X.; Tan, L. Optimization of water system pattern and connectivity in the Dongting Lake area. *Shuikexue Jinzhan/Adv. Water Sci.* **2019**, *30*, 661–672. [CrossRef]
46. Dou, M.; Jin, M.; Niu, X.T.; Ren, J. Analysis of evolution characteristics of urban water system form based on remote sensing data. *Eng. J. Wuhan Univ.* **2016**, *49*, 16–21. [CrossRef]
47. Freeman, L.C. A Set of Measures of Centrality Based on Betweenness. *Sociometry* **1977**, *40*, 35–41. [CrossRef]
48. Fronczak, A.; Fronczak, P.; Hołyst, J.A. Average path length in random networks. *Phys. Rev. E* **2004**, *70*, 056110. [CrossRef]
49. Watts, D.J.; Strogatz, S.H. Collective dynamics of 'small-world' networks. *Nature* **1998**, *393*, 440. [CrossRef]
50. Lin, Y.; Zhang, J.J. Centralization of complex networks. *Complex Syst. Complex. Sci.* **2006**, *3*, 13–20. [CrossRef]
51. Schick, R.S.; Lindley, S.T. Directed connectivity among fish populations in a riverine network. *J. Appl. Ecol.* **2007**, *44*, 1116–1126. [CrossRef]
52. Segurado, P.; Branco, P.; Ferreira, M.T. Prioritizing restoration of structural connectivity in rivers: A graph based approach. *Landsc. Ecol.* **2013**, *28*, 1231–1238. [CrossRef]
53. Bodin, O.; Saura, S. Ranking individual habitat patches as connectivity providers: Integrating network analysis and patch removal experiments. *Ecol. Model.* **2010**, *221*, 2393–2405. [CrossRef]
54. Liu, J.G.; Ren, Z.M.; Guo, Q.; Wang, B.H. Node importance ranking of complex networks. *Acta Phys. Sin.* **2013**, *62*, 9–18. [CrossRef]
55. Qi, X.; Duval, R.D.; Christensen, K. Terrorist Networks, Network Energy and Node Removal: A New Measure of Centrality Based on Laplacian Energy. *Soc. Netw.* **2013**, *2*, 19–31. [CrossRef]
56. Verbavatz, V.; Barthelemy, M. Betweenness centrality in dense spatial networks. *Phys. Rev. E* **2022**, *105*, 054303. [CrossRef] [PubMed]
57. Evans, T.S.; Chen, B. Linking the network centrality measures closeness and degree. *Commun. Phys.* **2022**, *5*, 172. [CrossRef]
58. Seidman, S.B. Network structure and minimum degree. *Soc. Netw.* **1983**, *5*, 269–287. [CrossRef]
59. Pandit, A.; Crittenden, J.C. Index of network resilience for urban water distribution systems. *Int. J. Crit. Infrastruct.* **2016**, *12*, 120–142. [CrossRef]
60. Yazdani, A.; Jeffrey, P. Applying Network Theory to Quantify the Redundancy and Structural Robustness of Water Distribution Systems. *J. Water Resour. Plan. Manag.* **2012**, *138*, 153–161. [CrossRef]
61. Medellín-Azuara, J.; MacEwan, D.; Howitt, R.E.; Koruakos, G.; Dogrul, E.C.; Brush, C.F.; Kadir, T.N.; Harter, T.; Melton, F.; Lund, J.R. Hydro-economic analysis of groundwater pumping for irrigated agriculture in California's Central Valley, USA. *Hydrogeol. J.* **2015**, *23*, 1205–1216. [CrossRef]
62. Pietrucha-Urbanik, K.; Rak, J. Water, Resources, and Resilience: Insights from Diverse Environmental Studies. *Water* **2023**, *15*, 3965. [CrossRef]
63. Wen, F.; Yang, M.; Guan, W.; Cao, J.; Zou, Y.; Liu, X.; Wang, H.; Dong, N. The Impact of Inter-Basin Water Transfer Schemes on Hydropower Generation in the Upper Reaches of the Yangtze River during Extreme Drought Years. *Sustainability* **2023**, *15*, 8373. [CrossRef]

MDPI AG
Grosspeteranlage 5
4052 Basel
Switzerland
Tel.: +41 61 683 77 34

Water Editorial Office
E-mail: water@mdpi.com
www.mdpi.com/journal/water

www.ingramcontent.com/pod-product-compliance
Lightning Source LLC
Chambersburg PA
CBHW061435210926
51850CB00008B/1517